Concise
Statistics

M G Godfrey
E M Roebuck
A J Sherlock

Edward Arnold

© M G Godfrey, E M Roebuck and A J Sherlock 1988

First published in Great Britain 1988 by
Edward Arnold (Publishers) Ltd, 41 Bedford Square, London WC1B 3DQ

Edward Arnold, 3 East Read Street, Baltimore, Maryland 21202, USA

Edward Arnold (Australia) Pty Ltd, 80 Waverley Road, Caulfield East, Victoria 3145, Australia

British Library Cataloguing in Publication Data

Godfrey, M.G.
  Concise statistics.
  1. Mathematical statistics—Examinations,
  questions, etc.
  I. Title   II. Roebuck, E.M.   III. Sherlock,
  A.J.
  519.5′076   QA276.2

ISBN 0-7131-3591-3

Text set in 10/12pt Times Compugraphic by Mathematical Composition Setters Ltd, Salisbury
Printed and bound in Great Britain by J. W. Arrowsmith Ltd, Bristol.

Preface

This course covers the Statistics for A level Mathematics syllabuses and the basic Statistics requirement for courses in computing, business studies, operational research and so on at Further Education Colleges, Polytechnics and Universities. It has been used successfully at Millfield over many years with students of widely differing mathematical ability. It can also be used as a self-teaching text.

It is intended that the book is worked through in order. The basic ideas of statistics are presented clearly with worked examples illustrating their application. Each section has a carefully structured exercise, and all the questions included are original. A particular feature of the book is the form of the Answers, which contain sufficient detail to indicate the method of solution. The authors (who are experienced teachers) consider this to be of great value to the serious student.

It is assumed that the student will be using an electronic calculator with statistics functions.

The book is complete in itself, and may be used in conjunction with past papers of the particular examination for which the student is preparing. However, the same authors have written a concise revision course entitled *Revision Statistics*, (also published by Edward Arnold), which can be used to reinforce the ideas contained in this book.

The Tables on pages 395–399 are reproduced with the kind permission of the Mathematics in Education and Industry Schools Project.

Maurice Godfrey
Elizabeth Roebuck
Alan Sherlock

Contents

1
Frequency Distributions

1.1 Histograms

A frequency distribution for a quantitative (i.e. numerical) variable can be illustrated by a histogram, which is a special type of bar chart where the *area* of the bars represents the frequency.

The variable of interest is represented on the horizontal axis, which is a continuous scale labelled in the usual manner. Bars are drawn vertically corresponding to the 'classes' into which the values have been grouped. There should be no gaps between the bars, and the precise dividing points, the *class boundaries*, are found as follows.

(a) For a continuous variable

A continuous variable is one which can take any value in a certain range, for example, lengths, weights, times and so on. For this we consider what exact values of the variable would fall into each class.

Example 1

Length (to nearest metre)	120–124	125–129	...
Frequency	8	17	

Any length between 119.5 m and 120.5 m will be recorded as 120 m (to the nearest metre), and similarly any length between 119.5 m and 124.5 m will fall into the first class. The class boundaries are

$$\mid 119.5 \text{ and } 124.5 \mid 124.5 \text{ and } 129.5 \mid$$

Note that a length of *exactly* 124.5 m is rounded up to 125 m (to the nearest metre), and therefore belongs to the second class. However any length which is just less than 124.5 m, however close (e.g. 124.499999 m) is rounded down to 124 m (to the nearest metre), and therefore belongs to the first class.

Example 2

Weight in kg (correct to 2 d.p.)	30.20–30.29	30.30–30.39
Frequency	15	27

A weight of 30.20 kg (correct to 2 d.p.) means that the exact weight is between 30.195 kg and 30.205 kg. The class boundaries are

$$|\,30.195 \text{ and } 30.295\,|\,30.295 \text{ and } 30.395\,|$$

Example 3

Age (in completed years)	0–5	6–12	13–18
Frequency	28	37	12

A child is said to be 5 years old when he is between his 5th and 6th birthdays, so his exact age is between 5 and 6 years.
The class boundaries are

$$|\,0 \text{ and } 6\,|\,6 \text{ and } 13\,|\,13 \text{ and } 19\,|$$

(b) For a discrete variable

Now suppose that the underlying variable is *discrete* (i.e. we can list its possible values). We can draw a histogram only if we are prepared to treat the variable as a continuous one.
    For example, if the possible values are the whole numbers $10, 11, 12, 13, \ldots$
the value 11 corresponds to the interval 10.5 to 11.5
the value 12 corresponds to the interval 11.5 to 12.5 and so on.
    If the possible values are (as in shoe sizes) $5, 5\frac{1}{2}, 6, 6\frac{1}{2}, \ldots$
the value $5\frac{1}{2}$ corresponds to the interval 5.25 to 5.75
the value 6 corresponds to the interval 5.75 to 6.25, and so on.

Example 4

Goals scored	0	1	2	...
Frequency	26	18	13	

The class boundaries are
$$|\,-0.5 \text{ and } 0.5\,|\,0.5 \text{ and } 1.5\,|\,1.5 \text{ and } 2.5\,|$$

Example 5

Number of enquiries	0–9	10–19	20–29
Frequency	10	22	17

The class boundaries are $\quad|\,-0.5 \text{ and } 9.5\,|\,9.5 \text{ and } 19.5\,|\,19.5 \text{ and } 29.5\,|$

Example 6

Shoe size	4–5½	6–7½	8–9½
Frequency	45	89	120

The class boundaries are

| 3.75 and 5.75 | 5.75 and 7.75 | 7.75 and 9.75 |

Note that, in all cases, the upper boundary of one class is equal to the lower boundary of the next class.

Sometimes it is necessary to make reasonable assumptions in order to determine the class boundaries.

Example 7

Height (cm)	150–	160–	170–
Frequency	4	16	24

Here '150–' clearly means the class beginning with 150; but height is a continuous variable and the accuracy to which the heights have been measured is not specified.

It could be to the nearest cm (in which case the lower class boundary is 149.5 cm)

or to the nearest 0.1 cm (in which case the lower class boundary is 149.95 cm)

and so on.

We shall assume that the heights are given exactly. The third class has no upper limit; it is reasonable to assume that it is the same size as the first two.

The class boundaries are | 150 and 160 | 160 and 170 | 170 and 180 |

Example 8

Height (cm)	150–159	160–169	170–179
Frequency	4	16	24

Again the accuracy of measurement is not specified, but since the first class ends with 159 cm and the second begins with 160 cm, it is clear that the measurements are to the nearest cm.

The class boundaries are

| 149.5 and 159.5 | 159.5 and 169.5 | 169.5 and 179.5 |

Example 9

Time (seconds) mid-interval value	35.5	55.5	75.5
Frequency	11	20	15

Here we are given the value at the centre of each class.
We assume that all the classes have the same width, and since the centres are 20 s apart, the width is 20 s. Hence each class extends 10 s each side of its centre.
The class boundaries are | 25.5 and 45.5 | 45.5 and 65.5 | 65.5 and 85.5 |

Frequency density

The **class width** is the difference between the class boundaries, and this is the width of the bar drawn on the histogram.

If the area of the bar is to be the frequency,

we have   height $\times$ width = frequency, and so   height $= \dfrac{\text{frequency}}{\text{width}}$

We define

$$\text{frequency density} = \frac{\text{frequency}}{\text{class width}}$$

and this gives the vertical scale on the histogram.

Since the area of the histogram is equal to the frequency, we can find frequencies and probabilities as shown in Example 10.

Example 10

The heights of the 400 trees in a small copse are as follows:

Height (nearest m)	5–9	10–11	12–13	14–16	17–19	20–22	23–26	27–36
Number of trees	18	58	62	72	57	42	36	55

Draw a histogram, and use it to find
(i) the number of trees with heights between 12 m and 25 m
(ii) the probability that one of these trees, chosen at random, is taller than 25 m.

Table 1.1 shows the same data with frequency densities calculated.

Table 1.1

Height	Frequency	Class boundaries	Class width	Frequency density
5–9	18	4.5 and 9.5	5	3.6
10–11	58	9.5 and 11.5	2	29
12–13	62	11.5 and 13.5	2	31
14–16	72	13.5 and 16.5	3	24
17–19	57	16.5 and 19.5	3	19
20–22	42	19.5 and 22.5	3	14
23–26	36	22.5 and 26.5	4	9
27–36	55	26.5 and 36.5	10	5.5
	400			

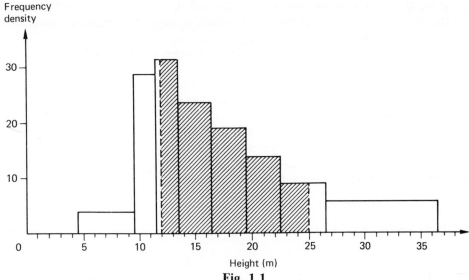

Fig. 1.1

(i) As height is a continuous variable, and the accuracy is not specified, we shall assume that 'between 12 m and 25 m' means between exactly 12 m and exactly 25 m. We require the area of the histogram between heights 12 m and 25 m, which is

$$(1.5 \times 31) + 72 + 57 + 42 + (2.5 \times 9) = 240$$

Hence there are 240 trees with heights between 12 m and 25 m.

(ii) The number of trees taller than 25 m is given by the area of the histogram to the right of 25 m, which is

$$(1.5 \times 9) + 55 = 68.5$$

There are 400 trees altogether, of which 68.5 are taller than 25 m. The probability that one tree chosen at random is taller than 25 m is therefore

$$\frac{68.5}{400} \approx 0.171.$$

Notes (1) These answers are only approximations, since we do not know how the heights are distributed within the given classes. In (ii) it is of course impossible to have 68.5 trees taller than 25 m.

(2) As the data are given to the nearest metre, we could have interpreted 'between 12 m and 25 m' to mean between 12 m and 25 m (inclusive) when measured to the nearest metre. This means that the exact height is between 11.5 m and 25.5 m, so we should find the area of the histogram between 11.5 m and 25.5 m.

The mode

The mode is the value which occurs most often; for a continuous variable it is the value with the highest frequency density.

For grouped data, the **modal class** is the class with the highest frequency density (i.e. having the tallest bar on the histogram). For the distribution of tree heights in Example 10, the modal class is 12–13 m.

Note that this is *not* necessarily the class with the highest frequency.

If we require a single value for the mode, we would clearly choose a value in the modal class (i.e. between 11.5 m and 13.5 m).

Since the next bar on the left is taller than that on the right, it is reasonable to choose a value closer to 11.5 m than to 13.5 m (see Fig. 1.2). Using the 'criss-cross' method on the histogram, we obtain

the mode is 11.9 m

This is, of course, only an approximate value.

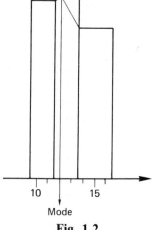

Fig. 1.2

The median

When the values are arranged in order, the median is the value in the centre. Since half the values are less than the median, and half are greater, the median can be found on a histogram by dividing the area in half.

For the distribution of tree heights in Example 10, the total area of the histogram is 400 and the area of the first three bars is $18 + 58 + 62 = 138$; we therefore need an area of 62 from the fourth bar (Fig. 1.2).

We have $24x = 62$

$x \approx 2.6$

and so the median is

$13.5 + 2.6 = 16.1$ m

Again this is only an approximate value.

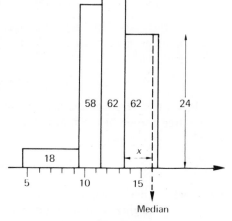

Fig. 1.3

Probability density

We define

$$\text{probability density} = \frac{\text{probability}}{\text{class width}}$$

$$= \frac{\text{frequency}}{(\text{total frequency}) \times (\text{class width})}$$

Table 1.2 shows the calculation for the distribution of tree heights in Example 10.

Table 1.2

Class boundaries	Frequency f	Class width w	Probability density $\dfrac{f}{400w}$
4.5 and 9.5	18	5	0.009
9.5 and 11.5	58	2	0.0725
11.5 and 13.5	62	2	0.0775
13.5 and 16.5	72	3	0.06
16.5 and 19.5	57	3	0.0475
19.5 and 22.5	42	3	0.035
22.5 and 26.5	36	4	0.0225
26.5 and 36.5	55	10	0.01375

A histogram may be drawn with probability density on the vertical scale.

Since probability density $= \dfrac{\text{frequency density}}{\text{total frequency}}$ this has exactly the same shape as the

original histogram (with frequency density on the vertical scale).
   The area of the histogram now gives the probability.
   Since the total probability is always one, the total area of this histogram is one.

Exercise 1.1   Histograms

1   The numbers of goals scored in 42 football matches were

3, 3, 1, 2, 2, 5, 1, 2, 2, 2, 1, 4, 4, 3, 3, 3, 6, 1, 2, 2, 1,
0, 2, 7, 1, 3, 2, 3, 3, 1, 4, 3, 0, 3, 5, 2, 1, 4, 1, 3, 2, 4.

Give a frequency table and draw a histogram.

2   For each of the following, state the limits on the horizontal scale between which the histogram bars should be drawn.

(a) *Number of wickets taken*	(b) *Golf scores*	(c) *Marks in a test*	(d) *Shoe size*
0	66–70	0–4	3–$5\frac{1}{2}$
1	71–75	5	6
2	76–80	6	$6\frac{1}{2}$

(e) *Heights (nearest cm)*	(f) *Resistance (nearest 0.1 ohm)*	(g) *Weights (nearest 5 kg)*	(h) *Ages (completed yrs)*
150–159	5.0–5.4	0–25	11–13
160–169	5.5–5.9	30–50	14–16
170–179	6.0–6.4	55–75	17–19

8 Frequency distributions

(i) Marks in an exam	(j) Price of 500 g butter (p)	(k) Lengths (nearest mm)	(l) Reaction time in s
0–10	38–39	110–	0.2–
11–20	40–41	120–	0.4–
21–30	42–43	130–	0.6–

(m) Diameters in cm (mid-interval value)	(n) Error (correct to 2 d.p.)
14.5	$-0.10 \rightarrow -0.05$
24.5	$-0.04 \rightarrow +0.04$
34.5	$+0.05 \rightarrow +0.10$

3 Draw a histogram for the following examination results

Marks	0–20	21–30	31–40	41–45	46–50	51–55	56–60	61–70	71–100
Number of candidates	14	9	15	11	18	14	10	16	24

Use your histogram to estimate the number of candidates who scored between 43 and 54 marks (inclusive).

4 The heights of 125 children were measured to the nearest 10 cm with the following results.

Height	50–70	80–100	110–120	130–140	150–170
Number of children	18	24	23	33	27

Draw a histogram.
If one of these children is selected at random, estimate the probability that his height is between 112 cm and 128 cm (when measured exactly).

5 (i) The ages of the people living in a village are as follows

Age (in completed years)	0–9	10–19	20–34	35–54	55–79
Number of people	440	480	630	440	150

Draw a histogram with probability density on the vertical scale.
(ii) The age distribution in a second village is

Age (in completed years)	0–3	4–23	24–38	39–48	49–58	59–73	74–88
Number of people	54	180	291	315	360	384	90

Why is it difficult to compare the two villages by simply looking at these tables? Draw a histogram (with probability density on the vertical scale) for the second village, and comment on the differences between the two villages.

1.2 Cumulative frequency diagrams

A histogram gives a clear 'picture' of a distribution, but when finding frequencies, the median, and so on, it is usually easier to use a cumulative frequency diagram.

Cumulative frequency is obtained by adding together all the frequencies so far, as shown below for the distribution of tree heights in the previous section.

Height (nearest m)	5–9	10–11	12–13	14–16	17–19	20–22	23–26	27–36
Frequency	18	58	62	72	57	42	36	55
Cumulative frequency	18	76	138	210	267	309	345	400

For example, the cumulative frequency 138 is calculated as $18 + 58 + 62$. This means that 138 trees have heights in the combined class 5–13, i.e. have heights less than 13.5 m. Similarly 267 trees have heights less than 19.5 m, and so on.

Thus on a cumulative frequency diagram, cumulative frequency (cf) is plotted against the **upper class boundary**. In this case we plot the points $(9.5, 18)$, $(11.5, 76)$, $(13.5, 138)$ and so on. We may also plot a cumulative frequency of zero against the lower class boundary of the first class—in this case the point $(4.5, 0)$.

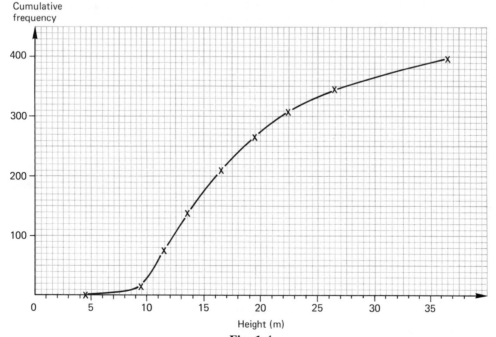

Fig. 1.4

The points may be joined by straight lines; this gives a **cumulative frequency polygon**. This assumes that the values are uniformly distributed within each class, and gives the same results as those obtained from the histogram.

However, slightly better results might be expected if the points are joined by a smooth curve; this gives a **cumulative frequency curve**.

The cumulative frequency curve for the above example is shown in Fig. 1.4.

The cumulative frequency curve can be used to find frequencies and probabilities: for any height x, the cumulative frequency gives the number of trees having height less than x.

Since half of the values are less than the median, the median can be found as the height for which the cumulative frequency is half of the total frequency.

We can also find other **percentiles**; for example the 65th percentile is the height for which the cumulative frequency is 65% of the total frequency.

Note that the median is the 50th percentile.

Example 1

For the distribution of tree heights given above, use the cumulative frequency curve to find

(i)   the number of trees taller than 20 m
(ii) the probability that a tree chosen at random has a height between 15 m and 30 m
(iii) the median height
(iv) the 15th percentile.

(i)   For a height of 20 m, the cumulative frequency is 275.
      Thus 275 trees are shorter than 20 m, and so the number of trees taller than 20 m is $400 - 275 = 125$.
(ii) For a height 30 m, cf = 369, i.e. 369 trees are shorter than 30 m.
      For a height 15 m, cf = 177, i.e. 177 trees are shorter than 15 m. So the number of trees with heights between 15 m and 30 m is $369 - 177 = 192$ and hence the probability is $192/400 = 0.48$.
(iii) For the median, cf = $\frac{1}{2} \times 400 = 200$;
      hence the median is 16.0 m.
(iv) For the 15th percentile, cf = $0.15 \times 400 = 60$;
      hence the 15th percentile is 11.1 m.
      This means that 15% of the trees are shorter than 11.1 m.

Linear interpolation

If we do not wish to draw an accurate cumulative frequency diagram, we can assume that the points are joined by straight lines and use similar triangles to *calculate* intermediate points.

This is *linear interpolation*.

Example 2

The marks obtained by 75 students in a test were as follows:

Mark	1–30	31–60	61–90	91–120	121–150	151–180
Number of students	3	9	20	22	13	8

Use linear interpolation to find:
(i) the median mark; (ii) the number of students who scored 140 marks or more.

It is clear that marks must be whole numbers, so
the first class has boundaries 0.5 and 30.5,
the second class has boundaries 30.5 and 60.5, and so on.
Cumulative frequency would be plotted as follows

Mark	30.5	60.5	90.5	120.5	150.5	180.5
Cumulative frequency	3	12	32	54	67	75

(i) For the median, $cf = \frac{1}{2} \times 75 = 37.5$ which comes between the points $(90.5, 32)$ and $(120.5, 54)$, see Fig. 1.5

We have $\dfrac{x}{30} = \dfrac{5.5}{22}$

$$x = 7.5$$

and so the median is $90.5 + 7.5 = 98.0$.

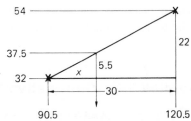

Fig. 1.5

(ii) Since marks are whole numbers, a score of 140 marks or more corresponds to 'greater than 139.5' on the continuous scale.
We consider the two points $(120.5, 54)$ and $(150.5, 67)$ (Fig. 1.6).

We have $\dfrac{y}{13} = \dfrac{19.0}{30.0}$

$$y \approx 8.2$$

so the cumulative frequency is

$$54 + 8.2 = 62.2.$$

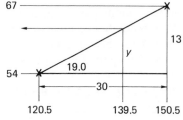

Fig. 1.6

Hence the number of students who scored 140 marks or more is $75 - 62.2 = 12.8$

These answers are only approximations, since we do not know the actual marks.
Note When the marks are arranged in order, the median is the 38th mark in the list. However, this is best estimated by interpolating to 37.5 on the cumulative frequency scale.

Cumulative probability

If we divide the cumulative frequencies by the total frequency, we obtain the **cumulative probabilities**.
For the distribution of tree heights given earlier, a cumulative probability diagram

would be plotted as follows:

Height (m)	9.5	11.5	13.5	16.5	19.5	22.5	26.5	36.5
Cumulative frequency (cf)	18	76	138	210	267	309	345	400
Cumulative probability $=\dfrac{cf}{400}$	0.045	0.19	0.345	0.525	0.6675	0.7725	0.8625	1

The cumulative probability scale always goes from 0 to 1.

The cumulative probability diagram can be used to obtain probabilities directly: for any height x, the cumulative probability gives the probability that a tree chosen at random has a height less than x.

The median has a cumulative probability 0.5.

The 65th percentile has cumulative probability 0.65, and so on.

Exercise 1.2   Cumulative frequency

1   The mid-day temperature at a seaside holiday resort was measured (to the nearest $^\circ$C) each day of the season, with the following results

Temperature ($^\circ$C)	10–11	12–13	14–15	16–17	18–19	20–21	22–23	24–25	26–27	28–29
Number of days	3	2	13	27	31	25	18	20	9	2

Draw a cumulative frequency curve, and use it to estimate:

(i)   the number of days on which the (exact) temperature was less than 19°C

(ii)   the number of days on which the temperature was between 16°C and 22°C

(iii)   the temperature which was exceeded on the hottest 25 days.

2   The populations of 220 British towns are given in the following table

Population (to nearest 100)	0–4900	5000–9900	10 000–19 900	20 000–29 900	
Number of towns	28	35	39	28	
Population	30 000–39 900	40 000–59 900	60 000–79 900	80 000–99 900	100 000–
Number of towns	20	23	18	9	20

Draw a cumulative frequency curve, and use it to estimate

(i)   the median   (ii)   the 65th percentile

(iii)   the percentage of towns having a population of more than 45 000

(iv)   the probability that one of these towns, selected at random, will have a population between 12 000 and 36 000

(v)   the population which is exceeded by 15% of these towns.

3   The following table gives the ages of 240 people.

Age	Number of people	Age	Number of people
0–	31	50–	28
10–	39	60–	24
20–	34	70–	17
30–	33	80–	7
40–	27		

Draw a cumulative probability curve, and use it to find
(i)   the median   (ii) the 20th percentile
(iii) the proportion of people aged between 18 and 65.

4   The marks obtained by 350 candidates in an examination were:

Marks	0–30	31–40	41–45	46–50	51–55	56–60	61–70	71–100
Number of candidates	51	38	26	39	37	44	68	47

Without drawing an accurate cumulative frequency curve, use linear interpolation to find
(i)   the median
(ii)   the number of candidates who scored between 55 and 64 marks (inclusive)
(iii) the pass mark, if 60% of the candidates passed the examination.

5   The engine capacities of cars produced by a large car manufacturer are as follows

Engine capacity (c.c.)	0–1000	1000–1600	1600–2800	2800–
Number of cars for export	64	126	38	24
Number of cars for the home market	126	312	90	13

Use linear interpolation to find the median engine capacity
(i)   of the cars produced for export
(ii)   of the cars produced for the home market
(iii) of all the cars produced.

6   The following information relates to the annual salaries of manual workers employed in the public sector in 1980:
10% earned less than £3148; 25% earned less than £3721; 50% earned less than £4449; 25% earned more than £5202; 10% earned more than £6036.

Write down the coordinates of the five points on the cumulative probability curve which are given by the above figures.

State the median salary, and use linear interpolation to find
(i) the 35th percentile
(ii) the percentage of these employees who earn more than £5500.

1.3   The mean

The (arithmetic) mean is obtained by adding together all the values, then dividing by the number of values.

For n values $x_1, x_2, ..., x_n$, the mean is

$$\bar{x} = \frac{x_1 + x_2 + \cdots + x_n}{n} = \frac{\sum\limits_{i=1}^{n} x_i}{n}$$

which we write simply as

$$\boxed{\bar{x} = \frac{\sum x}{n}}$$

If we have a frequency table, and the value x_1 occurs f_1 times
the value x_2 occurs f_2 times, and so on,

then the sum of the values is

$$x_1 f_1 + x_2 f_2 + \cdots = \sum x_i f_i, \quad \text{or simply } \sum xf$$

and the numbers of values is

$$f_1 + f_2 + \cdots = \sum f_i, \quad \text{or simply } \sum f \quad \text{(the total frequency)}$$

so the mean is

$$\boxed{\bar{x} = \frac{\sum xf}{\sum f}}$$

If the values are grouped into classes, we use the same formula, but for x we take the **mid-interval value** of each class; this is the average of the upper and lower class boundaries.

We cannot calculate the mean exactly, because we do not know the actual values, but this method gives a reasonable approximation, assuming that the values are uniformly distributed within the classes.

Example 1
Find the mean height of the following distribution of tree heights:

Height (nearest m)	5–9	10–11	12–13	14–16	17–19	20–22	23–26	27–36
Number of trees	18	58	62	72	57	42	36	55

The first class has boundaries 4.5 and 9.5, so the mid-interval value is $\dfrac{4.5 + 9.5}{2} = 7.0$.

The remainder of the calculation is set out in Table 1.3.

Table 1.3

Height	Frequency f	Mid-interval value x	xf
5–9	18	7.0	126
10–11	58	10.5	609
12–13	62	12.5	775
14–16	72	15.0	1080
17–19	57	18.0	1026
20–22	42	21.0	882
23–26	36	24.5	882
27–36	55	31.5	1732.5
$\sum f = 400$		$\sum xf = 7112.5$	

The mean height, $\bar{x} = \dfrac{7112.5}{400} = 17.78$ m

Linear functions

Example 2
Calculate the mean of the numbers 3, 7, 8, 21, 35, and deduce the mean of
(i)   203, 207, 208, 221, 235
(ii) 0.6, 1.4, 1.6, 4.2, 7
(iii) 5.03, 5.07, 5.08, 5.21, 5.35

The mean is $\dfrac{\sum x}{n} = \dfrac{74}{5} = 14.8$

(i)   We have added 200 to all the numbers;
      so the mean is now $14.8 + 200 = 214.8$
(ii) We have multiplied all the numbers by 0.2
      so the mean is now $0.2 \times 14.8 = 2.96$
(iii) We have multiplied all the numbers by 0.01, then added 5,
      so the mean is now $0.01 \times 14.8 + 5 = 5.148$.

In general, if the numbers $u_1, u_2, ..., u_n$ have mean $\bar{u}$, and the numbers $x_1, x_2, ..., x_n$ are defined by $x = au + b$ (i.e. $x_1 = au_1 + b, x_2 = au_2 + b, ...,$ where a and b are constants), then the mean of $x_1, x_2, ..., x_n$ is $\bar{x} = a\bar{u} + b$.

To prove this, we have

$$\bar{x} = \frac{\sum_{i=1}^{n} x_i}{n} = \frac{\sum_{i=1}^{n}(au_i + b)}{n}$$

$$= \frac{a\sum_{i=1}^{n} u_i + bn}{n} = \frac{a\sum_{i=1}^{n} u_i}{n} + b$$

$$= a\bar{u} + b$$

Coding

Example 3

Find the mean mark for the following distribution:

Mark	1–30	31–60	61–90	91–120	121–150	151–180
Number of students	3	9	20	22	13	8

The first class has boundaries 0.5 and 30.5, so the mid-interval value is $\dfrac{0.5 + 30.5}{2} = 15.5$; similarly the other mid-interval values are 45.5, 75.5, 105.5, 135.5, 165.5.

We could now calculate the mean from the formula $\bar{x} = \sum xf / \sum f$, but as the values 15.5, 45.5, 75.5, 105.5, 135.5, 165.5 are equally spaced, we can simplify the work as follows.

If we subtract one of these values, say 75.5, from all of them, we obtain

$$-60, \quad -30, \quad 0, \quad 30, \quad 60, \quad 90$$

and these are clearly easier numbers to work with.

We can make a further simplification by dividing each number by 30 to obtain

$$-2, \quad -1, \quad 0, \quad 1, \quad 2, \quad 3$$

We have 'coded' the original mid-interval values (x) by $u = \dfrac{x - 75.5}{30}$.

We now calculate the mean of the coded values(u) as $\bar{u} = \dfrac{\sum uf}{\sum f}$

This is an easy calculation since the values of u are simple.

Since $u = \dfrac{x - 75.5}{30}$, we have $x = 30u + 75.5$;

so the mean of the original distribution is then $\bar{x} = 30\bar{u} + 75.5$

Mark	Frequency f	Mid-interval value x	$u = \dfrac{x - 75.5}{30}$	uf
1–30	3	15.5	−2	−6
31–60	9	45.5	−1	−9
61–90	20	75.5	0	0
91–120	22	105.5	1	22
121–150	13	135.5	2	26
151–180	8	165.5	3	24
	$\sum f = 75$			$\sum uf = 57$

We have $\bar{u} = \dfrac{\sum uf}{\sum f} = \dfrac{57}{75}$

so the mean mark is

$$\bar{x} = 30\bar{u} + 75.5 = 30 \times \frac{57}{75} + 75.5$$

$$= 98.3.$$

In general, if we 'code' the original values (x) by

$$u = \frac{x - b}{a},$$

then we have

$$\bar{x} = a\bar{u} + b$$

Central measures

The mode, the median, or the mean can be used as an indication of the *central value* of a distribution.

Figure 1.7 shows the histogram for the tree heights given earlier, with the positions of the mode, median and mean marked on it.

This distribution is not symmetrical, and is said to be *positively skew*. Clearly the median and the mean are to the right of the mode; but note that the *extreme values* (say the trees with height more than 25 m) have a greater effect on the mean than on the median.

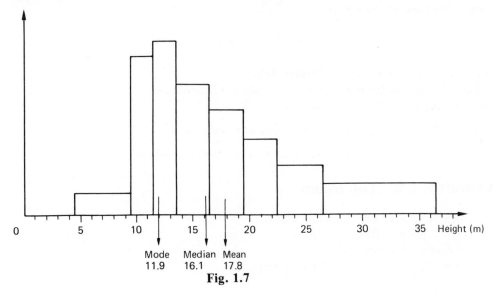

Mode 11.9   Median 16.1   Mean 17.8

Fig. 1.7

For a negatively skew distribution, the median and the mean are on the left of the mode.

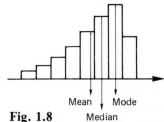

Mean · Mode

Median

Fig. 1.8

For a symmetrical distribution as shown, the mode, median and mean all coincide.

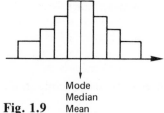

Mode
Median
Fig. 1.9   Mean

The mode is perhaps the simplest idea: for example, a shoe manufacturer would clearly wish to know the most popular shoe size. However, the mode does not always give a good indication of the centre of the distribution; it can be right at an end of the distribution (Fig. 1.10), and sometimes there is more than one mode (Fig. 1.11).

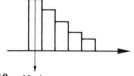

Fig. 1.10   Mode

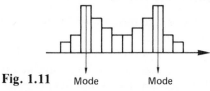

Fig. 1.11   Mode   Mode

The median very often gives the best indication when a 'typical' value is required.

The mean can be affected by the presence of a few extreme values. Nevertheless the mean is by far the most commonly used of the central measures. This is because it is a simple function of the values

$$\bar{x} = \frac{x_1 + x_2 + \cdots + x_n}{n}$$

and it is easy to deal with mathematically.

For a linear transformation $x = au + b$, the mode and the median behave in exactly the same way as the mean, for example,

if $u_1, u_2, ..., u_n$ have median M

then $x_1, x_2, ..., x_n$ have median $aM + b$

Exercise 1.3   The mean

1   Draw accurately on graph paper the histogram sketched in Fig. 1.12.

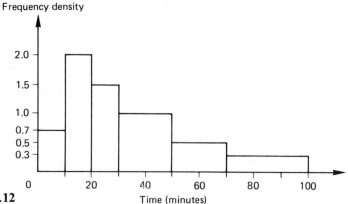

Fig. 1.12

(i) Use your histogram to find the mode and the median, and mark in these values on your histogram.

(ii) Derive the frequency table, and hence calculate the mean. Mark in the mean value on your histogram, and comment on the relative positions of the mode, median and mean.

2 The following *cumulative* frequency table refers to the ages of the members of an 'Under 30's' social club:

Age	Cumulative frequency	Age	Cumulative frequency
under 10	0	under 20	80
14	5	21	91
16	15	22	112
17	26	23	156
18	48	24	190
19	70	30	200

(i)   Draw a cumulative frequency curve and hence find the median.

(ii) Derive the frequency table, and hence calculate the mean.

(iii) Draw a histogram, and mark on it the mode, the median and the mean.

3 Estimate the mean of the following distribution of test marks:

Marks	1–10	11–20	21–30	31–40	41–50
Number of pupils	3	17	28	21	11

Why is your answer only an estimate?

What is (i) the greatest value, and (ii) the least value, that the mean could possibly have?

4 For the following distribution of weights

Weight (kg)	10–	20–	30–	40–	50–
Frequency	27	38	16	5	0

a student calculates the mean as $\dfrac{10 \times 27 + 20 \times 38 + 30 \times 16 + 40 \times 5}{86} = 19.88$

Criticise this calculation, and state the value of the mean which he should have obtained.

5   The following table gives the ages of 200 people:

Age	0–20	20–40	40–60	over 60
Number of people	60	55	45	40

(a) Estimate the mean age
    (i)   assuming that the mean age for those over 60 is 68 years
    (ii) assuming that the mean age for those over 60 is 78 years.
(b) Given that the mean age for these 200 people is 36.7 years, estimate the mean age
of those who are over 60.

6   Calculate the mean value of u from the table:

u	Frequency
−2	7
−1	19
0	37
1	26
2	15
3	6

Hence find the means of x, y, z and w given in the tables below:

(i)

x	Frequency
51	7
52	19
53	37
54	26
55	15
56	6

(ii)

y	Frequency
51	14
52	38
53	74
54	52
55	30
56	12

(iii)

z	Frequency
40	7
43	19
46	37
49	26
52	15
55	6

(iv)

w	Frequency
75.5	7
85.5	19
95.5	37
105.5	26
115.5	15
125.5	6

7 Calculate the mean mark from the following table

Marks in an Exam	10–19	20–29	30–39	40–49	50–59	60–69	70–79	80–89
Number of candidates	2	5	16	8	13	11	5	1

8 Calculate the mean length of the following

Length (nearest 0.01 mm)	35.26–35.30	35.31–35.35	35.36–35.40	35.41–35.45	35.46–35.50	35.51–35.55
Number of components	17	44	104	86	35	14

9 Calculate the mean age of the following 228 people

Age (in completed years)	Number of people	Age	Number
16–18	10	31–33	37
19–21	13	34–36	28
22–24	22	37–39	20
25–27	38	40–42	18
28–30	35	43–45	7

2
Measures of Spread

2.1 Interquartile range and mean deviation

If we are told, for example, only that the mean of a distribution is 36, this does not tell us very much; all the values might be close to 36, but they could be well spaced out.

We now consider ways of measuring how 'spread out' the values are.

Range

This is defined by

$$\text{range} = (\text{highest value}) - (\text{lowest value})$$

This is a very simple idea, but when there are a large number of values it is of little use because it depends only on 'extreme' values and not on 'typical' ones. For example, if a public mathematics examination (marked out of 100) is taken by several thousand candidates, the marks usually have range 100, because there is almost always somebody who scores 100, and somebody else who scores zero.

Interquartile range

The **lower quartile** is the value for which the cumulative frequency is one quarter of the total frequency (i.e. the 25th percentile).

The **upper quartile** is the value for which the cumulative frequency is three quarters of the total frequency (i.e. the 75th percentile).

> Interquartile range = (upper quartile) − (lower quartile)

We ignore the top quarter and the bottom quarter, and find the range of the 'middle half' of the distribution.

The quartiles may be found from a cumulative frequency diagram, by linear interpolation, or by calculating areas on a histogram.

Example 1
Find the interquartile range for the distribution of the tree heights given in Example 10 (p. 4).

We shall use the cumulative frequency curve drawn on p. 9. For the lower quartile $cf = \frac{1}{4} \times 400 = 100$, so the lower quartile is 12.2 m. For the upper quartile $cf = \frac{3}{4} \times 400 = 300$, so the upper quartile is 22.0 m.

Hence the interquartile range is

$$22.0 - 12.2 = 9.8 \text{ m}$$

Example 2

Using linear interpolation, find the interquartile range for the following distribution of marks:

Mark	1–30	31–60	61–90	91–120	121–150	151–180
Number of students	3	9	20	22	13	8
Cumulative frequency	3	12	32	54	67	75

Remember that cumulative frequency is plotted at the upper class boundaries, which are 30.5, 60.5, and so on.

For the lower quartile (Fig. 2.1), cf $= \frac{1}{4} \times 75 = 18.75$

$$\frac{x}{30} = \frac{6.75}{20}$$

$$x = 10.125$$

the lower quartile is $60.5 + 10.125 = 70.625$

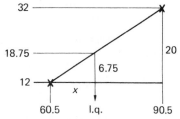

Fig. 2.1

For the upper quartile (Fig. 2.2), cf $= \frac{3}{4} \times 75 = 56.25$

$$\frac{y}{30} = \frac{2.25}{13}$$

$$y = 5.192$$

the upper quartile is $120.5 + 5.192 = 125.692$.
The interquartile range is $125.692 - 70.625 \approx 55.1$ marks

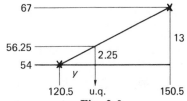

Fig. 2.2

Most measures of spread give an indication of the 'typical' distance of a value from the 'centre'.

For this reason, the **semi-interquartile range**, which is half of the interquartile range, is sometimes used (Fig. 2.3).

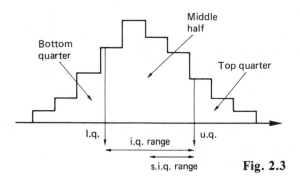

Fig. 2.3

For the marks given in Example 2, the semi-interquartile range is

$$\tfrac{1}{2} \times 55.1 \approx 27.5 \text{ marks.}$$

Mean deviation

Consider the five numbers (x): $3, 7, 8, 21, 35$

The mean is $\bar{x} = \dfrac{\sum x}{n} = \dfrac{74}{5} = 14.8$

If we take the mean as the 'centre', the 'deviations' of the five numbers from the mean are

$$(x - \bar{x}): \ -11.8, \ -7.8, \ -6.8, \ +6.2, \ +20.2$$

Some of these are positive and some are negative, and in fact the average of these deviations is always zero.

If we ignore the signs we obtain

$$|x - \bar{x}|: \ 11.8, \ 7.8, \ 6.8, \ 6.2, \ 20.2.$$

The **mean deviation** is the average of these:

$$\frac{11.8 + 7.8 + 6.8 + 6.2 + 20.2}{5} = 10.56$$

We have

$$\boxed{\text{mean deviation} = \frac{\sum |x - \bar{x}|}{n} \ \text{or} \ \frac{\sum |x - \bar{x}| f}{\sum f}}$$

Example 3
Find the mean deviation for the distribution of marks given in Example 2.

We found earlier that the mean is $\bar{x} = 98.3$. This enables us to construct Table 2.1.

Table 2.1

Marks	Frequency f	Mid-interval value x	$x - \bar{x}$	$|x - \bar{x}| f$
1–30	3	15.5	−82.8	248.4
31–60	9	45.5	−52.8	475.2
61–90	20	75.5	−22.8	456
91–120	22	105.5	7.2	158.4
121–150	13	135.5	37.2	483.6
151–180	8	165.5	67.2	537.6
	$\sum f = 75$			$\sum |x - x| f = 2359.2$

$$\text{mean deviation} = \frac{2359.2}{75} \approx 31.5 \text{ marks}$$

Linear functions

Example 4
The five numbers 3, 7, 8, 21, 35 have mean deviation 10.56
Deduce the mean deviation of
(i)   203, 207, 208, 221, 235
(ii)   0.6, 1.4, 1.6, 4.2, 7
(iii) 5.03, 5.07, 5.08, 5.21, 5.35.

(i)   Adding 200 to all the numbers does not affect their 'spread' (the deviation of each value from the mean remains the same); so the mean deviation is 10.56.
(ii)   We have multiplied all the numbers by 0.2; so their 'spread' will also be multiplied by 0.2; the mean deviation is $0.2 \times 10.56 = 2.112$
(iii) We have multiplied the numbers by 0.01 (which multiplies the mean deviation by 0.01), and then added 5 (which does not affect the mean deviation). So the mean deviation is $0.01 \times 10.56 = 0.1056$.

In general if numbers (u) have mean deviation d, and numbers (x) are defined by $x = au + b$ (where a and b are constants, and a is positive), then the mean deviation of (x) is ad.
If values (x) are 'coded' as $u = \dfrac{x - b}{a}$ (so that $x = au + b$) and the mean deviation of the u-values is calculated as $d = \dfrac{\sum |u - \bar{u}| f}{\sum f}$ then the mean deviation of the original values (x) is ad.
Similar results hold for *all* measures of spread (for example, the interquartile range).

Exercise 2.1   Interquartile range and mean deviation

1 Two dice were thrown repeatedly, and the following total scores were obtained:
   8, 2, 7, 3, 10, 8, 8, 5, 12.
   (i)   State the range of these scores.
   (ii)   Calculate the mean, and find the deviation of each value from the mean.
   (iii) Calculate the mean deviation.

2 The histogram in Fig. 2.4 illustrates the growth of some young trees over one year.
   (i)   State the range of these growths.
   (ii)   Find the quartiles. Copy the histogram and mark the quartiles on it. State the interquartile range.

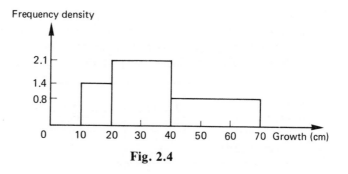

Fig. 2.4

   (iii) How many trees have grown an amount which is between the lower quartile and the upper quartile?

3   For the distribution of ages given in Exercise 1.3, Question 2, use your cumulative frequency curve to find the upper and lower quartiles, and mark them on the histogram. State the interquartile range.

4   For the temperatures given in Exercise 1.2, Question 1, use your cumulative frequency curve to find the interquartile range.

5   For the ages given in Exercise 1.2, Question 3, use your cumulative probability curve to find the interquartile range.

6   The following table shows the gross earnings of all men in the UK in 1980.

Gross earnings less than	£2000	£3000	£3500	£4000	£4500	£5000	£5500	£6000	£7000	£8000	£10000
Percentage of men	0.8	7.8	17.1	29.6	44.2	58.0	69.4	77.8	85.0	93.1	97.1

Draw a curve showing these cumulative percentages, and use it to estimate the median earnings, and the interquartile range.

7   The 128 competitors in a golf tournament had the following total scores for four rounds. Use linear interpolation to find the interquartile range.

Score	261–264	265–268	269–272	273–276	277–280	281–284	285–288	289–292	293–296	297–300	301–304	305–308	309–312	313–316
Number of competitors	1	0	2	1	5	10	16	24	23	28	14	3	0	1

8   Nominal 13 A fuses are tested by passing a gradually increasing electric current through them until they blow. The current which caused them to blow is shown in the following table. Use linear interpolation to find the interquartile range.

Current (nearest 0.1 A)	12.5–12.9	13.0–13.4	13.5–13.9	14.0–14.4	14.5–14.9
Number of fuses	3	15	36	44	57
Current	15.0–15.4	15.5–15.9	16.0–16.4	16.5–16.9	17.0–
Number	49	21	7	2	9

9 30 people were each asked 5 general knowledge questions, and the numbers of correct answers they gave are shown below

No of correct answers (x)	Number of people (f)	xf	$x - \bar{x}$	$\lvert x - \bar{x} \rvert f$
0	3			
1	5			
2	8			
3	7			
4	5			
5	2			
	$\sum f = 30$	$\sum xf =$		$\sum \lvert x - \bar{x} \rvert f =$

Complete the xf column and calculate the mean, $\bar{x} = \dfrac{\sum xf}{\sum f}$

Complete the $(x - \bar{x})$ and $\lvert x - \bar{x} \rvert f$ columns, and hence calculate the mean deviation, $\dfrac{\sum \lvert x - \bar{x} \rvert f}{\sum f}$

10 For the following distribution of lengths, complete the blank columns and calculate the mean and the mean deviation of the coded values (u). Deduce the mean and the mean deviation of the lengths (x).

Lengths (nearest 0.01 mm)	Number of components f	Mid-interval value x	$u = \dfrac{x - 35.38}{0.05}$	uf	$u - \bar{u}$	$\lvert u - \bar{u} \rvert f$
35.26–35.30	17	35.28	−2			
35.31–35.35	44	35.33	−1			
35.36–35.40	104	35.38	0			
35.41–35.45	86	35.43	1			
35.46–35.50	35	35.48	2			
35.51–35.55	14	35.53	3			

2.2 Standard deviation

The mean deviation may appear to be a good measure of the spread of a distribution, but in fact it is rarely used, because the modulus function $\lvert x - \bar{x} \rvert$ is difficult to handle mathematically.

Consider again the five numbers (x): 3, 7, 8, 21, 35, with mean $\bar{x} = 14.8$ and deviations $(x - \bar{x})$: -11.8, -7.8, -6.8, 6.2, 20.2

Another way of dealing with the negative signs is to square the deviations, to obtain $(x - \bar{x})^2$: 139.24, 60.84, 46.24, 38.44, 408.04.

We define the *variance*, or mean squared deviation, to be the average of these:

$$\frac{139.24 + \cdots + 408.04}{5} = 138.56$$

As the deviations were squared, we should now take the square root to obtain a reasonable measure of the spread in the right units; we define the *standard deviation*, or root mean squared deviation, to be the square root of the variance.

The standard deviation of the five numbers above is thus

$$s = \sqrt{138.56} = 11.77$$

We have

$$\text{Standard deviation, } s = \sqrt{\frac{\sum (x - \bar{x})^2}{n}} \quad \text{or} \quad \sqrt{\frac{\sum (x - \bar{x})^2 f}{\sum f}}$$

$$\text{variance} = s^2$$

Example 1

Using the formula $s = \sqrt{\dfrac{\sum (x - \bar{x})^2 f}{\sum f}}$, find the standard deviation for the following distribution of marks

Mark	1–30	31–60	61–90	91–120	121–150	151–180
Number of students	3	9	20	22	13	8

We know that the mean is $\bar{x} = 98.3$.

Mark (mid-interval value) x	Frequency f	$x - \bar{x}$	$(x - \bar{x})^2 f$
15.5	3	−82.8	20 567.52
45.5	9	−52.8	25 090.56
75.5	20	−22.8	10 396.8
105.5	22	7.2	1 140.48
135.5	13	37.2	17 989.92
165.5	8	67.2	36 126.72
	$\sum f = 75$		$\sum (x - \bar{x})^2 f = 111\ 312$

The standard deviation is $s = \sqrt{\dfrac{111\ 312}{75}} = 38.525 \approx 38.5$ marks

Alternative formula for calculating the standard deviation

For numbers $x_1, x_2, ..., x_n$, with mean $\bar{x} = \dfrac{\sum\limits_{i=1}^{n} x_i}{n}$ the variance is

$$s^2 = \frac{\sum\limits_{i=1}^{n} (x_i - \bar{x})^2}{n} = \frac{\sum\limits_{i=1}^{n} (x_i^2 - 2\bar{x}x_i + \bar{x}^2)}{n} = \frac{\sum\limits_{i=1}^{n} x_i^2}{n} - 2\bar{x}\frac{\sum\limits_{i=1}^{n} x_i}{n} + \frac{n\bar{x}^2}{n}$$

$$= \frac{\sum\limits_{i=1}^{n} x_i^2}{n} - 2\bar{x}^2 + \bar{x}^2 = \frac{\sum\limits_{i=1}^{n} x_i^2}{n} - \bar{x}^2$$

Hence

$$\text{Standard deviation } s = \sqrt{\frac{\sum x^2}{n} - \bar{x}^2}$$

$$\text{or} \quad \sqrt{\frac{\sum x^2 f}{\sum f} - \bar{x}^2}$$

$$= \sqrt{\frac{\sum x^2 f}{\sum f} - \left(\frac{\sum xf}{\sum f}\right)^2}$$

$$\text{variance} = s^2$$

This alternative formula simplifies the calculation of the standard deviation for a set of data, since there is no longer any need to find the deviations $(x - \bar{x})$.

Example 2
Calculate the mean and the standard deviation of the five numbers 3, 7, 8, 21, 35.

$$\text{We have } \sum x = 74 \text{ and } \sum x^2 = 3^2 + 7^2 + \cdots + 35^2 = 1788$$

$$\text{Mean } \bar{x} = \frac{74}{5} = 14.8$$

$$\text{Standard deviation } s = \sqrt{\frac{1788}{5} - \left(\frac{74}{5}\right)^2} = 11.77 \text{ (as before)}$$

Example 3
Find the mean and the standard deviation for the distribution of marks given in Example 1.

Mark	Frequency f	Mid-interval value x	xf	$x^2 f$
1–30	3	15.5	46.5	720.75
31–60	9	45.5	409.5	18 632.25
61–90	20	75.5	1 510	114 005
91–120	22	105.5	2 321	244 865.5
121–150	13	135.5	1 761.5	238 683.25
151–180	8	165.5	1 324	219 122
	$\sum f = 75$		$\sum xf = 7\ 372.5$	$836\ 028.75 = \sum x^2 f$

$$\text{Mean } \bar{x} = \frac{7372.5}{75} = 98.3; \text{ standard deviation } s = \sqrt{\frac{836028.75}{75} - \left(\frac{7372.5}{75}\right)^2}$$

$$= 38.525 \text{ (as before)}$$

When the formula

$$s = \sqrt{\frac{\sum x^2}{n} - \bar{x}^2}$$

is used to calculate the standard deviation, the result can be seriously affected by rounding errors, especially if the mean is large compared with the standard deviation. It is important to use the most accurate value available for the mean $\bar{x}$.

Example 4
A set of 28 numbers has $\sum x = 1893$

$$\text{and } \sum x^2 = 127994$$

Calculate the mean and the standard deviation.

The mean is $\bar{x} = \dfrac{1893}{28} \approx 67.6$

The standard deviation is $s = \sqrt{\dfrac{127994}{28} - \left(\dfrac{1893}{28}\right)^2} \approx 0.699$

The accurate value $\bar{x} = \dfrac{1893}{28}$ was used in the calculation of s.

Note that

$$\sqrt{\frac{127994}{28} - 67.6^2} \approx 1.206$$

$$\sqrt{\frac{127994}{28} - 67.61^2} \approx 0.320$$

$$\sqrt{\frac{127994}{28} - 67.607^2} \approx 0.713.$$

Use of calculators

A calculator having statistics functions may be used to find the mean and standard deviation of a set of numbers quickly.

Calculators vary, but typically the procedure is as follows:
For a set of numbers such as 3, 7, 8, 21, 35,
put the calculator into 'STAT' or 'SD' mode;
enter the numbers

3 | DATA | 7 | DATA | 8 | DATA | 21 | DATA | 35 | DATA

press the mean key (usually labelled $\bar{x}$) to obtain the mean (14.8). Most calculators have two standard deviation keys (with labels such as s, σ, s', σ_n, σ_{n-1}). Discover which key gives the correct standard deviation (11.77115...), and always use this key when finding a standard deviation.

If you now wish to find the mean and standard deviation of a different set of numbers, remember to clear the calculator first (otherwise the new set of numbers will be treated as a continuation of the previous set), and reset into 'STAT' mode (if necessary).

For values given in a frequency table, for example

Mark (mid-interval value)	15.5	45.5	75.5	105.5	135.5	165.5
Frequency	3	9	20	22	13	8

enter the values and frequencies by a sequence such as

$$15.5 \quad \boxed{\times} \quad 3 \quad \boxed{\text{DATA}}$$

$$45.5 \quad \boxed{\times} \quad 9 \quad \boxed{\text{DATA}}$$

$$\cdots\cdots$$

$$165.5 \quad \boxed{\times} \quad 8 \quad \boxed{\text{DATA}}$$

then obtain the mean (98.3) and the standard deviation (38.52479...) as before.
The calculator uses the formulae

$$\bar{x} = \frac{\sum xf}{\sum f}, \quad s = \sqrt{\frac{\sum x^2 f}{\sum f} - \left(\frac{\sum xf}{\sum f}\right)^2}$$

(as in Example 3) and the totals $\sum f = 75$, $\sum xf = 7372.5$, $\sum x^2 f = 836\,028.75$ can usually be obtained by pressing keys labelled 'n', '$\sum x$' and '$\sum x^2$'.

When a calculator is used in an examination, the expression being calculated should be stated. We could write it out as follows:

$$\sum f = 3 + 9 + \cdots = 75$$

$$\sum xf = 15.5 \times 3 + 45.5 \times 9 + \cdots = 7372.5$$

$$\sum x^2 f = 15.5^2 \times 3 + 45.5^2 \times 9 + \cdots = 836\,028.75$$

$$\bar{x} = \frac{7372.5}{75} = 98.3 \qquad s = \sqrt{\frac{836\,028.75}{75} - \left(\frac{7372.5}{75}\right)^2} = 38.525$$

or alternatively as

$$\bar{x} = \frac{15.5 \times 3 + 45.5 \times 9 + \cdots}{75} = 98.3$$

$$s = \sqrt{\frac{15.5^2 \times 3 + 45.5^2 \times 9 + \cdots}{75} - \left(\frac{15.5 \times 3 + 45.5 \times 9 + \cdots}{75}\right)^2} = 38.525$$

Interpretation of the standard deviation

Example 5
Find the mean and the standard deviation of the following tree heights, and estimate the percentage of trees having heights
(i)   within one standard deviation of the mean height

(ii) within two standard deviations of the mean height
(iii) within three standard deviations of the mean height.

Height (nearest metre)	5–9	10–11	12–13	14–16	17–19	20–22	23–26	27–36
Mid-interval value (x)	7.0	10.5	12.5	15.0	18.0	21.0	24.5	31.5
Frequency (f)	18	58	62	72	57	42	36	55

We have $\sum f = 18 + 58 + \cdots = 400$

$$\sum xf = 7.0 \times 18 + 10.5 \times 58 + \cdots = 7112.5$$

$$\sum x^2 f = 7.0^2 \times 18 + 10.5^2 \times 58 + \cdots = 146\,336.75$$

thus mean $\bar{x} = \dfrac{7112.5}{400} = 17.781 \approx 17.8$ m

$$\text{standard deviation } s = \sqrt{\frac{146\,336.75}{400} - \left(\frac{7112.5}{400}\right)^2} = 7.048 \approx 7.0 \text{ m}$$

(i) Heights which are one standard deviation each side of the mean are 17.781 ± 7.048, i.e. 10.73 and 24.83 m.
To estimate the number of trees between these heights, we can use cumulative frequencies, or we can work directly from the frequency table as follows
10.73 lies in the class 9.5 to 11.5, which contains 58 trees
so between 10.73 and 11.5 m there are approximately $\dfrac{11.5 - 10.73}{2} \times 58 = 22.3$ trees.
Between 11.5 and 22.5 m there are $62 + 72 + 57 + 42 = 233$ trees.
Between 22.5 and 24.83 m there are approximately $\dfrac{24.83 - 22.5}{4} \times 36 = 21.0$ trees.
Hence the number of trees with heights between 10.73 and 24.83 m is approximately $22.3 + 233 + 21.0 = 276.3$, and the percentage of trees is $\dfrac{276.3}{400} \times 100 = 69.1\%$.

(ii) Heights which are two standard deviations each side of the mean are 17.781 ± 14.096, i.e. 3.69 and 31.88 m.
The only trees *outside* these limits are those with heights more than 31.88 m; their number is approximately $\dfrac{36.5 - 31.88}{10} \times 55 = 25.4$.
Hence the number of trees with heights within two standard deviations of the mean is $400 - 25.4 = 374.6$, or 93.6%.

(iii) Heights which are three standard deviations each side of the mean are 17.781 ± 21.144, i.e. -3.36 and 38.92 m.
Hence 100% of the trees have heights within three standard deviations of the mean.

For most distributions which are likely to be encountered in practice, it is found that
About $\frac{2}{3}$ of the values lie within one standard deviation of the mean

i.e. between $\bar{x} - s$ and $\bar{x} + s$

About 95% of the values lie within two standard deviations of the mean

i.e. between $\bar{x} - 2s$ and $\bar{x} + 2s$

Almost all the values lie within three standard deviations of the mean

i.e. between $\bar{x} - 3s$ and $\bar{x} + 3s$

The exact proportions depend on the shape of the distribution; the guidelines given above are very reliable for approximately symmetrical distributions having the usual 'bell' shape sketched in Fig 2.5, and Example 5 shows that they are approximately correct even for the markedly skew distribution of tree heights.

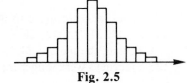

Fig. 2.5

If we are told, for example, that a distribution has mean 36 and standard deviation 5, we can reasonably expect that

about $\frac{2}{3}$ of the values are between 31 and 41
about 95% of the values are between 26 and 46
and almost all the values are between 21 and 51.

However, it is possible to construct distributions for which the proportions differ considerably from those given above.

Exercise 2.2   Standard deviation

1   Complete the following table, and hence calculate the mean and standard deviation of the number of letters received per day.

No of letters x	No of days f	xf	$x - \bar{x}$	$(x - \bar{x})^2$	$(x - \bar{x})^2 f$
0	17				
1	12				
2	6				
3	4				
4	1				

2   You are given the set of numbers 3, 11, 7, 5, 15, 12, 9, 4.
(i)   Calculate the mean $\bar{x}$.
(ii) Calculate the deviations from the mean, and use the formula

$$\sqrt{\frac{\sum (x - \bar{x})^2}{n}}$$

to calculate the standard deviation.
(iii) Find the sum of squares $\sum x^2$, and use the formula

$$\sqrt{\frac{\sum x^2}{n} - \bar{x}^2}$$

to calculate the standard deviation.

You may use your calculator statistics functions in the following questions.

3   Calculate the mean and the standard deviation of the following weights (in kg):

3.5, 1.8, 4.3, 1.7, 2.9, 6.2, 4.0, 3.3, 2.2, 4.1.

How many of these weights are
(i) within one standard deviation of the mean?
(ii) within two standard deviations of the mean?

4   The times taken to drive to work on seven consecutive mornings were (in minutes):
11, 20, 14, 31, 22, 12, 28.
     Calculate the mean and standard deviation of these journey times. How many of these journeys were longer than the mean time by more than one standard deviation?

5   At a race meeting I bet on six races and my winnings were:

$$£5.50, \ -£2, \ -£5, \ £23.30, \ -£10, \ -£7$$

(where a negative amount indicates a loss). Calculate the mean and standard deviation of the amounts I won.

6   A set of 120 numbers has $\sum x = 3460$, $\sum x^2 = 121\ 440$.
Calculate the mean and the standard deviation.

7   A bridge player recorded the number of spades he had in 50 hands as follows:

3, 3, 1, 2, 6, 2, 2, 2, 4, 4, 2, 3, 4, 3, 3, 7, 5, 4, 3, 5, 2, 5, 6, 4, 4
5, 1, 3, 2, 3, 0, 3, 3, 6, 6, 2, 4, 1, 6, 2, 2, 2, 2, 5, 5, 5, 2, 6, 7, 2

Form a frequency table, and calculate the mean and the standard deviation.

8   The following table shows the length of time for which 250 young people have been unemployed.

Time (in weeks)	0–2	2–4	4–8	8–13	13–26	26–52	52–104
No of people	22	18	31	30	47	46	56

Calculate the mean time, and the standard deviation. Draw a histogram to illustrate the data, and mark on it the mean, and the times which are one standard deviation each side of the mean.
     Estimate the percentage of these people who have been unemployed for a time which is within one standard deviation of the mean time.

9   Three methods are available for finding the alcohol level in samples of blood. 15 samples, each known to contain exactly 80 mg per 100 ml were sent for analysis, five by each method, with the following results:

Method A: 78.4, 85.9, 87.0, 82.3, 89.6
Method B: 100.8, 82.0, 68.4, 60.0, 88.8
Method C: 72.1, 72.6, 74.0, 73.4, 72.9

Calculate the mean and standard deviation for each of the Methods A, B, and C. Which method would you recommend, and why?

10 For the following distribution show that 5% of the values are more than four standard deviations from the mean. What is unusual about this?

x	1	2	3	4	5
f	2	0	76	0	2

2.3 Linear functions and coding

If numbers $(x_1, x_2, ..., x_n)$ are related to numbers $(u_1, u_2, ..., u_n)$ by $x = au + b$, then, as with other measures of spread, the standard deviations s_x and s_u are related by $s_x = as_u$.

To prove this, we have $\bar{x} = a\bar{u} + b$, and so

$$s_x^2 = \frac{\sum_{i=1}^{n}(x_i - \bar{x})^2}{n} = \frac{\sum_{i=1}^{n}\{(au_i + b) - (a\bar{u} + b)\}^2}{n}$$

$$= \frac{\sum_{i=1}^{n}\{a(u_i - \bar{u})\}^2}{n} = a^2\frac{\sum_{i=1}^{n}(u_i - \bar{u})^2}{n} = a^2 s_u^2$$

Hence $s_x = as_u$

> If $x = au + b$, then mean $\bar{x} = a\bar{u} + b$
> variance $s_x^2 = a^2 s_u^2$
> standard deviation $s_x = as_u$

Note We have assumed that a is positive.

For any a (positive or negative), the results $\bar{x} = a\bar{u} + b$ and $s_x^2 = a^2 s_u^2$ are still true, but when we take the square root we obtain $s_x = |a|s_u$

For example, if $x = 7 - 2u$, then $\bar{x} = 7 - 2\bar{u}$

$$s_x^2 = 4 s_u^2$$

$$s_x = 2 s_u$$

Coding

If the original values (x) are coded as $u = \dfrac{x - b}{a}$, (so that $x = au + b$) then

$$\bar{x} = a\bar{u} + b = a\frac{\sum uf}{\sum f} + b$$

and

$$s_x = a s_u = a\sqrt{\frac{\sum u^2 f}{\sum f} - \left(\frac{\sum uf}{\sum f}\right)^2}$$

The use of coding can greatly simplify the arithmetic. It can also reduce the effects of rounding errors.

Example 1

Find the mean and the standard deviation for the distribution of marks given in Example 1 on page 28.

Mark	Frequency f	Mid-interval value x	$u = \dfrac{x - 75.5}{30}$	uf	u^2f
1–30	3	15.5	-2	-6	12
31–60	9	45.5	-1	-9	9
61–90	20	75.5	0	0	0
91–120	22	105.5	1	22	22
121–150	13	135.5	2	26	52
151–180	8	165.5	3	24	72
$\sum f = 75$				$\sum uf = 57$	$167 = \sum u^2f$

$$\text{mean } \bar{x} = 30 \times \frac{57}{75} + 75.5 = 98.3$$

$$\text{standard deviation } s = 30 \sqrt{\frac{167}{75} - \left(\frac{57}{75}\right)^2} = 38.525 \text{ (as before)}$$

Measures of dispersion

For the distribution of marks given in Example 1, we have now calculated three different measures of the 'spread' of the marks:

the semi-interquartile range is 27.5
the mean deviation is   31.5
the standard deviation is   38.5

The standard deviation is always greater than (or equal to) the mean deviation, and the mean deviation is usually greater than the semi-interquartile range.

These are sometimes called 'measures of dispersion'. The standard deviation is by far the most commonly used.

Variance

The variance is simply the square of the standard deviation, but take care not to confuse the two.

Variance is measured in units which are the square of those in which the original variable is measured.

Remember that if the values are multiplied by a constant k, then the variance will be multiplied by k^2.

Example 2

The amounts of oil consumed by a central heating system in six successive weeks were 14.2, 9.8, 9.5, 16.0, 12.2, 10.4 gallons.

(i) Find the mean and variance of these amounts.

(ii) Deduce the mean and the variance, if the amount is measured in litres (1 gallon = 4.546 litres).

(i) We have $\sum x = 72.1$, $\sum x^2 = 900.93$

Mean $\bar{x} = \dfrac{72.1}{6} = 12.02$ gal

$$\text{Variance} = \dfrac{900.93}{6} - \left(\dfrac{72.1}{6}\right)^2 = 5.755 \text{ gal}^2$$

(ii) To convert the amounts to litres, we multiply by 4.546

Hence,   mean $= 12.02 \times 4.546 = 54.6\,\ell$

variance $= 5.755 \times 4.546^2 = 118.9\,\ell^2$

Exercise 2.3   Linear functions and coding

1   The numbers 1, 2, 3, ..., 100 have mean 50.5 and standard deviation 28.87. State the mean and the standard deviation of:

(i)   the numbers 51, 52, 53, ..., 150
(ii) the numbers 3, 6, 9, ..., 300
(iii) the numbers 3, 5, 7, ..., 201
(iv) the numbers 5.6, 15.6, 25.6, ..., 995.6
(v)   the numbers 4.01, 4.02, 4.03, ..., 5.00.

2   Calculate the mean and standard deviation of u from the table:

u	-2	-1	0	1	2	3
Frequency	7	19	37	26	15	6

Hence find the mean and standard deviation of $w, x, y,$ and z given in the following tables:

(i)

w	114	115	116	117	118	119
Frequency	7	19	37	26	15	6

(ii)

x	114	115	116	117	118	119
Frequency	21	57	111	78	45	18

(iii)

y	62	67	72	77	82	87
Frequency	7	19	37	26	15	6

(iv)

z	254.5	274.5	294.5	314.5	334.5	354.5
Frequency	7	19	37	26	15	6

3 Using the coding $u = \dfrac{x - 45}{10}$, calculate the mean and standard deviation of the ages given below.

Age x (in completed yrs)	0–9	10–19	20–29	30–39	40–49	50–59	60–69	70–79	80–89
Number of people	5	13	24	35	40	36	22	16	9

Estimate the number of people whose ages are within two standard deviations of the mean age.

4 Using a suitable coding, calculate the mean and standard deviation of the following lengths.

Lengths (nearest 0.01 mm)	35.26–35.30	35.31–35.35	35.36–35.40	35.41–35.45	35.46–35.50	35.51–35.55
Number of components	17	44	104	86	35	14

5 Calculate the mean and standard deviation of the following weights

Weight of pebble (grams)	10–15	15–20	20–25	25–30	30–35	35–40	40–45	45–50
Number of pebbles	7	21	38	46	50	54	18	2

Later it is discovered that the balance was faulty, and all the above weights are 3 g too high. Estimate the true mean and standard deviation of the pebble weights.

6 For the following distribution

No of job applications	1–3	4–6	7–9	10–12	13–15	16–18	19–21
Number of people	24	22	8	4	1	0	1

find (i)   the semi-interquartile range
 (ii) the mean deviation
 (iii) the standard deviation,
and comment on the relative sizes of these three measures of spread.

7 Calculate the mean and variance of the following lengths:

4.2, 5.3, 4.8, 6.2, 3.7, 5.0 metres

and deduce the mean and variance of the following:
(i)   42, 53, 48, 62, 37, 50 km
(ii)   34.2, 35.3, 34.8, 36.2, 33.7, 35.0 cm
(iii) 0.84, 1.06, 0.96, 1.24, 0.74, 1.00 m
(iv) 6420, 6530, 6480, 6620, 6370, 6500 mm

8 Calculate the values of the mean and variance of the following speeds

Speed (m s^{-1})	11–15	16–20	21–25	26–30	31–35	36–40
Number of cars	12	34	42	40	24	8

State the values of the mean and variance when the speed is measured in terms of km and hours.

2.4 Standard scores

Consider a set of values (x) with mean $\bar{x}$ and standard deviation s. The deviation $(x - \bar{x})$ gives the distance of a value x from the mean $\bar{x}$. The *standard score*, defined by $z = \dfrac{x - \bar{x}}{s}$ gives the number of standard deviations from the mean for a particular value x.
   For example, if a set of examination marks has mean 54 and standard deviation 15, then a mark of 36 has standard score $z = \dfrac{36 - 54}{15} = -1.2$, indicating that this mark is 1.2 standard deviations *below* the mean.
   Standard scores are useful when comparing values from different distributions, and when scaling values so as to have a new mean and standard deviation.

Example 1

The third form pupils at a school sat examinations in mathematics and English. For mathematics, the marks had mean 46 and standard deviation 18. For English, the marks had mean 62 and standard deviation 7.

One pupil scored 91 in mathematics and 80 in English. Which of these is the better mark, relative to the overall performance of the pupils in that school?

91 marks in mathematics has standard score $z = \dfrac{91 - 46}{18} = 2.5$

80 marks in English has standard score $z = \dfrac{80 - 62}{7} = 2.571$

Hence the English mark is better (since it has the higher standard score).

Example 2

The marks in an examination had mean 61.8 and standard deviation 9.5. All the marks are to be re-scaled so as to have a mean of 50 and a standard deviation of 15.

What would the scaled mark be for a candidate who scored 44 in the examination?

The original mark 44 has standard score $z = \dfrac{44 - 61.8}{9.5} = -1.874$

This mark is 1.874 standard deviations below the mean, so the re-scaled mark would be $50 - 1.874 \times 15 = 21.9$.

Usually (for about 95% of values) the standard scores are between -2 and $+2$. It is very unusual for a standard score to be outside the range -3 to $+3$.

Note that if the values (x) have mean $\bar{x}$ and standard deviation s, the deviations ($x - \bar{x}$) have mean $\bar{x} - \bar{x} = 0$, and standard deviation s, and so $z = \dfrac{x - \bar{x}}{s}$ have mean $\dfrac{0}{s} = 0$, and standard deviation $\dfrac{s}{s} = 1$.

Hence the standard scores themselves (z) have mean 0 and standard deviation 1.

The process of converting values to standard scores is called *standardising*.

Exercise 2.4   Standard scores

1   A girl obtained the following marks in her end-of-term examinations. The mean and standard deviation of the marks of her class are given in brackets for each subject:

$$
\begin{aligned}
&\text{English} &&: 53 &&(\text{mean } 47,\ \text{standard deviation } 8)\\
&\text{history} &&: 62 &&(\text{mean } 55,\ \text{standard deviation } 12)\\
&\text{French}   &&: 73 &&(\text{mean } 50,\ \text{standard deviation } 13)\\
&\text{maths}   &&: 87 &&(\text{mean } 51,\ \text{standard deviation } 21)\\
&\text{physics} &&: 59 &&(\text{mean } 64,\ \text{standard deviation } 10)
\end{aligned}
$$

(i) Standardise each of her marks. Comment on her general performance (in comparison with the others in her class). Which is her best subject, and which is her worst?

(ii) A boy in the same class scored 33 in English and 46 in physics. Which is the better of these two marks?

2   The marks in a mathematics examination had mean 42.3 and standard deviation 15.8. All the marks are to be scaled so as to have a mean of 50 and a standard deviation of 20.

Calculate the scaled marks for candidates having the following original marks:

(a) 64   (b) 29   (c) 44   (d) 92   (e) 10

3   Find the mean and standard deviation of the six numbers

$$3, 14, 15, 9, 26, 5$$

Express each number as a standard score (z).
Find $\sum z$ and $\sum z^2$, and deduce the mean and standard deviation of the z-scores. Rescale the six numbers so as to have mean 40 and standard deviation 15.

4   When a fair coin is tossed 100 times, the number of heads obtained has mean 50 and standard deviation 5.

Would you suspect that a coin might be biased if 100 tosses produced

(a) 58 heads   (b) 64 heads   (c) 29 heads   (d) 40 heads?

5   In a talent competition, the three judges awarded a mark (out of 100) for each act. The five finalists were A, B, C, D, E, and the results were as follows.

	1st judge	2nd judge	3rd judge	Total	Position
A	50	82	70	202	2nd
B	72	75	40	187	4th
C	67	79	52	198	3rd
D	68	80	25	173	5th
E	48	74	92	214	1st

(i)   How did the 3rd judge ensure that the final placings coincided with his own views?
(ii) For the 1st judge, calculate the mean and standard deviation of his marks, and hence standardise his marks. Repeat for the 2nd and 3rd judges.
(iii) Revise the results of the competition, using the standardised marks. Explain why this is fairer.

2.5 Combining samples

If we know the mean and the standard deviations of two (or more) separate samples, we can calculate the mean and the standard deviation of the single sample obtained by combining them.

We use the formulae

$$\bar{x} = \frac{\sum x}{n} \quad \text{and} \quad s = \sqrt{\frac{\sum x^2}{n} - \bar{x}^2}$$

together with their rearrangements $\sum x = n\bar{x}$, $\sum x^2 = n(s^2 + \bar{x}^2)$

Example 1

For a sample of 60 men, the mean weight is 72 kg with standard deviation 5 kg. For a sample of 90 women, the mean weight is 58 kg with standard deviation 7 kg. Calculate the mean and the standard deviation of the weights of the combined sample of 150 people.

For the 60 men we have $\qquad \sum x = 60 \times 72 = 4320$

and $\qquad \sum x^2 = 60(5^2 + 72^2) = 312\,540$

For the 90 women we have $\qquad \sum x = 90 \times 58 = 5220$

$\qquad \sum x^2 = 90(7^2 + 58^2) = 307\,170$

Hence for all 150 people, $\qquad \sum x = 4320 + 5220 = 9540$

$\qquad \sum x^2 = 312\,540 + 307\,170 = 619\,710$

so the mean is $\dfrac{9540}{150} = 63.6$ kg

and the standard deviation is $\sqrt{\dfrac{619\,710}{150} - \left(\dfrac{9540}{150}\right)^2} = 9.30$ kg,

Note that the mean of the combined sample is simply $\dfrac{60 \times 72 + 90 \times 58}{150}$, and this necessarily lies between the means of the two separate samples. However, the calculation of the standard deviation is more complicated, and (as in this case), it does not necessarily lie between the standard deviations of the separate samples.

Exercise 2.5   Combining samples

1   A set of 10 numbers (x) has $\sum x = 53$, $\sum x^2 = 330$. Calculate the mean and standard deviation of these 10 numbers.

Two more numbers, 12 and 8, are added to the set. State the new values of $\sum x$ and $\sum x^2$ and calculate the mean and standard deviation of the new set of 12 numbers.

2   The weights (w kg) of 20 children have mean 28 kg and standard deviation 8 kg. Find the values of $\sum w$ and $\sum w^2$. Given that one more child of weight 35 kg is added, find the mean and standard deviation of the 21 children.

3   In a mathematics test, the marks of the top set (15 pupils) had mean 79.2 and standard deviation 6.4 and the marks of the second set (10 pupils) had mean 59.0 and standard deviation 11.8.

Calculate the mean and standard deviation of the marks of the two sets combined.

4   The following table gives details of the weekly earnings of the employees of a small manufacturing company.

Category	Number of employees	Mean earnings	Standard deviation
Management	8	£260	£25
Office staff	32	£145	£12
Factory staff	110	£170	£30

Calculate the mean and standard deviation of the earnings of all the employees.

5   A group of 8 children have mean age 14.2 years and standard deviation 1.2 years. Given that one girl, whose age is 15.0 years, leaves the group, find the mean age and standard deviation for the remaining 7 children.

6   In a village there are 100 men and 150 women. The heights of all 250 people have mean 174 cm and standard deviation 12 cm. Given that the heights of the men have mean 178 cm and standard deviation 10 cm, calculate the mean and standard deviation of the heights of the women.

7   You are given 16 rock samples, and you are told that their gold content (x mg) satisfies $\sum x = 496$, $\sum x^2 = 15\,386$. The first sample which you examine has a gold content of 27 mg. Are you surprised by this value?

Show that, in fact, a mistake must have been made.
(*Hint: attempt to calculate the standard deviation for the other 15 rock samples.*)

3
Discrete Random Variables

3.1 Probability distributions

A *random variable* is a numerical variable whose value depends to some extent on chance.
  We shall use capital letters $X, Y, \ldots$, to represent random variables. For example:

(1) Throw a die; X is the score.
(2) Throw two dice: X is the total score
    $\qquad\qquad\qquad$ Y is the number of sixes
    $\qquad\qquad\qquad$ Z is the higher of the two scores
(3) Toss three coins; X is the number of heads.
(4) Deal a hand of 13 cards from a pack; X is the number of aces.
(5) Select 100 people at random; X is the number of people having blue eyes.
(6) Select one person at random; X is the height of that person.
(7) Select 20 potatoes from a large batch; X is the mean weight of these 20 potatoes.
(8) A gambler stakes £1 on number 21 at roulette; X is the amount he wins.
(9) A bag contains 9 white discs and 1 black disc; discs are removed (without replacement)
    until the black disc is taken; X is the number of discs removed.

  A random variable is *discrete* if we can list its possible values; or *continuous* if it can take any value within a certain range.
  Of the examples above, (6) and (7) are continuous, and all the others are discrete.
  In this chapter we shall consider only discrete random variables. The (probability) *distribution* of a discrete random variable is a list of its possible values, together with the probability that it takes each value.

Example 1
A bag contains three discs numbered $1, 2, 3$, and a second bag contains four discs numbered $4, 5, 6, 8$. One disc is taken from the first bag, and one disc from the second bag; X is the sum of the numbers on these two discs.
(i)   Tabulate the distribution of X.
(ii) State the mode.
(iii) Find the probability that X is greater than 8.
(iv) Draw a histogram.
(v)   When this procedure (taking a disc from each bag, and then replacing them) is repeated 100 times, give the expected frequency distribution of X.

(i) When a disc is taken from each bag, there are 12 possible outcomes:

$$
\begin{array}{cccc}
(1, 4) & (1, 5) & (1, 6) & (1, 8) \\
(2, 4) & (2, 5) & (2, 6) & (2, 8) \\
(3, 4) & (3, 5) & (3, 6) & (3, 8)
\end{array}
$$

and all of these are equally likely.

Thus X (the sum of the two numbers) has possible values 5, 6, 7, 8, 9, 10, 11.

$X = 5$ occurs for only one outcome $(1, 4)$; so the probability is $\dfrac{1}{12}$

$X = 6$ occurs for $(1, 5)$ and $(2, 4)$;   so the probability is $\dfrac{2}{12}$

$X = 7$ occurs for $(1, 6)$, $(2, 5)$ and $(3, 4)$;   so the probability is $\dfrac{3}{12}$

$X = 8$ occurs for $(2, 6)$ and $(3, 5)$;   so the probability is $\dfrac{2}{12}$

$X = 9$ occurs for $(1, 8)$ and $(3, 6)$;   so the probability is $\dfrac{2}{12}$

$X = 10$ occurs for $(2, 8)$;   so the probability is $\dfrac{1}{12}$

$X = 11$ occurs for $(3, 8)$;   so the probability is $\dfrac{1}{12}$

Hence the distribution is

X	5	6	7	8	9	10	11
Probability	$\dfrac{1}{12}$	$\dfrac{2}{12}$	$\dfrac{3}{12}$	$\dfrac{2}{12}$	$\dfrac{2}{12}$	$\dfrac{1}{12}$	$\dfrac{1}{12}$

Note that the probabilities add up to one.

(ii) The mode, or most likely value, is the value with the highest probability. Hence the mode is 7.

(iii) The probability that $X > 8$, written $P(X > 8)$, is the probability that X is 9, 10 or 11;

so $P(X > 8) = P(X = 9) + P(X = 10) + P(X = 11) = \dfrac{2}{12} + \dfrac{1}{12} + \dfrac{1}{12}$

$$= \dfrac{1}{3}$$

(iv) Since X takes only integer (i.e. whole number) values, we adopt the usual convention when drawing the histogram (Fig. 3.1), so that, for example, the probability that $X = 8$ is given by the area of the histogram between 7.5 and 8.5. Since the bars have width 1, the probability density $\left(\text{which is } \dfrac{\text{probability}}{\text{class width}}\right)$ is equal to the probability. The total area of the histogram is one.

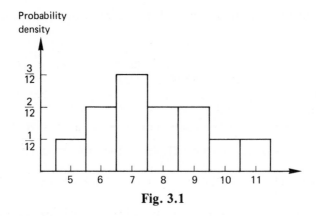

Fig. 3.1

(v)   If we repeat the process 100 times, we would expect the value $X = 5$ to occur about $100 \times \frac{1}{12} \approx 8.3$ times.

We obtain the 'expected frequencies' by multiplying the probabilities by 100, so the expected frequency distribution is as shown below.

X	5	6	7	8	9	10	11
Expected frequency	8.3	16.7	25	16.7	16.7	8.3	8.3

Note that the expected frequencies need not be whole numbers.

Example 2

A coin is tossed 3 times. Obtain the distribution of the number of heads, and find the mode.

There are 8 equally likely sequences HHH,   HHT,   HTH,   HTT
                                    THH,   THT,   TTH,   TTT.
The distribution of X, the number of heads, is thus as shown below.

X	0	1	2	3
Probability	$\frac{1}{8}$	$\frac{3}{8}$	$\frac{3}{8}$	$\frac{1}{8}$

There are two modes, 1 and 2.

Example 3

From a group of 6 boys and 4 girls, three children are selected by drawing names out of a hat.

Given that X is the number of girls selected, tabulate the distribution of X.

We can draw a probability tree for this situation (Fig. 3.2)

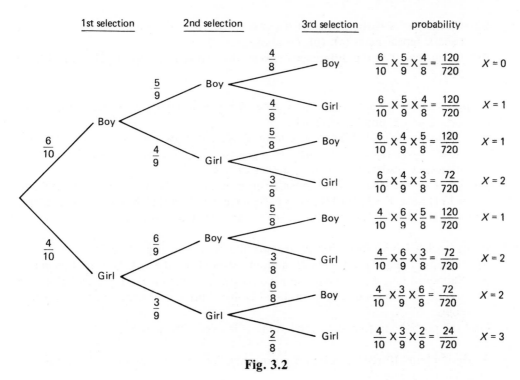

Fig. 3.2

The possible values of X are 0, 1, 2, 3.

$$P(X=0) = \frac{120}{720} = \frac{5}{30}, \qquad P(X=1) = \frac{120}{720} \times 3 = \frac{15}{30}$$

$$P(X=2) = \frac{72}{720} \times 3 = \frac{9}{30}, \qquad P(X=3) = \frac{24}{720} = \frac{1}{30}$$

The distribution is shown below

X	0	1	2	3
Probability	$\frac{5}{30}$	$\frac{15}{30}$	$\frac{9}{30}$	$\frac{1}{30}$

Exercise 3.1 Probability distributions

1 Two ordinary dice are thrown. Write down (in any suitable form) all 36 possible outcomes.
   (i) Given that X is the total score
      (a) tabulate the probability distribution
      (b) state the mode
      (c) find the probability that X is greater than 2.
   (ii) Given that X is the number of sixes repeat parts (a), (b), (c) above

(iii) Given that X is the higher of the two scores (or the common score if both are equal), repeat parts (a), (b), (c) above.

(iv) Given that X is the difference between the two scores, repeat parts (a), (b), (c) above.

2   A coin is tossed 4 times. Write down the 16 possible sequences of heads and tails.
   (i)   Let X be the number of heads.
        Tabulate the probability distribution of X, and draw a histogram. If the experiment (tossing a coin 4 times) is repeated 32 times, write down the expected frequency distribution of the number of heads.
   (ii) Let Y be the length of the longest run of either heads or tails (e.g. HHHT and TTTH have $Y = 3$; HTTH and TTHH have $Y = 2$; HTHT has $Y = 1$, and so on).
   Tabulate the probability distribution of Y, and draw a histogram.

3   In my pocket I have three 50p coins, two 20p coins and five 10p coins.
   (i)   Suppose I take out one coin at random. Write down the distribution of its value.
   (ii) Suppose I take out two coins at random (the first coin is *not* replaced before the second is taken). Obtain the probability distribution of the total value of the two coins.
   Find the probability that their total value is between 25p and 65p.

4   A bag contains 10 red discs and 6 green ones. Three discs are taken from the bag (without replacement).
   (ii) Given that X is the number of red discs taken, obtain the distribution of X.
   (ii) Given that each red disc is worth 5 points and each green one is worth 8 points, and that Y is the total number of points for the three discs taken, obtain the distribution of Y.

5   When a person takes the driving test, the probability that he passes is $\frac{1}{3}$. If he fails the test, he takes it again, and so on until he passes. Given that X is the number of times he takes the test, state the possible values of X, and find the probabilities that

$$X = 1, \ X = 2, \ X = 3, \ X = 4.$$

Show that the probabilities form a geometric progression, and hence verify that the probabilities add up to 1.
   In a sample of 243 people, how many would you expect to
   (i)   pass first time   (ii) pass second time
   (iii) pass third time   (iv) need more than three attempts?

3.2   Mean and Variance

Expectation

Suppose, in general, that a discrete random variable X has
possible values x_1 x_2 $\cdots$
with probabilities p_1 p_2 $\cdots$

$$\text{(where } p_1 + p_2 + \cdots = 1)$$

This corresponds to a frequency table (values x_i, with frequencies f_i) so the mean $\dfrac{\sum x_i f_i}{\sum f_i}$ becomes $\dfrac{\sum x_i p_i}{\sum p_i} = \sum x_i p_i$ since $\sum p_i = 1$.

Thus the mean is $\sum x_i p_i = x_1 p_1 + x_2 p_2 + \cdots$; i.e., we multiply each value by its probability, and add up.

We define the *mean*, or *expectation*, or *expected value*, of X, written $E[X]$ to be

$$E[X] = \sum x_i p_i$$

Example 1
Find the mean of the following distribution (defined in Example 1, p. 44).

X	5	6	7	8	9	10	11
Probability	$\dfrac{1}{12}$	$\dfrac{2}{12}$	$\dfrac{3}{12}$	$\dfrac{2}{12}$	$\dfrac{2}{12}$	$\dfrac{1}{12}$	$\dfrac{1}{12}$

The mean is $E[X] = 5 \times \dfrac{1}{12} + 6 \times \dfrac{2}{12} + 7 \times \dfrac{3}{12} + 8 \times \dfrac{2}{12} + 9 \times \dfrac{2}{12} + 10 \times \dfrac{1}{12} + 11 \times \dfrac{1}{12}$

$$= \dfrac{93}{12} = \dfrac{31}{4} = 7.75$$

Note that the 'expected value' of a random variable is the mean (and *not* the most likely value). In Example 1 above, the expected value of X is 7.75.

Variance

Suppose that a random variable X has mean $E[X] = \mu$
The deviation from the mean is $(X - \mu)$
The *variance* of X, written var (X), is the mean squared deviation, i.e. the average value of $(X - \mu)^2$,
so we define

$$\text{var}(X) = E[(X - \mu)^2]$$

For a frequency distribution, the variance is usually calculated by the alternative formula $\dfrac{\sum x_i^2 f_i}{\sum f_i} - \bar{x}^2$ which suggests that var $(X) = \sum x_i^2 p_i - \mu^2$

Note that, since X^2 has possible values $\quad x_1^2 \quad x_2^2 \quad \cdots$
with probabilities $\quad p_1 \quad p_2 \quad \cdots$

$\sum x_i^2 p_i$ is the mean of X^2, i.e. $E[X^2]$

We shall now prove that var $(X) = E[X^2] - \mu^2$

We have, by definition, var $(X) = E[(X - \mu)^2]$
Now $(X - \mu)^2$ has possible values $(x_1 - \mu)^2 \quad (x_2 - \mu)^2 \quad \cdots$
with probabilities $\quad p_1 \quad p_2 \quad \cdots$

so

$$\begin{aligned}
\text{var}\,(X) = \text{E}\,[\,(X - \mu)^2\,] &= \sum \,(x_i - \mu)^2 p_i \\
&= \sum \,(x_i^2 - 2\mu x_i + \mu^2) p_i \\
&= \sum \,x_i^2 p_i - 2\mu \sum \,x_i p_i + \mu^2 \sum \,p_i \\
&= \sum \,x_i^2 p_i - 2\mu\,\mu + \mu^2 \\
&= \sum \,x_i^2 p_i - \mu^2 \\
&= \text{E}\,[\,X^2\,] - \mu^2
\end{aligned}$$

We have

$$\text{variance var}\,(X) = \text{E}\,[\,X^2\,] - \mu^2 = \sum \,x_i^2 p_i - \mu^2$$

$$\text{standard deviation } \sigma = \sqrt{\text{var}\,(X)}$$

Note The variance of a random variable is quite often used (in preference to the standard deviation), since it does not involve a square root, and there are some useful theoretical properties.

Example 2

Find the variance and the standard deviation of the following distribution (given in Example 1).

X	5	6	7	8	9	10	11
Probability	$\dfrac{1}{12}$	$\dfrac{2}{12}$	$\dfrac{3}{12}$	$\dfrac{2}{12}$	$\dfrac{2}{12}$	$\dfrac{1}{12}$	$\dfrac{1}{12}$

We know that $\text{E}\,[\,X\,] = \dfrac{31}{4}$

$$\text{E}\,[\,X^2\,] = 25 \times \frac{1}{12} + 36 \times \frac{2}{12} + 49 \times \frac{3}{12} + 64 \times \frac{2}{12}$$

$$+ \, 81 \times \frac{2}{12} + 100 \times \frac{1}{12} + 121 \times \frac{1}{12}$$

$$= \frac{755}{12}$$

Thus $\text{var}\,(X) = \dfrac{755}{12} - \left(\dfrac{31}{4}\right)^2$

$$= \frac{137}{48}$$

The standard deviation is $\sigma = \sqrt{\dfrac{137}{48}} \approx 1.689.$

Example 3
A single die is thrown.
Find the mean and variance of the score.

If X is the score, its distribution is as given below

X	1	2	3	4	5	6
Probability	$\frac{1}{6}$	$\frac{1}{6}$	$\frac{1}{6}$	$\frac{1}{6}$	$\frac{1}{6}$	$\frac{1}{6}$

$$E[X] = 1 \times \frac{1}{6} + 2 \times \frac{1}{6} + \cdots + 6 \times \frac{1}{6} = \frac{21}{6} = \frac{7}{2} = 3.5$$

$$E[X^2] = 1 \times \frac{1}{6} + 4 \times \frac{1}{6} + \cdots + 36 \times \frac{1}{6} = \frac{91}{6}$$

$$\text{var}(X) = \frac{91}{6} - \left(\frac{7}{2}\right)^2 = \frac{35}{12} \approx 2.917$$

Example 4
Three children are selected at random from a group of 6 boys and 4 girls.
Given that X is the number of girls selected, find the mean and standard deviation of X.
Find the probability that X is within one standard deviation of its mean.

The distribution of X is shown below, (see Example 3, p. 46).

X	0	1	2	3
Probability	$\frac{5}{30}$	$\frac{15}{30}$	$\frac{9}{30}$	$\frac{1}{30}$

We have

$$E[X] = 0 \times \frac{5}{30} + 1 \times \frac{15}{30} + 2 \times \frac{9}{30} + 3 \times \frac{1}{30} = 1.2$$

$$E[X^2] = 0 \times \frac{5}{30} + 1 \times \frac{15}{30} + 4 \times \frac{9}{30} + 9 \times \frac{1}{30} = 2$$

so

$$\text{var}(X) = 2 - (1.2)^2 = 0.56.$$

Thus X has mean $\mu = 1.2$ and standard deviation $\sigma = \sqrt{0.56} \approx 0.748$.
$\mu - \sigma = 0.452$ and $\mu + \sigma = 1.948$; so the probability that X is within one standard deviation of its mean is

$$P(0.452 < X < 1.948) = P(X = 1) = \frac{15}{30} = \frac{1}{2}$$

Example 5
A bag contains 4 white discs and 3 black discs. Two players, R and S, take it in turns (with R starting) to remove one disc from the bag, without replacing it. The game ends

when a player removes a black disc; that player then pays the other one 5p for each disc which has been removed.

Calculate the expected amount which R wins in a single game. If they play this game 150 times, how much should R expect to win or lose?

Consider a single game, and let X pence be the amount which R wins. The probability tree is shown in Fig. 3.3.

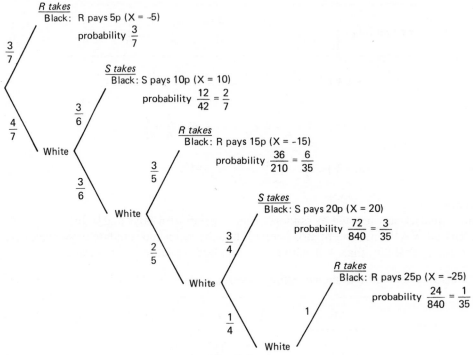

Fig. 3.3

The distribution of X is

X	-25	-15	-5	10	20
Probability	$\dfrac{1}{35}$	$\dfrac{6}{35}$	$\dfrac{15}{35}$	$\dfrac{10}{35}$	$\dfrac{3}{35}$

The expected amount which R wins in a single game is

$$E[X] = \left(-25 \times \frac{1}{35}\right) + \left(-15 \times \frac{6}{35}\right) + \left(-5 \times \frac{15}{35}\right) + \left(10 \times \frac{10}{35}\right) + \left(20 \times \frac{3}{35}\right)$$

$$= -\frac{30}{35} = -\frac{6}{7}$$

Thus, on average, R loses $\frac{6}{7}$p per game.

If they play 150 times, R should expect to lose $150 \times \frac{6}{7}$ pence, i.e. about £1.29

Example 6

A die is thrown repeatedly until the first 6 is observed; X is the number of throws required. Obtain the distribution of X, and hence calculate the mean number of throws required.

There are infinitely many possible values for X: 1, 2, 3, ...

If $X = 1$, the first throw is a six; so $P(X = 1) = \dfrac{1}{6}$

If $X = 2$, the first throw is not a six, and the second throw is a six,

$$\text{so } P(X = 2) = \frac{5}{6} \times \frac{1}{6}$$

If $X = 3$, the first two throws are not sixes, and the third throw is a six;

$$\text{so } P(X = 3) = \frac{5}{6} \times \frac{5}{6} \times \frac{1}{6}, \text{ and so on.}$$

The distribution is

X	1	2	3	4	...
Probability	$\dfrac{1}{6}$	$\dfrac{5}{6} \times \dfrac{1}{6}$	$\left(\dfrac{5}{6}\right)^2 \times \dfrac{1}{6}$	$\left(\dfrac{5}{6}\right)^3 \times \dfrac{1}{6}$	...

Note that the sum of the probabilities $\dfrac{1}{6} + \dfrac{5}{6} \times \dfrac{1}{6} + \left(\dfrac{5}{6}\right)^2 \times \dfrac{1}{6} + \cdots$ is an infinite geometric series, with first term $\dfrac{1}{6}$ and common ratio $\dfrac{5}{6}$, so the sum is

$$\frac{\dfrac{1}{6}}{1 - \dfrac{5}{6}} = 1,$$

as expected.

This is an example of a 'geometric' distribution.

$$E[X] = 1 \times \frac{1}{6} + 2 \times \frac{5}{6} \times \frac{1}{6} + 3 \times \left(\frac{5}{6}\right)^2 \times \frac{1}{6} + \cdots$$

$$= \frac{1}{6} \left\{ 1 + 2 \times \frac{5}{6} + 3 \times \left(\frac{5}{6}\right)^2 + \cdots \right\}$$

Comparing this with the binomial series

$$(1 - x)^{-2} = 1 + 2x + 3x^2 + 4x^3 + \cdots$$

we see that

$$E[X] = \frac{1}{6} \left(1 - \frac{5}{6}\right)^{-2} = \frac{1}{6} \times 36 = 6$$

The mean number of throws required is 6.

Linear functions

If a random variable X has mean μ and standard deviation σ (and therefore variance σ^2), then the random variable $Y = aX + b$ has mean $a\mu + b$ and standard deviation $a\sigma$ (or $|a|\sigma$ if a is negative) and hence variance $a^2\sigma^2$

i.e.
$$E[aX + b] = aE[X] + b$$
$$\text{var}(aX + b) = a^2 \text{ var}(X)$$

To prove these results, suppose X has possible values x_i with probabilities p_i, then $aX + b$ has values $ax_i + b$ with probabilities p_i,

so $E[aX + b] = \sum (ax_i + b)p_i = a \sum x_i p_i + b \sum p_i$

$$= aE[X] + b$$

If $E[X] = \mu$, then $(aX + b)$ has mean $aE[X] + b = a\mu + b$, and thus

$$\text{var}(aX + b) = E[\{(aX + b) - (a\mu + b)\}^2] = E[\{aX - a\mu\}^2]$$

$$= E[a^2(X - \mu)^2]$$

$$= a^2 E[(X - \mu)^2] = a^2 \text{ var}(X)$$

Example 7

A die is thrown. The random variable Y is obtained by doubling the score and adding three.

Find the mean and variance of Y.

Let X be the score on the die.
We know (see Example 3) that $E[X] = \dfrac{7}{2}$ and $\text{var}(X) = \dfrac{35}{12}$

We have $Y = 2X + 3$, so $E[Y] = 2 \times \dfrac{7}{2} + 3 = 10$

$$\text{var}(Y) = 4 \times \frac{35}{12} = \frac{35}{3}$$

Other functions

Example 8

The random variable X has the distribution given below.

X	0	1	2	3	4
Probability	0.1	0.2	0.3	0.25	0.15

Obtain the distribution of the random variable $Y = X^3 - 4X^2 + 10$, and deduce $E[Y]$ and $\text{var}(Y)$.

The table below shows the values of Y.

X	0	1	2	3	4
Probability	0.1	0.2	0.3	0.25	0.15
$Y = X^3 - 4X^2 + 10$	10	7	2	1	10

The possible values of Y are 1, 2, 7, 10, and its distribution is given below.

Y	1	2	7	10
Probability	0.25	0.3	0.2	0.25

$$E[Y] = 1 \times 0.25 + 2 \times 0.3 + 7 \times 0.2 + 10 \times 0.25 = 4.75$$

$$E[Y^2] = 1 \times 0.25 + 4 \times 0.3 + 49 \times 0.2 + 100 \times 0.25 = 36.25$$

Hence var $(Y) = 36.25 - (4.75)^2 = 13.6875$.

Note We use the Greek letters μ and σ for the theoretical mean and standard deviation of a random variable.

Roman letters $\bar{x}$ and s are used for the mean and standard deviation of actual numerical data.

Exercise 3.2   Mean and variance

1   Two dice are thrown (see Exercise 3.1, Question 1)
Calculate the mean of each of the following variables
(i)   the total score
(ii) the number of sixes
(iii) the higher of the two scores
(iv) the difference between the two scores.

2   A coin is tossed 4 times (see Exercise 3.1, Question 2)
(i)   Calculate the mean and the standard deviation of the number of heads.
(ii) Calculate the mean and the standard deviation of the length of the longest run of either heads or tails.

3   In my pocket I have three 50p coins, two 20p coins and five 10p coins (see Exercise 3.1, Question 3)
(i)   Given that I take out one coin at random, find the mean and variance of its value.
(ii) Given that I take out two coins at random, calculate the mean and variance of their total value.
Find the probability that this value exceeds the mean by more than one standard deviation.

4   A bag contains 10 red discs and 6 green ones. Three discs are taken from the bag without replacement (see Exercise 3.1, Question 4).
   (i)   Given that X is the number of red discs taken, calculate the mean and variance of X.
   (ii) Given that each red disc is worth 5 points and each green one is worth 8 points and that Y is the total number of points for the three discs taken, use the distribution of Y to calculate its mean and variance.
   (iii) Explain why $Y = 5X + 8(3 - X) = 24 - 3X$, and hence check your values for the mean and variance of Y.

5   Three letters are placed at random into three envelopes which have already been addressed. Write down the 6 possible arrangements of the letters in the envelopes. Given that X is the number of letters which are in their correct envelopes, deduce the probability distribution of X.
   Calculate $E[X]$ and var (X).

6   From a bag containing 4 white discs and 1 black disc, discs are removed one at a time (without replacement) until the black disc is taken. Given that X is the number of discs removed, calculate the mean and variance of X.
   Find the probability that X is within one standard deviation of its mean.

7   From a bag containing 4 white and 2 black discs, discs are removed one at a time (without replacement) until the first black disc is taken. Given that X is the number of discs removed, find the mean of X.

8   A and B play a game as follows: 3 coins are tossed; if there are 3 heads, B pays 30p to A; if there are just 2 heads, no money changes hands; if there are less than 2 heads, A pays 10p to B. Given that the winnings of A are X pence (and a loss is regarded as a negative winning), calculate the mean and variance of X.
   If they play this game 50 times, how much should A expect to win or lose?

9   (i) X and Y are each chosen randomly from the set $\{0, 1, 2, 3\}$, and repeats (e.g. $X = 2$ and $Y = 2$) are allowed. By considering all 16 possibilities, tabulate the probability distributions of X, Y, and their sum $X + Y$, and their product XY. Is it true that
         (a) $E[X + Y] = E[X] + E[Y]$
         (b) $E[XY] = E[X] \times E[Y]$?
   (ii) Repeat part (i) for the case where repeats are *not* allowed. (There are now 12 possible ways of selecting X and Y).

3.3   Probabilities from Binomial distributions

Example 1
A coin is tossed 3 times; X is the number of heads. Give the distribution of X.

The distribution of X (see Example 2, p. 46) is

X	0	1	2	3
Probability	$\dfrac{1}{8}$	$\dfrac{3}{8}$	$\dfrac{3}{8}$	$\dfrac{1}{8}$

Now $(q+p)^3 = q^3 + 3q^2p + 3qp^2 + p^3$

and so $(\frac{1}{2}+\frac{1}{2})^3 = (\frac{1}{2})^3 + 3(\frac{1}{2})^2(\frac{1}{2}) + 3(\frac{1}{2})(\frac{1}{2})^2 + (\frac{1}{2})^3$

$$= \frac{1}{8} + \frac{3}{8} + \frac{3}{8} + \frac{1}{8}$$

Note that the probabilities in the above distribution are the terms in the expansion of $(\frac{1}{2}+\frac{1}{2})^3$.

Example 2
Four dice are thrown; X is the number of sixes. Find the distribution of X and show that the probabilities are the terms in the expansion of $\left(\dfrac{5}{6} + \dfrac{1}{6}\right)^4$.

X has possible values $0, 1, 2, 3, 4$.

$X = 0$ when all four are not sixes; so $P(X=0) = \dfrac{5}{6}\times\dfrac{5}{6}\times\dfrac{5}{6}\times\dfrac{5}{6} = \left(\dfrac{5}{6}\right)^4$

$X = 1$ when one die shows a six and the other three do not.

$\qquad$ $P(6...) = \frac{1}{6}\times\frac{5}{6}\times\frac{5}{6}\times\frac{5}{6}$ (where . denotes any score other than a six)

$\qquad$ $P(.6..) = \frac{5}{6}\times\frac{1}{6}\times\frac{5}{6}\times\frac{5}{6}$

$\qquad$ $P(..6.) = \frac{5}{6}\times\frac{5}{6}\times\frac{1}{6}\times\frac{5}{6}$

$\qquad$ $P(...6) = \frac{5}{6}\times\frac{5}{6}\times\frac{5}{6}\times\frac{1}{6}$

Thus $P(X=1) = 4 \times\dfrac{1}{6}\times\left(\dfrac{5}{6}\right)^3$

$X = 2$ for the following sequences: $\quad$ 66.. $\quad$ 6.6. $\quad$ 6..6

$\qquad\qquad\qquad\qquad\qquad\qquad\qquad\quad$ .66. $\quad$ .6.6 $\quad$ ..66

and each of these has probability $\frac{1}{6}\times\frac{1}{6}\times\frac{5}{6}\times\frac{5}{6}$

Thus $P(X=2) = 6 \times \left(\dfrac{1}{6}\right)^2 \times \left(\dfrac{5}{6}\right)^2$

$X = 3$ for the sequences: $\quad$ 666. $\quad$ 66.6 $\quad$ 6.66 $\quad$ .666

Thus $P(X=3) = 4\times\frac{1}{6}\times\frac{1}{6}\times\frac{1}{6}\times\frac{5}{6} = 4 \times (\frac{1}{6})^3 \times \frac{5}{6}$

$X = 4$ when all four dice show sixes; so $P(X=4) = \frac{1}{6}\times\frac{1}{6}\times\frac{1}{6}\times\frac{1}{6} = (\frac{1}{6})^4$

The distribution is therefore as shown below.

X	0	1	2	3	4
Probability	$\left(\dfrac{5}{6}\right)^4$	$4\times\dfrac{1}{6}\times\left(\dfrac{5}{6}\right)^3$	$6\times\left(\dfrac{1}{6}\right)^2\times\left(\dfrac{5}{6}\right)^2$	$4\times\left(\dfrac{1}{6}\right)^3\times\dfrac{5}{6}$	$\left(\dfrac{1}{6}\right)^4$
	$\dfrac{625}{1296}$	$\dfrac{500}{1296}$	$\dfrac{150}{1296}$	$\dfrac{20}{1296}$	$\dfrac{1}{1296}$

Now $(q+p)^4 = q^4 + 4q^3p + 6q^2p^2 + 4qp^3 + p^4$

and thus $\left(\dfrac{5}{6}+\dfrac{1}{6}\right)^4 = \left(\dfrac{5}{6}\right)^4 + 4\times\left(\dfrac{5}{6}\right)^3\times\dfrac{1}{6} + 6\times\left(\dfrac{5}{6}\right)^2\times\left(\dfrac{1}{6}\right)^2 + 4\times\dfrac{5}{6}\times\left(\dfrac{1}{6}\right)^3 + \left(\dfrac{1}{6}\right)^4$ so the

probabilities in the distribution of X above are the terms in the expansion of $\left(\dfrac{5}{6}+\dfrac{1}{6}\right)^4$

Examples 1 and 2 above are 'Binomial distributions'. In general, consider a sequence of independent 'trials' where each trial has two possible outcomes, called 'success' and 'failure', and for any one trial

the probability of success is p
and the probability of failure is $q = 1 - p$.

If the trial is repeated n times, and X is the number of successes, then X is said to have the *Binomial Distribution* B(n, p)
   For example:
(1) A coin is tossed 3 times; X is the number of heads (see Example 1 above).
   Here a 'trial' is to toss the coin once, and 'success' is to get a head; at each trial, the probability of success is $\frac{1}{2}$ (thus $p = \frac{1}{2}$).
   The trial is repeated 3 times (thus $n = 3$). Hence the number of heads X has the Binomial distribution B($3, \frac{1}{2}$).
(2) Four dice are thrown; X is the number of sixes (see Example 2 above).
   Here a 'trial' is to throw one die; this is repeated 4 times.
   At each trial, 'success' is to get a six, so the probability of success is $\frac{1}{6}$.
   Hence the number of sixes, X, has the Binomial distribution B($4, \frac{1}{6}$).

(3) X is the number of girls in a family of 6 children.
   Here a 'trial' is to have a child; this is repeated 6 times. The probability that a child is a girl is $\frac{1}{2}$; hence the number of girls X has the Binomial distribution B($6, \frac{1}{2}$).
(4) Suppose that 15% of people are regular church-goers. 80 people are selected at random, and X is the number of regular church-goers selected.
   Here a 'trial' is to select one person at random; this is repeated 80 times. Each time that a person is selected, the probability that he is a regular church-goer is 0.15; hence the number of church-goers selected, X, has the Binomial distribution B($80, 0.15$).
(5) 1% of the light-bulbs produced by a factory are defective. X is the number of defective bulbs in a batch of 250 bulbs.
   Here a 'trial' is to produce one bulb; this is repeated 250 times. The probability that

a bulb is defective is 0.01 (note that 'success', is to produce a defective bulb!); hence the number of defective bulbs, X, has the Binomial distribution B(250, 0.01).

Probabilities

For the distribution B$(3, \frac{1}{2})$, given in Example 1, the possible values are 0, 1, 2, 3 and the probabilities are given by the terms in the expansion of $(\frac{1}{2} + \frac{1}{2})^3$. For the distribution B$(4, \frac{1}{6})$ given in Example 2, the possible values are 0, 1, 2, 3, 4 and the probabilities are given by the terms in the expansion of $(\frac{5}{6} + \frac{1}{6})^4$

In general, if X has the distribution B(n, p), then X is the number of successes in n trials, so the possible values of X are 0, 1, 2, ..., n and the probabilities are given by the terms in the expansion of $(q + p)^n$ (where $q = 1 - p$).

Now
$$(q + p)^1 = q + p$$
$$(q + p)^2 = q^2 + 2qp + p^2$$
$$(q + p)^3 = q^3 + 3q^2 p + 3qp^2 + p^3$$
$$(q + p)^4 = q^4 + 4q^3 p + 6q^2 p^2 + 4qp^3 + p^4$$
$$\vdots$$
$$(q + p)^n = q^n + \binom{n}{1} q^{n-1} p + \binom{n}{2} q^{n-2} p^2 + \cdots + \binom{n}{r} q^{n-r} p^r + \cdots + p^n$$

where $\binom{n}{r} = \dfrac{n!}{(n-r)! \, r!}$ is the usual Binomial coefficient.

Thus for the general Binomial distribution B(n, p), we have the distribution given below.

X	0	1	2	...	r	...	n
Probability	q^n	$\binom{n}{1} q^{n-1} p$	$\binom{n}{2} q^{n-2} p^2$	...	$\binom{n}{r} q^{n-r} p^r$	...	p^n

i.e.
$$P(X = r) = \binom{n}{r} p^r q^{n-r}$$

Example 3
A driving examiner finds that he passes 40% of the candidates. For a particular day on which he examines 9 people, find the probability that
(i)   he passes exactly 6 people
(ii) he passes at least one person
(iii) he passes more than 7 people.

If X is the number of people he passes on that day, then X has the Binomial distribution B(9, 0.4)

(i)   P(he passes 6 people) $= P(X = 6) = \binom{9}{6}(0.4)^6 (0.6)^3 = 84 \times (0.4)^6 \times (0.6)^3$

$$= 0.0743$$

(ii)   P(he passes at least one person) $= 1 - $ P(he passes none)

$$= 1 - \text{P}(X = 0)$$
$$= 1 - (0.6)^9$$
$$= 0.9899$$

(iii) P(he passes more than 7 people) $=$ P$(X > 7) =$ P$(X = 8) +$ P$(X = 9)$

$$= (^9_8)(0.4)^8(0.6) + (0.4)^9$$
$$= 0.00354 + 0.00026$$
$$= 0.0038$$

Example 4

Batteries are dispatched from a factory in boxes of 120, and 3% of the batteries produced are defective. Find the probability that a box contains
(i)   exactly 5 defective batteries
(ii) at least 3 defective batteries.

If X is the number of defective batteries in a box, then X has the Binomial distribution B(120, 0.03).
The Binomial coefficients are

$$\binom{120}{1} = \frac{120!}{119!\ 1!} = 120, \qquad \binom{120}{2} = \frac{120!}{118!\ 2!} = \frac{120 \times 119}{2 \times 1}$$

$$\binom{120}{3} = \frac{120!}{117!\ 3!} = \frac{120 \times 119 \times 118}{3 \times 2 \times 1} \quad \text{and so on.}$$

(i) $\text{P}(X = 5) = \binom{120}{5}(0.03)^5(0.97)^{115} = \frac{120 \times 119 \times 118 \times 117 \times 116}{5 \times 4 \times 3 \times 2 \times 1}(0.03)^5(0.97)^{115}$

$$= 0.1395$$

(ii) $\text{P}(X \geqslant 3) = 1 - \{\text{P}(X = 0) + \text{P}(X = 1) + \text{P}(X = 2)\}$

$$= 1 - \left\{(0.97)^{120} + \binom{120}{1}(0.03)(0.97)^{119} + \binom{120}{2}(0.03)^2(0.97)^{118}\right\}$$

$$= 1 - \{0.02586 + 0.09597 + 0.17661\}$$
$$= 0.7016$$

Note If we are asked to 'state the distribution' of a random variable, it is sufficient to say, for example, that the distribution is B(4, $\frac{1}{6}$), because the possible values and their probabilities can then be obtained using standard formulae.

However, if we are asked to 'tabulate the distribution' we should list the possible values and compute the probabilities.

Exercise 3.3   Probabilities from Binomial distributions

1   A family has 4 children. Assuming that each child is equally likely to be a boy or a girl, give the probability distribution of X, the number of boys. Draw a histogram.

If 64 families, each having 4 children, are considered, what is the expected frequency distribution for the number of boys in a family?

2   When the seeds of a certain flower are sown, one quarter of the seeds grow into new plants. When 6 seeds are sown, give the distribution of the number of new plants which grow, and draw a histogram.
Find the probability that less than three new plants will grow.

3   Five dice are thrown. Find the probability that there are
(i)   exactly 3 sixes   (ii) at least 3 sixes.

4   Given that 10% of eggs are bad, find the probabilities that a box of 6 eggs contains
(i)   no bad eggs   (ii) exactly one bad egg
(iii) at least one bad egg.

5   In a certain town, 65% of the people are in favour of a controversial development scheme. Given that 12 people are interviewed, find the probability that exactly 5 of them are in favour of the scheme.

6   A large crate of oranges is classified as follows:
20 oranges are selected at random and examined; if this sample contains no damaged fruit, the crate is sold as Class I; if there are one or two damaged fruit, it is sold as Class II; if there are three or more damaged fruit, it is sold as Class III.
In fact 3% of the oranges in the crate are damaged. Find the probabilities that the crate is sold as

(i) Class I   (ii) Class II   (iii) Class III

7   90% of drivers wear seat-belts. Find the probability that, in the next 10 cars to pass, less than 8 drivers will be wearing seat-belts.

8   Random number tables consist of independent digits chosen from $0, 1, 2, ..., 9$ (each having probability $\frac{1}{10}$). If 5 digits are taken from such a table find the probability that at least 3 of them are less than 2.

9   A drunkard takes 8 steps. Each step is either one metre forwards, or one metre backwards, with equal probability, and the steps may be assumed to be independent. By considering all the possibilities:

8 steps forward;   7 steps forward (and 1 backwards);
6 steps forward (and 2 backwards); and so on,

list the possible finishing positions.
Find the probabilities that, after 8 steps,
(i) he is back at his starting point
(ii) he is more than 4 m from his starting point.

10   A multiple choice test consists of 12 questions and each question gives a choice of 5 answers. An ill-prepared candidate guesses the answer to each question. Given that each correct answer scores 4 marks and each wrong answer scores minus one mark, find the probabilities that the candidate scores

(i) 13 marks   (ii) less than zero marks.

3.4 Binomial distributions

Conditions for a Binomial distribution

If a 'trial', which can result in 'success' or failure', is repeated n times, then the number X of successes has the Binomial distribution B(n, p) *provided that*

(i)   the probability of success, p, is the same for each trial

and (ii) the outcomes of different trials are independent (for example, if it is known that the first trial is a success, this does not affect the probability that any subsequent trial is a success).

These two conditions should be carefully checked before we assert that a random variable has a Binomial distribution. (Discuss whether the conditions are satisfied for each example in the previous section).

For example, consider taking three cards (without replacement) from an ordinary pack of 52 cards, and let X be the number of aces taken. Here a 'trial' is to take one card, and 'success' is taking an ace.

However, the three trials are not independent. If the first card is an ace, the probability that the second card is an ace is $\frac{3}{51}$ but if the first card is not an ace, the probability that the second card is an ace is $\frac{4}{51}$; i.e. the probability that the second trial is a success is affected by the outcome of the first trial.

Hence X does *not* have a Binomial distribution.

The mean and variance of the Binomial distribution B(n, p)

Example 1

Calculate the mean and variance of the distribution $B\left(4, \frac{1}{6}\right)$ (given in Example 2, p. 57).

The distribution is as shown below.

X	0	1	2	3	4
Probability	$\frac{625}{1296}$	$\frac{500}{1296}$	$\frac{150}{1296}$	$\frac{20}{1296}$	$\frac{1}{1296}$

$$E[X] = 0 \times \frac{625}{1296} + 1 \times \frac{500}{1296} + 2 \times \frac{150}{1296} + 3 \times \frac{20}{1296} + 4 \times \frac{1}{1296}$$

$$= \frac{864}{1296} = \frac{2}{3}$$

$$E[X^2] = 0 \times \frac{625}{1296} + 1 \times \frac{500}{1296} + 4 \times \frac{150}{1296} + 9 \times \frac{20}{1296} + 16 \times \frac{1}{1296}$$

$$= \frac{1296}{1296} = 1$$

Thus var $(X) = 1 - \left(\frac{2}{3}\right)^2 = \frac{5}{9}$

Consider the distribution B(1, p):

X	0	1
Probability	q	p

We have

$$E[X] = 0 \times q + 1 \times p = p$$

$$E[X^2] = 0 \times q + 1 \times p = p$$

and so

$$\text{var }(X) = p - p^2 = p(1-p) = pq$$

Now consider the distribution B(2, p):

X	0	1	2
Probability	q^2	$2qp$	p^2

We have

$$E[X] = 0 \times q^2 + 1 \times 2qp + 2 \times p^2 = 2p(q+p) = 2p \quad (\text{since } q+p=1)$$

$$E[X^2] = 0 \times q^2 + 1 \times 2qp + 4 \times p^2 = 2qp + 4p^2$$

and so

$$\text{var }(X) = 2qp + 4p^2 - (2p)^2 = 2pq$$

Now consider the distribution B(3, p):

X	0	1	2	3
Probability	q^3	$3q^2p$	$3qp^2$	p^3

We have

$$E[X] = 3q^2p + 6qp^2 + 3p^3 = 3p(q^2 + 2qp + p^2)$$
$$= 3p(q+p)^2 = 3p$$

$$E[X^2] = 3q^2p + 12qp^2 + 9p^3 = 3p(q^2 + 4qp + 3p^2)$$
$$= 3p(q+3p)(q+p) = 3p(q+3p) = 3pq + 9p^2$$

and so

$$\text{var }(X) = 3pq + 9p^2 - (3p)^2 = 3pq$$

The results

Distribution	Mean	Variance
B(1, p)	p	pq
B(2, p)	$2p$	$2pq$
B(3, p)	$3p$	$3pq$

suggest the general result

> If X has the Binomial distribution B(n, p)
> then   mean $E[X] = np$
> variance var $(X) = npq$

It is of course very reasonable that if the probability of success is p, and the trial is repeated n times, the average number of successes will be np.

A formal proof of these formulae direct from the general Binomial probability distribution is possible, but difficult. An easier indirect proof is given on p. 235.

As an example, if X has the distribution $B\left(4, \dfrac{1}{6}\right)$, then

$$E[X] = 4 \times \frac{1}{6} = \frac{2}{3} \text{ and var } (X) = 4 \times \frac{1}{6} \times \frac{5}{6} = \frac{5}{9}$$

(agreeing with the values calculated in Example 1, p. 62).

Exercise 3.4   Binomial distributions

1   For each of the following situations, state whether the random variable X has a Binomial distribution B(n, p). If it does, state the values of n and p; if not, explain why.
   (i)   Two dice are thrown 15 times. X is the number of doubles.
   (ii) Five cards are dealt from a normal pack. X is the number of aces.
   (iii) Two cards are dealt from a normal pack; they are then replaced and the pack shuffled. This is repeated 8 times. X is the number of times that both cards were of the same suit.
   (iv) One third of days are rainy. X is the number of rainy days in a week.
   (v)   A class consists of 10 boys and 10 girls. A group of 4 children is selected at random; X is the number of girls selected.
   (vi) 1% of the items produced by a machine are faulty. X is the number of faulty items in a random sample of 50 items.

2   (i)   For the Binomial distribution B($3, \frac{1}{4}$)
         (a) tabulate the probability distribution, and state the mode
         (b) calculate the mean and the variance, and verify that these are as given by the formulae np and npq.
   (ii) Repeat part (i) for the Binomial distribution B(5, 0.7).

3   In a certain city, 10% of the people are civil servants.
    Given that 12 people are selected at random, state the mean and the standard deviation of the number of civil servants in this group. Calculate the probability that the number of civil servants is within one standard deviation of its mean.

4   A die is thrown 150 times. Find the mean and standard deviation of the number of sixes.

5   An industrial engine has 16 sparking plugs. For each plug there is a probability of 0.05 that it will fail during a day's operation. State the distribution of X, the number of plugs which fail during a certain day.
    By calculating $P(X \leqslant 0)$, $P(X \leqslant 1)$, $P(X \leqslant 2)$, ..., determine how many new plugs need to be available if the probability that all the plugs which fail on that day can be replaced is to be more than 0.99. (The plugs may be assumed to fail independently. If a failed plug is replaced by a new one, you may assume that the new one does not fail on the same day.)

6   The probability that a lobster pot left overnight will contain a lobster in the morning is 0.2. Given that a fisherman leaves n lobster pots, write down the probability that next morning
    (i) none will contain a lobster
    (ii) at least one will contain a lobster.
    Deduce the lowest number of pots which should be left if the probability that at least one will contain a lobster next morning is to be greater than 0.95.

7   100 army recruits were each asked 8 simple questions about a lecture which they had just attended, and the number of correct responses given by the recruits were as follows:

Number of correct responses	0	1	2	3	4	5	6	7	8
Number of recruits	7	2	0	6	22	28	19	10	6

It is claimed that if p is the probability that any one question is answered correctly, then the number of correct responses should have the Binomial distribution B(8, p) so the above is a sample of 100 values from this Binomial distribution.
    (i)   Write down the mean of B(8, p). Calculate the mean of the above sample and hence estimate the value of p.
    (ii) Compare the variance of the sample with the variance of B(8, p) using the value of p found in (i).
    (iii) Using the value of p found in (i) calculate the expected frequencies for a sample of 100 values from B(8, p) and compare these with the actual frequencies observed.
    (iv) What conditions are necessary to ensure that the number of correct answers really does have the Binomial distribution B(8, p)? Comment on whether these conditions are likely to be satisfied in practice.

4
Probability Density Functions

4.1 Probability density functions

Suppose we have a large sample of values of a continuous variable (such as heights, weights, times and so on).

We can group the values into classes, and draw a histogram, with probability density on the vertical scale.

In Fig. 4.1 the class width is 10. The probability that X lies between 18 and 36 is given by the shaded area, and the total area of the histogram is 1.

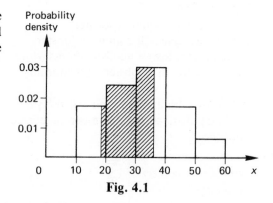

Fig. 4.1

We could now use smaller classes. In Fig. 4.2 the class width is 2.5. As the total area is still 1, the histogram has the same overall size and shape as before.

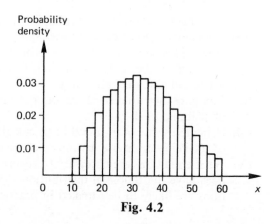

Fig. 4.2

Considering smaller and smaller classes, it seems reasonable to replace the histogram by a smooth curve (Fig. 4.3). The total area under the curve is still 1, and P$(18 < X < 36)$ is given by the shaded area.

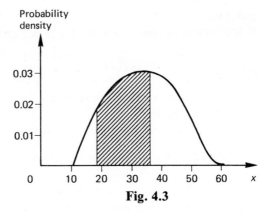

Fig. 4.3

A continuous random variable is specified by expressing the probability density in terms of x (i.e. by giving the equation of this curve). If the probability density is given by the function $f(x)$, then $f(x)$ is the *probability density function* (pdf).

Conditions for a pdf

We must have (i) $f(x) \geqslant 0$

and (ii) the total area under the curve is 1, i.e. $\displaystyle\int_{\text{all } x} f(x)\,\mathrm{d}x = 1$

Calculation of probabilities

The probability that X lies between a and b is given by the area under the pdf curve between a and b, i.e.

$$P(a < X < b) = \int_a^b f(x)\,\mathrm{d}x$$

Note that, for a continuous variable, we always find the probability that X lies in some *interval*; the probability that X takes any particular value is always zero, i.e. $P(X = a) = 0$

$P(a \leqslant X \leqslant b)$ is also given by $\displaystyle\int_a^b f(x)\,\mathrm{d}x$

The mode

This is a value of x for which the pdf $f(x)$ is maximum. It can often be found by inspection of a sketch of the pdf curve; sometimes it is necessary to solve $f'(x) = 0$.

The median

This is the value of x which divides the area into two equal halves.

Example 1

A continuous random variable X has pdf given by

$$f(x) = \begin{cases} kx^2(1-x) & \text{for} \quad 0 \leqslant x \leqslant 1 \\ 0 & \text{otherwise} \end{cases}$$

Find (i)   the value of the constant k
(ii)   the mode
(iii)   the probability that X is between 0.4 and 0.6.

(i)   Since the total area under the curve must be 1, we have

$$1 = \int_0^1 kx^2(1-x)\,dx = k\left[\frac{1}{3}x^3 - \frac{1}{4}x^4\right]_0^1 = \frac{1}{12}k$$

Hence $k = 12$

(ii)   The pdf is $f(x) = 12x^2(1-x) = 12(x^2 - x^3)$

$$f'(x) = 12(2x - 3x^2)$$

$$= 12x(2-3x), \text{ which is zero when } x = 0 \quad \text{or} \quad x = \frac{2}{3}.$$

$x = 0$ gives a minimum value, and $x = \frac{2}{3}$ gives a maximum.

Hence the mode is $\frac{2}{3}$.

(iii) $P(0.4 < X < 0.6) = \displaystyle\int_{0.4}^{0.6} 12x^2(1-x)\,dx$

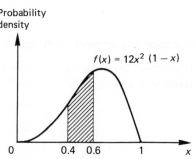

Probability density

$f(x) = 12x^2\,(1-x)$

$$= \left[4x^3 - 3x^4\right]_{0.4}^{0.6}$$

$$= \{4 \times (0.6)^3 - 3 \times (0.6)^4\}$$

$$- \{4 \times (0.4)^3 - 3 \times (0.4)^4\}$$

$$= 0.296$$

Fig. 4.4

Example 2

A random variable X has pdf given by

$$f(x) = \begin{cases} kx^3 & \text{for} \quad 0 \leqslant x \leqslant 3 \\ 0 & \text{otherwise} \end{cases}$$

Find (i)   the constant k
(ii)   the mode
(iii)   the median
(iv)   the value of a if $P(X > a) = 0.1$

(i)   We have $1 = \displaystyle\int_0^3 kx^3\,dx = k\left[\frac{1}{4}x^4\right]_0^3 = \frac{81}{4}k$

Thus $k = \dfrac{4}{81}$.

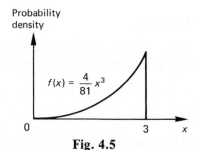

Probability density

$f(x) = \dfrac{4}{81}x^3$

Fig. 4.5

(ii)   The pdf $f(x) = \dfrac{4}{81} x^3$ is clearly a maximum when $x = 3$

The mode is 3.

(iii) If the median is M, then $\displaystyle\int_0^M \frac{4}{81} x^3 \, dx = \frac{1}{2}$

$$\left[\frac{1}{81} x^4 \right]_0^M = \frac{1}{2}$$

$$\frac{1}{81} M^4 = \frac{1}{2}$$

so the median is $M = \left(\dfrac{81}{2} \right)^{1/4} \approx 2.523$

(iv) $\mathrm{P}(X > a) = \displaystyle\int_a^3 \frac{4}{81} x^3 \, dx,$

so we require $0.1 = \displaystyle\int_a^3 \frac{4}{81} x^3 \, dx = \left[\frac{1}{81} x^4 \right]_a^3 = 1 - \frac{1}{81} a^4$

$$a^4 = 72.9$$

$$a \approx 2.922$$

Sometimes the area under the pdf can be found without integrating.

Example 3
A random variable X has pdf (Fig. 4.6) given by

$$f(x) = \begin{cases} k & \text{for} \quad 0 \leqslant x \leqslant 4 \\ 0 & \text{otherwise} \end{cases}$$

Find (i)   the constant k
       (ii) the probability $\mathrm{P}(2 \leqslant X \leqslant 3)$

(i)   Since the total area must be 1 we have $4k = 1$

$$k = \tfrac{1}{4}$$

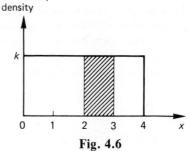

Fig. 4.6

(ii) $\mathrm{P}(2 \leqslant X \leqslant 3)$ is the area between $x = 2$ and $x = 3$;

Hence $\mathrm{P}(2 \leqslant X \leqslant 3) = 1 \times \tfrac{1}{4} = \tfrac{1}{4}$

This is an example of a *rectangular*, or *uniform* distribution. It may be said that 'X is uniformly distributed between 0 and 4' or 'X is equally likely to take any value between 0 and 4'.

Example 4

A random variable X has a 'triangular distribution', with pdf as shown in Fig. 4.7

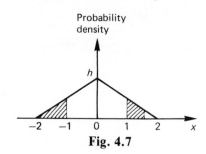

Fig. 4.7

Calculate the probabilities
(i) $P(1 < X < 1\frac{1}{2})$ (ii) $P(X \leq -1)$

The total area of the triangle is $\frac{1}{2} \times 4 \times h = 2h$
Hence $2h = 1$, and so $h = \frac{1}{2}$.
(i) $P(1 < X < 1\frac{1}{2})$ is the area of the shaded trapezium in Fig. 4.7.

We have $f(1) = \frac{1}{4}$ and $f\left(1\frac{1}{2}\right) = \frac{1}{8}$,

thus $P\left(1 < X < 1\frac{1}{2}\right) = \frac{1}{2} \times \left(\frac{1}{4} + \frac{1}{8}\right) \times \frac{1}{2} = \frac{3}{32}$.

(ii) $P(X \leq -1)$ is the area of the shaded triangle in Fig 4.7.

$f(-1) = \frac{1}{4}$, thus $P(X \leq -1) = \frac{1}{2} \times 1 \times \frac{1}{4} = \frac{1}{8}$.

Exercise 4.1   Probability density functions.

1   The time taken to perform a certain task is a random variable X which may be assumed to be uniformly distributed between 2 and 6 minutes. The probability density function (pdf) is shown in Fig. 4.8.

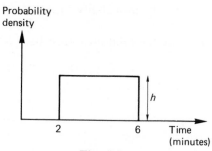

Fig. 4.8

(i)   State the height h of the rectangle.
(ii) Find the probability that the task takes between 2.5 and 5 minutes.
(iii) In group of 75 people, how many would you expect to take longer than $4\frac{1}{2}$ minutes to complete the task?
(iv) Find the time by which 95% of people will have completed the task.

2   In a nursery plantation, apple trees are grown from seed until they reach a height of 1 m, when they are transplanted elsewhere. The height of trees in the nursery plantation is a random variable X metres, whose pdf is shown in Fig. 4.9.

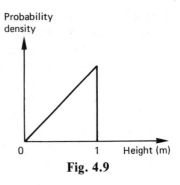

Fig. 4.9

(i)   State the mode of this distribution.
(ii) Find the equation of the pdf $f(x)$.
(iii) Given that one tree is chosen at random in the nursery find the probability that it is (a) shorter than 0.2 m (b) taller than 0.8 m.
(iv) Find the median height of the trees in the nursery.
(v)   If the nursery contains 800 trees, how many would you expect to have heights between 0.1 and 0.9 m?
(vi) If two trees are chosen at random, find the probability that
      (a) they are both taller than 0.8 m
      (b) one is taller then 0.8 m, and the other is shorter than 0.2 m.

3   A lorry and its trailer are weighed separately, and each of these two measured weights is subject to an error of ± 0.1 tonne.
The error in the total weight is a random variable X tonnes, whose pdf is shown in Fig. 4.10

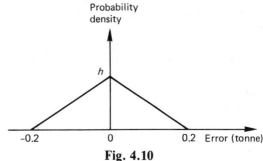

Fig. 4.10

(i)   Find the height h of the triangle
(ii) Find the probability that the magnitude of the error is less than 0.1 tonne (i.e. X is between -0.1 and $+0.1$)
(iii) Given that the measured weights of the lorry and its trailer are 6.7 and 21.6 tonnes, find the probability that the actual total weight is between 28.25 and 28.45 tonnes.

4   Currants are randomly distributed in a spherical bun of radius 3 cm. The distance of a currant from the centre is a random variable X cm with pdf $f(x) = kx^2$ for $0 \leqslant x \leqslant 3$
(i)   Sketch this pdf. Find (by integration) the area under the pdf in terms of k, and deduce the value of k.
(ii) Find the probability that a currant is more than 2 cm from the centre.
(iii) Find the median and the interquartile range of the distance of a currant from the centre.
(iv) If the bun contains 250 currants, how many currants would you expect to find within 1 cm of the centre of the bun?

5   For mass-produced resistors, the resistance is a random variable X k ohm with pdf
    $f(x) = \frac{3}{4}x(2 - x)$, for $0 \leqslant x \leqslant 2$.
    Sketch the distribution, and verify that the area under the pdf is 1.
    What proportion of resistors have resistances
    (i) more than 1.5 k ohm   (ii) between 0.9 and 1.1 k ohm?

6   The lifetime of a certain insect is a random variable X months with pdf
    $f(x) = kx^2(4 - x)$ for $0 \leqslant x \leqslant 4$.
    (i)   Sketch the distribution, and find the value of k.
    (ii) Find the mode.
    (iii) Find the probability that an insect dies before it is 1 month old.

7   On a certain remote island, the bearing of the wind direction is a random variable X
    (measured clockwise from north, between 0 and 2π radians) with pdf

    $$f(x) = \frac{1}{2\pi}(1 - \cos x) \qquad \text{for} \qquad 0 \leqslant x \leqslant 2\pi$$

    (i)   Sketch the pdf, and verify that the total area is 1.
    (ii) State the mode (i.e. the prevailing wind direction).
    (iii) Find the probability that the wind is blowing within an angle $\pi/6$ of due south.

4.2   Mean and variance

Consider a continuous random variable X with
pdf $f(x)$. Divide the range of possible values
into small intervals of width δx. The probability that X lies in a typical interval is given by
the area of the shaded strip, which is approximately $p = f(x)\,\delta x$.

The mean of X,
$E[X] = \sum xp = \sum xf(x)\,\delta x$,
which as $\delta x \to 0$ becomes the

integral $\int_{\text{all } x} xf(x)\,dx$.

Similarly $E[X^2] = \sum x^2 p = \sum x^2 f(x)\,\delta x$,

which becomes the integral $\int_{\text{all } x} x^2 f(x)\,dx$.

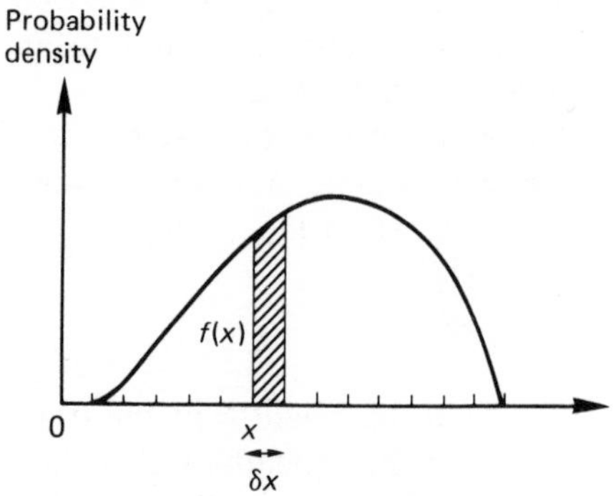

Hence we have

Probability
density

$f(x)$

x

δx

Fig. 4.11

$$\text{Mean, } \mu = E[X] = \int_{\text{all } x} xf(x)\,dx.$$

$$\text{Variance, var}(X) = E[(X - \mu)^2]$$

$$= E[X^2] - \mu^2$$

$$= \int_{\text{all } x} x^2 f(x)\,dx - \mu^2$$

$$\text{Standard deviation, } \sigma = \sqrt{\text{var}(X)}$$

Note The result $p = f(x) \, \delta x$ can be written in the form

$$P(x < X < x + \delta x) \approx f(x) \, \delta x$$

which is sometimes useful for finding the probability that X lies in a very small interval (for example, $P(0.4 < X < 0.401) \approx f(0.4) \times 0.001$).

Also, probability density $= \dfrac{\text{probability}}{\text{width of class}}$

$$= \frac{P(x < X < x + \delta x)}{\delta x}$$

This leads to the formal definition of the pdf as

$$f(x) = \lim_{\delta x \to 0} \frac{P(x < X < x + \delta x)}{\delta x}$$

We shall now find the mean, variance and standard deviation of the four random variables studied earlier.

Example 1

The random variable X with pdf (Fig. 4.12) $f(x) = \begin{cases} 12x^2(1 - x) & \text{for} \quad 0 \leqslant x \leqslant 1 \\ 0 & \text{otherwise} \end{cases}$

Mean, $E[X] = \displaystyle\int_0^1 x \, 12x^2(1 - x) \, dx$

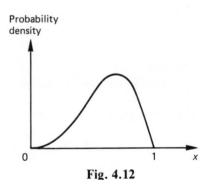

Probability density

Fig. 4.12

$$= 12 \left[\frac{1}{4} x^4 - \frac{1}{5} x^5 \right]_0^1$$

$$= 12 \left(\frac{1}{4} - \frac{1}{5} \right)$$

$$= 0.6$$

$E[X^2] = \displaystyle\int_0^1 x^2 \, 12x^2(1 - x) \, dx = 12 \left[\frac{1}{5} x^5 - \frac{1}{6} x^6 \right]_0^1$

$$= 12 \left(\frac{1}{5} - \frac{1}{6} \right)$$

$$= 0.4$$

Variance, $\text{var}(X) = 0.4 - (0.6)^2$

$$= 0.04$$

Standard deviation $\sigma = \sqrt{0.04}$

$$= 0.2$$

Example 2
The random variable X with pdf

$$f(x) = \begin{cases} \dfrac{4}{81}\, x^3 & \text{for} \quad 0 \leqslant x \leqslant 3 \\ 0 & \text{otherwise} \end{cases}$$

Find also the probability that X is within one standard deviation of its mean.

$$E[X] = \int_0^3 x\, \frac{4}{81}\, x^3\, dx = \frac{4}{81}\left[\frac{1}{5}\, x^5\right]_0^3$$

$$= 2.4$$

$$E[X^2] = \int_0^3 x^2 \frac{4}{81}\, x^3\, dx = \frac{4}{81}\left[\frac{1}{6}\, x^6\right]_0^3$$

$$= 6.0$$

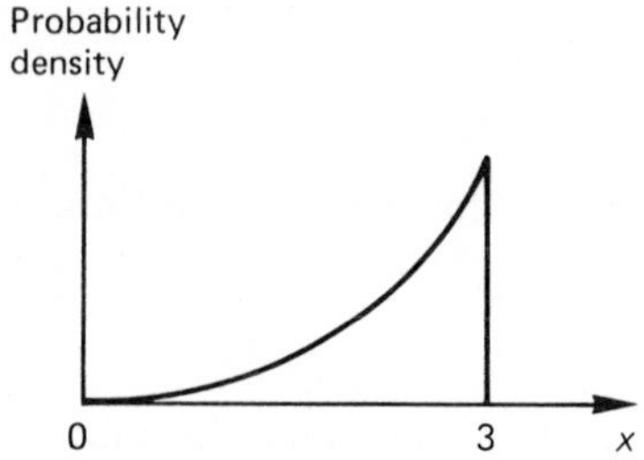

Variance, var $(X) = 6 - (2.4)^2$

$$= 0.24$$

$$\sigma \approx 0.4899$$

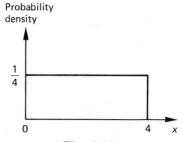

Probability density

Fig. 4.13

The probability that X is within one standard deviation of its mean,

$$P(1.9101 < X < 2.8899) = \int_{1.9101}^{2.8899} \frac{4}{81}\, x^3\, dx = \left[\frac{1}{81}\, x^4\right]_{1.9101}^{2.8899}$$

$$= \frac{1}{81}\, \{(2.8899)^4 - (1.9101)^4\}$$

$$= 0.6967$$

Example 3
The random variable X, having the rectangular distribution between 0 and 4.

The pdf is $f(x) = \frac{1}{4}$ for $0 \leqslant x \leqslant 4$

$$E[X] = \int_0^4 x\, \frac{1}{4}\, dx = \left[\frac{1}{8}\, x^2\right]_0^4$$

$$= 2.$$

This is also clear from the symmetry of the pdf.

$$E[X^2] = \int_0^4 x^2 \frac{1}{4}\, dx = \left[\frac{1}{12}\, x^3\right]_0^4 = \frac{16}{3}$$

Variance, var $(X) = \frac{16}{3} - 2^2$

$$= \frac{4}{3}$$

$$\sigma = \sqrt{\frac{4}{3}} \approx 1.155$$

Probability density

Fig. 4.14

Example 4

The random variable X having the triangular distribution shown in Fig. 4.15.

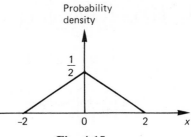

We must first find the equations of the pdf.
Between $x = -2$ and $x = 0$, the gradient is $\frac{1}{2}/2 = \frac{1}{4}$, so the pdf is $f(x) = \frac{1}{4}x + \frac{1}{2}$
Between $x = 0$ and $x = 2$, the gradient is $-\frac{1}{2}/2 = -\frac{1}{4}$, so the pdf is $f(x) = -\frac{1}{4}x + \frac{1}{2}$

Fig. 4.15

$$\text{Thus } f(x) = \begin{cases} \frac{1}{4}x + \frac{1}{2} & \text{for} \quad -2 \leqslant x \leqslant 0 \\ -\frac{1}{4}x + \frac{1}{2} & \text{for} \quad 0 \leqslant x \leqslant 2 \\ 0 & \text{otherwise} \end{cases}$$

The mean, $E[X] = 0$, by symmetry.

$$E[X^2] = \int_{-2}^{2} x^2 f(x) \, dx = \int_{-2}^{0} x^2 f(x) \, dx + \int_{0}^{2} x^2 f(x) \, dx$$

$$= \int_{-2}^{0} x^2 \left(\frac{1}{4}x + \frac{1}{2}\right) dx + \int_{0}^{2} x^2 \left(-\frac{1}{4}x + \frac{1}{2}\right) dx$$

$$= \left[\frac{1}{16} x^4 + \frac{1}{6} x^3\right]_{-2}^{0} + \left[-\frac{1}{16} x^4 + \frac{1}{6} x^3\right]_{0}^{2}$$

$$= 0 - \left(1 - \frac{4}{3}\right) + \left(-1 + \frac{4}{3}\right) - 0$$

$$= \frac{2}{3}$$

Variance, $\text{var}(X) = \frac{2}{3} - 0^2 = \frac{2}{3}$

$$\sigma = \sqrt{\frac{2}{3}} \approx 0.8165$$

Example 5

The random variable X has pdf given by

$$f(x) = \begin{cases} kx & \text{for} \quad 0 \leqslant x \leqslant 1 \\ k & \text{for} \quad 1 \leqslant x \leqslant 4 \\ 0 & \text{otherwise} \end{cases}$$

Find (i) the mean, (ii) the variance, (iii) the median.

Since the total area must be one, we have

$$1 = \frac{1}{2} \times 1 \times k + 3k = \frac{7}{2} k$$

Hence $k = \frac{2}{7}$.

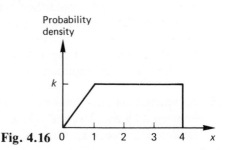

Fig. 4.16

(i) $E[X] = \int_0^4 x\, f(x)\, dx = \int_0^1 x\, f(x)\, dx + \int_1^4 x\, f(x)\, dx$

$= \int_0^1 x \frac{2}{7} x\, dx + \int_1^4 x \frac{2}{7}\, dx$

$= \left[\frac{2}{21} x^3 \right]_0^1 + \left[\frac{1}{7} x^2 \right]_1^4$

$= \frac{2}{21} + \frac{15}{7} = \frac{47}{21}$

≈ 2.238

(ii) $E[X^2] = \int_0^4 x^2\, f(x)\, dx = \int_0^1 x^2 \frac{2}{7} x\, dx + \int_1^4 x^2 \frac{2}{7}\, dx$

$= \left[\frac{1}{14} x^4 \right]_0^1 + \left[\frac{2}{21} x^3 \right]_1^4 = \frac{1}{14} + 6 = \frac{85}{14}$

var $(X) = \frac{85}{14} - \left(\frac{47}{21} \right)^2 = \frac{937}{882} \approx 1.062$

(iii) If the median is M, then the area to the right of M is $\frac{1}{2}$.

Thus $(4 - M) \times \frac{2}{7} = \frac{1}{2}$, giving $M = \frac{9}{4} = 2.25$

Exercise 4.2   Mean and variance

1   Shots are fired at random towards a circular target of radius 2 cm. For shots which land on the target, their distance from the centre is a random variable X cm with pdf $f(x) = kx$, for $0 \leqslant x \leqslant 2$
   (i)   Sketch the pdf, and find the value of k.
   (ii) Find the mean and variance of X.
   (iii) Find the probability that the distance from the centre is within one standard deviation of the mean distance.

2   A method for estimating the breaking strain of a rope (without actually breaking it) gives an estimate which is subject to an error of ± 1.5 kN.
   A stockpile of ropes have all had their breaking strains estimated as 2.5 kN. Between what limits could their actual breaking strains lie?
   Assume that the breaking strains are uniformly distributed between these two values.
   (i)   Sketch the pdf of the actual breaking strains X kN
   (ii) Calculate the mean and standard deviation of the breaking strains.

3   A supermarket buys chickens in bulk and classifies them by weight. Those weighing between 2 kg and 3 kg are sold as 'medium'. The weight of a medium chicken is a random variable X kg with pdf $f(x) = k(x^2 - 1)$ for $2 \leqslant x \leqslant 3$.

(i)   Sketch the distribution and find the value of k.

(ii) Calculate the mean and variance of the weights of medium chickens.

4 The area of single leaves of a certain plant is a random variable X cm^2 with pdf
$f(x) = kx^2(x-2)^2$ for $0 \leqslant x \leqslant 2$.

(i)   Sketch the distribution and show that $k = \dfrac{15}{16}$.

(ii) Find the mean and variance of the area of a leaf.

5 The time X minutes which I have to wait for a train in the morning has the pdf shown
in Fig. 4.17.

(i)   Find the maximum time I have to wait.

(ii) Find the equation of the pdf for
    (a) $0 \leqslant x \leqslant 5$ (b) $5 \leqslant x \leqslant a$

(iii) Calculate the mean waiting time.

(iv) Find the median waiting time.

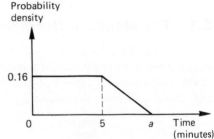

Fig. 4.17

5
Normal Distributions

5.1 The standard Normal curve

When measuring natural objects (for example, the weights of babies at birth, or the width of oak leaves), we might obtain a symmetrical histogram similar to Fig. 5.1. Distributions of this shape are very common, and they can often be modelled by a theoretical distribution called a *Normal distribution*.

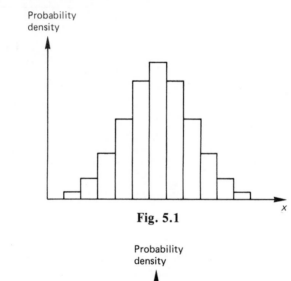

Fig. 5.1

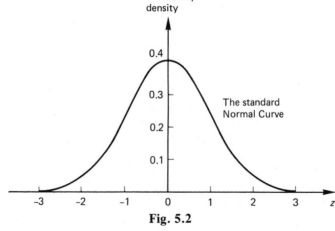

Fig. 5.2

The *standard Normal distribution* has mean 0 and standard deviation 1. Its pdf is shown in Fig. 5.2. Note that

(1) the curve is symmetrical about $z = 0$;
(2) the points of inflexion occur at $z = \pm 1$; these are one standard deviation on each side of the mean.

The equation of this curve will be given on p. 284. Unfortunately the pdf cannot be integrated analytically, so it is not possible to find areas under this curve (corresponding to probabilities) in the same way as for the other continuous distributions we met in Chapter 4.

Instead, areas under this curve (which have been obtained by numerical integration) are tabulated in *Normal distribution tables* (see p. 395).
These tables give the area under the curve to the left of a value z (Fig. 5.3). This area is usually written $\Phi(z)$. (Φ is the Greek letter capital 'phi'.)
Note that, as expected, $\Phi(0) = 0.5$, and as z increases, $\Phi(z)$ increases towards 1 (the total area under the curve).

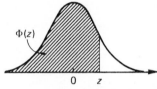

Fig. 5.3

The tables list only positive values of z. For negative values, we use the symmetry of the curve (Fig. 5.4). $\Phi(-z)$ is the area to the left of $-z$. By symmetry, this is equal to the area to the right of z, which is $1 - \Phi(z)$.

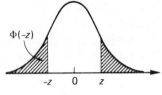

Fig. 5.4

Thus

$$\Phi(-z) = 1 - \Phi(z)$$

Example 1

$$\Phi(-0.62) = 1 - \Phi(0.62) = 1 - 0.7324$$
$$= 0.2676$$

Finding probabilities

Suppose Z is a random variable having the standard Normal distribution. The probability that Z is less than a, $P(Z < a)$, is the area under the curve to the left of a, which is $\Phi(a)$.

The probability that Z is greater than a, $P(Z > a)$, is the area under the curve to the right of a, which is $1 - \Phi(a)$. Figure 5.5 shows both of these areas.

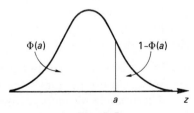

Fig. 5.5

The probability that Z is between a and b, $P(a < Z < b)$, is the area under the curve between a and b (Fig. 5.6), which is the area to the left of b minus the area to the left of a, i.e. $\Phi(b) - \Phi(a)$.

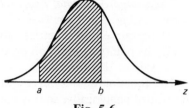

Fig. 5.6

We have, therefore,

$$P(Z < a) = \Phi(a)$$

$$P(Z > a) = 1 - \Phi(a)$$

$$P(a < Z < b) = \Phi(b) - \Phi(a)$$

Example 2

$P(Z < 1.645) = \Phi(1.645) = 0.9500$

$P(Z > 1.239) = 1 - \Phi(1.239) = 1 - 0.8923 = 0.1077$

$P(0.693 < Z < 1.493) = \Phi(1.493) - \Phi(0.693) = 0.9323 - 0.7559 = 0.1764$

$P(-0.497 < Z < 1.864) = \Phi(1.864) - \Phi(-0.497) = 0.9689 - (1 - 0.6904) = 0.6593$

$$\begin{aligned}P(-1.314 < Z < -0.628) &= \Phi(-0.628) - \Phi(-1.314) \\ &= (1 - 0.7350) - (1 - 0.9055) \\ &= 0.1705\end{aligned}$$

Note that

$$P(-1 < Z < 1) = \Phi(1) - \Phi(-1) = 0.8413 - (1 - 0.8413) = 0.6826$$

Similarly

$$P(-2 < Z < 2) = 0.9544$$

and

$$P(-3 < Z < 3) = 0.9974$$

This means that

   68.3% of values lie within one standard deviation of the mean

   95.4% of values lie within two standard deviations of the mean

and   99.7% of values lie within three standard deviations of the mean, agreeing with the results discovered in Chapter 2 (see p. 32).

Example 3

(i)   Find the value of a if $P(Z < a) = 0.893$

(ii) Find the value of b if $P(Z < b) = 0.369$

(iii) Find the value of c if $P(-c < Z < c) = 0.95$

(iv) Find the value of d if $P(-d < Z < d) = 0.99$

(i) $P(Z < a) = \Phi(a)$, so we have $\Phi(a) = 0.893$.

   We could use the tables 'backwards' to find which value of z gives $\Phi(z) = 0.893$.

   Alternatively we simply look up 0.893 in *inverse Normal tables* (see pp. 396–397).

   We obtain $a = 1.243$.

(ii) Now $\Phi(b) = 0.369$. Clearly b is negative
      and we cannot use the tables directly,
      (Fig. 5.7).
      However, $\Phi(-b) = 1 - \Phi(b)$
      $= 1 - 0.369 = 0.631$
      Using inverse Normal tables,
      $-b = 0.3345$ and so $b = -0.3345$.

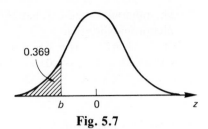

Fig. 5.7

(iii) The area to the left of c is 0.975, (Fig. 5.8)
      i.e. $\Phi(c) = 0.975$
      Thus $c = 1.960$

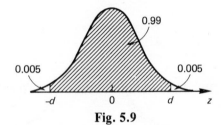

Fig. 5.8

(iv) The area to the left of d is 0.995, (Fig. 5.9)
      i.e. $\Phi(d) = 0.995$
      Thus $d = 2.576$

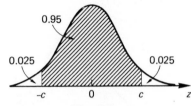

Fig. 5.9

Exercise 5.1   The standard Normal distribution

2   Figure 5.10 a–d shows the pdf of the standard Normal distribution. Use the table of
    areas under the standard Normal curve to find the shaded areas.

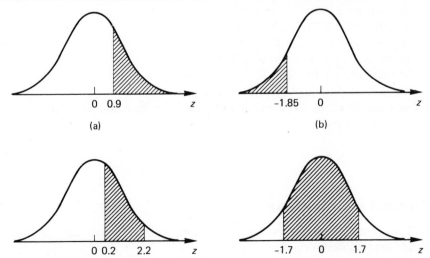

Fig. 5.10

2   The random variable Z has the standard Normal distribution. Find
    the probability that
    (i) $Z < 1.287$ (ii) $Z > 0.6$ (iii) $Z > -0.6$ (iv) $2.3 < Z < 2.6$
    (v) $-1.35 < Z < -1.05$ (vi) $-0.5 < Z < 1.3$ (vii) $-1.823 < Z < 1.823$
    (viii) $-2.008 < Z < -0.426$.

3   Find the values of a, b, c, d, e on the sketches of the standard Normal curve, shown
    in Fig. 5.11.

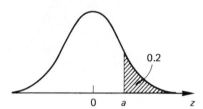
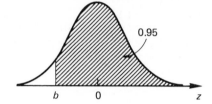
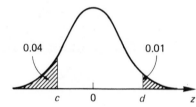
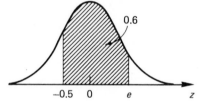

Fig. 5.11

4   The random variable Z has the standard Normal distribution. Find
    (i)   the value of a if $P(Z < a) = 0.9$
    (ii)   the value of b if $P(Z < b) = 0.02$
    (iii) the value of c if $P(Z > c) = 0.65$
    (iv)   the value of d if $P(-d < Z < d) = 0.7$
    (v)   the value of e if $P(-e < Z < e) = 0.98$
    (vi)   the value of f if $P(0 < Z < f) = 0.32$
    (vii) the upper quartile, the lower quartile, and the interquartile range
    (viii) the 88th percentile.

5.2   Normal distributions

Suppose Z is a random variable having the standard Normal distribution, so that Z has
mean 0 and standard deviation 1.

   Consider, for example, the random variable $X = 15Z + 50$, then X
has mean $15 \times 0 + 50 = 50$, and standard deviation $15 \times 1 = 15$.

The pdf of X has a similar shape to that of Z; we say that X has the Normal distribution with mean 50 and standard deviation 15.

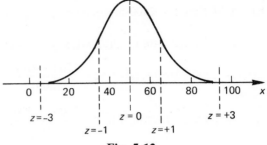

Fig. 5.12

Notes (1) the curve is symmetrical about the mean (i.e. $x = 50$);
(2) the points of inflexion on the curve are at one standard deviation on each side of the mean (i.e. at $x = 35$ and $x = 65$), and
(3) almost all of the distribution (in fact 99.7%) lies within three standard deviations of the mean (i.e. between $x = 5$ and $x = 95$).

Generally, the random variable $X = \sigma Z + \mu$ has the Normal distribution with mean μ and standard deviation σ

Thus $Z = \dfrac{X - \mu}{\sigma}$ has the standard Normal distribution

We can therefore use the table of areas under the standard Normal curve to find probabilities associated with X, by first converting the values x to their corresponding standard values $z = \dfrac{x - \mu}{\sigma}$.

Note that this is the *standardisation* process studied in Chapter 2 (see p. 39); z is the number of standard deviations from the mean.

Notation

The Normal distribution with mean μ and standard deviation σ, and hence variance σ^2, may be written as $N(\mu, \sigma^2)$.

The first random variable X above is thus $N(50, 15^2)$, and we may write $X \sim N(50, 15^2)$ or $X \sim N(50, 225)$.

The standard Normal distribution can thus be written as $N(0, 1)$.

Example 1

After extensive testing, it was found that a particular brand of light bulb had a mean life of 2100 hours with a standard deviation of 200 hours. Assuming that the life of the bulbs is Normally distributed, find
(i)   the probability that one of these bulbs will last more than 2400 hours
(ii) the percentage of bulbs expected to fail between 1700 and 2200 hours
(iii) the time after which all but 3% of the bulbs will have failed.

The lifetime X hours is Normally distributed with mean 2100 and standard deviation 200 hours; therefore $Z = \dfrac{X - 2100}{200}$ has the standard Normal distribution.

(i)   A lifetime $x = 2400$ hours corresponds to $z = \dfrac{2400 - 2100}{200} = 1.5$

Thus $P(X > 2400) = P(Z > 1.5) = 1 - \Phi(1.5) = 1 - 0.9332 = 0.0668$

(ii) $x = 1700$ corresponds to $z = \dfrac{1700 - 2100}{200} = -2.0$

$x = 2200$ corresponds to $z = \dfrac{2200 - 2100}{200} = 0.5$

Thus $P(1700 < X < 2200) = P(-2.0 < Z < 0.5) = \Phi(0.5) - \Phi(-2.0)$

$= 0.6915 - (1 - 0.9772) = 0.6687$

Hence the percentage expected to fail is $0.6687 \times 100 = 66.87\%$.

(iii) The time x after which all but 3% of the bulbs will have failed corresponds to the standard value z, where $\quad \Phi(z) = 0.97$, (Fig. 5.13)

so $\quad z = 1.881$

Thus $\dfrac{x - 2100}{200} = 1.881$

$x = 2476.2$

Hence all but 3% of the bulbs will have failed in the first 2476.2 hours.

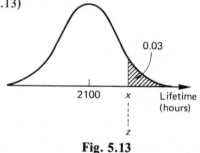

Fig. 5.13

Example 2

The mean weight of 1000 packets of sugar is found to be 1012 g. Of the 1000 packets, 70 were found to have a weight in excess of 1015 g. Assuming that the weights are Normally distributed about the given mean, estimate the standard deviation of the weights.

Estimate also the number of packets that may be expected to weigh less than 1008 g.

Suppose the weight X g is Normally distributed with mean 1012 and standard deviation σ.

We know that the probability that the weight exceeds 1015 g is $\dfrac{70}{1000} = 0.07$, (Fig. 5.14).

Therefore 1015 g corresponds to the standard value z

where $\quad \Phi(z) = 0.93$

$z = 1.476$

So $\dfrac{1015 - 1012}{\sigma} = 1.476$

$\sigma = \dfrac{3}{1.476}$

$= 2.0325$ g

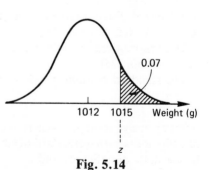

Fig. 5.14

Then $P(X < 1008) = P\left(Z < \dfrac{1008 - 1012}{2.0325}\right)$

$$= P(Z < -1.968)$$

$$= \Phi(-1.968)$$

$$= 1 - 0.9755 = 0.0245$$

Hence of the 1000 packets, we expect $1000 \times 0.0245 = 24.5$ packets to weigh less than 1008 g.

Example 3

As a result of a survey of traffic speed on a motorway, it was found that 87% of the vehicles exceed 90 km h^{-1}, whilst 9% of vehicles exceed 120 km h^{-1}. Assuming that the speeds are Normally distributed, find their mean and standard deviation.

What percentage of vehicles are exceeding the speed limit of 112 km h^{-1}?

Suppose the speeds $X \text{ km h}^{-1}$ are Normally distributed with mean μ and standard deviation σ.

90 km h^{-1} corresponds to z_1, where

$\Phi(z_1) = 0.13$, $z_1 = -1.126$

120 km h^{-1} corresponds to z_2, where

$\Phi(z_2) = 0.91$, $z_2 = 1.341$

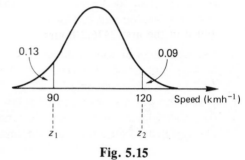

Fig. 5.15

Thus $\dfrac{90 - \mu}{\sigma} = -1.126,$ i.e. $90 = \mu - 1.126\sigma$

and $\dfrac{120 - \mu}{\sigma} = 1.341,$ i.e. $120 = \mu + 1.341\sigma$

subtracting $30 = 2.467\sigma$

$$\sigma = 12.16$$

then substituting $\mu = 103.69$

Hence the speeds have mean 103.69 km h^{-1} and standard deviation 12.16 km h^{-1}

Then $P(X > 112) = P\left(Z > \dfrac{112 - 103.69}{12.16}\right) = P(Z > 0.683)$

$$= 1 - \Phi(0.683) = 1 - 0.7527 = 0.2473$$

so 24.73% of the vehicles are exceeding the speed limit.

Exercise 5.2 Normal distributions

1 Figure 5.16(a, b) shows the pdfs of Normal variables. State approximately the mean and standard deviation of each one.

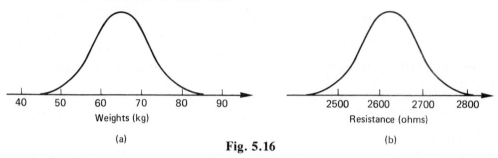

Fig. 5.16

2 Sketch the pdf for a Normal variable having mean 125 cm and standard deviation 20 cm.

3 The weights of cows are Normally distributed with mean 250 kg and standard deviation 40 kg. Find the probability that a cow weighs
(i) more than 300 kg   (ii) less than 175 kg
(iii) between 260 and 270 kg   (iv) between 235 and 265 kg.

4 The diameters of rods produced by a machine are Normally distributed with mean 4.74 cm and standard deviation 0.03 cm. Only rods with diameters between 4.7 cm and 4.8 cm are acceptable.
   If 6000 rods are produced in a day, how many of these would you expect to be
(i) rejected because they are too large
(ii) rejected because they are too small?

5 The temperature at which an individual bacterium (of a certain type) is killed is a Normal variable having mean $120°$C and standard deviation $18°$C.
(i) What percentage of the bacteria will be killed at a temperature of $100°$C?
(ii) Find the temperature at which 99% of the bacteria will be killed.

6 Cars travelling along a certain stretch of road have speeds which are Normally distributed with mean $86 \, \text{km h}^{-1}$ and standard deviation $14 \, \text{km h}^{-1}$.
(i) Find the speed which is exceeded by 60% of the cars.
(ii) Find two speeds u and v (equally spaced about the mean speed) such that 95% of cars have speeds between u and v.
(iii) Find the upper quartile, the lower quartile, and the semi-interquartile range of the speeds.

7 The useful life of a communications satellite is a Normal variable having mean 4.2 years and standard deviation 1.2 years.
(a) Find the probability that a satellite will last more than 6 years.
(b) If two satellites are launched, find the probability that
   (i) both will last for more than 6 years
   (ii) at least one of them will last for more than 6 years.

8 The heights of men may be assumed to be Normally distributed with mean 176 cm and standard deviation 10 cm. Given that 5 men are selected at random, find the probability that
(i) all five of them are taller than 170 cm
(ii) three of them are taller than 170 cm and two of them are shorter than 170 cm.

9 A machine packs flour into bags. The weights of the bags have standard deviation 0.08 kg, but the mean weight can be adjusted. You may assume that the weights are Normally distributed. To what value should the mean be set if the probability that a bag weighs less than 3 kg is to be 0.02?
   With the mean set at this value, find the probability that a bag weighs between 3.1 and 3.2 kg.

10 The current at which a certain type of fuse will blow is a Normal variable with mean 14.5 A, and 70% of the fuses will blow at a current of less than 15.0 A.
(i) Find the standard deviation of the current at which a fuse will blow.
(ii) Find the percentage of fuses which will blow at a current of less than 13.0 A.

11 The volumes of lead pellets are Normally distributed with mean 20 mm^3, and 90% of the pellets have volumes between 18 and 22 mm^3.
(i) Find the standard deviation of the volumes.
(ii) Give two values between which the volumes of 99% of the pellets will lie.

12 The time taken by a girl to travel to school has a Normal distribution. The journey takes longer than 20 minutes on 65% of days, and it takes longer than 30 minutes on 8% of days.
(i)   Find the mean and standard deviation of the journey time.
(ii) If she allows 25 minutes for the journey, what is the probability that she will be late?
(iii) How long should she allow for the journey if the probability that she will be late is to be 0.02?

13 There are 640 trees in a wood, and their heights may be assumed to be Normally distributed. 25 trees are shorter than 18 m and 110 trees are taller than 24 m.
(i) Find the mean and standard deviation of the heights.
(ii) Estimate the number of trees in the wood which have heights between 16 and 20 m.

14 Find the mean and standard deviation of the following distribution:

Concentration (mg per litre)	31–35	36–40	41–45	46–50	51–55
Frequency	6	13	28	43	55
Concentration (mg per litre)	56–60	61–65	66–70	71–75	76–80
Frequency	40	26	16	10	3

For a Normal distribution having the same mean and standard deviation, find P(60.5 < X < 65.5), and hence find the expected frequency for the class 61–65.
Similarly find the expected frequency for the class 36–40.

15 The length of a certain component is critical, and when it is manufactured the length is Normally distributed with mean 120.1 mm and standard deviation 0.25 mm.

Components with lengths between 119.8 and 120.2 mm are sold for £25.
Components with lengths between 119.5 and 119.8 mm, or between 120.2 and 120.5 mm, are sold for £20.
Components with lengths less than 119.5 mm or greater than 120.5 mm are sold for scrap at £2.

Find the probabilities that a component is sold for £25, £20 and £2, and hence find the mean selling price of these components.

5.3 The Normal approximation to the Binomial distribution

We studied the Binomial distribution $B(n, p)$ in Chapter 3. Remember that the mean is np, and the standard deviation is $\sqrt{npq}$ (where $q = 1 - p$).

If we keep p fixed, then as n increases, the histogram of the Binomial distribution $B(n, p)$ becomes more and more like the histogram of a Normal distribution.
Figure 5.17 illustrates the case $p = 0.7$, with $n = 5, 15, 25$ and 50.

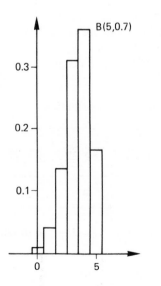

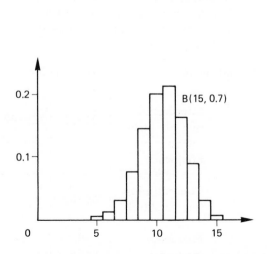

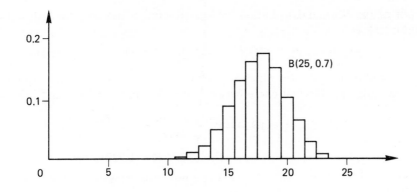

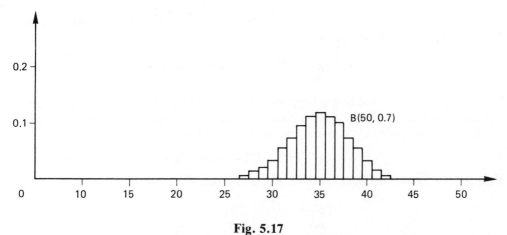

Fig. 5.17

Notice that, if X is B(50, 0.7) the possible values of X range from 0 to 50, but it is very unlikely that X is close to these extremes (for example $P(X = 0) = 7.2 \times 10^{-27}$, and $P(X = 50) = 1.8 \times 10^{-8}$). Practically all the distribution lies between 25 and 45.

Figure 5.18 shows the histogram of the Binomial distribution B(20, 0.7). Also shown

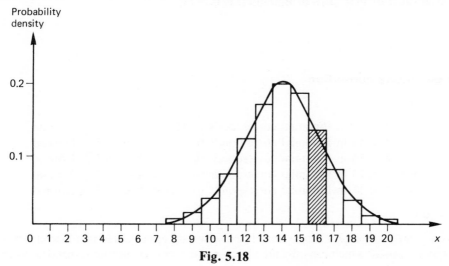

Fig. 5.18

is the pdf of the Normal distribution having the same mean and standard deviation as $B(20, 0.7)$; that is

$$\mu = 20 \times 0.7 = 14, \quad \text{and} \quad \sigma = \sqrt{20 \times 0.7 \times 0.3} = \sqrt{4.2} = 2.0494$$

The probability that $X = 16$, which is given by the shaded area, is approximately equal to the area under the Normal curve between $x = 15.5$ and $x = 16.5$, and this is

$$P(15.5 < X < 16.5) = P\left(\frac{15.5 - 14}{2.0494} < Z < \frac{16.5 - 14}{2.0494}\right)$$

$$= P(0.732 < Z < 1.220)$$

$$= \Phi(1.220) - \Phi(0.732)$$

$$= 0.8888 - 0.7679 = 0.1209$$

This compares with the true value

$$P(X = 16) = \binom{20}{16}(0.7)^{16}(0.3)^4 = 0.1304$$

The great advantage of the Normal approximation is that a probability such as $P(X \geqslant 16)$ can be found as a single area, in this case the area under the Normal curve for $x > 15.5$—instead of having to sum the individual probabilities:

$$P(X = 16) + P(X = 17) + P(X = 18) + \cdots$$

Thus $P(X \geqslant 16 \text{ in Binomial}) \approx P(X > 15.5 \text{ in Normal})$

$$= P\left(Z > \frac{15.5 - 14}{2.0494}\right)$$

$$= P(Z > 0.732)$$

$$= 1 - \Phi(0.732) = 1 - 0.7679$$

$$= 0.2321$$

The true value of $P(X \geqslant 16 \text{ in Binomial})$ is 0.2375.

The continuity correction

Notice that

$$X = 16 \text{ in Binomial corresponds to } 15.5 < X < 16.5 \text{ in Normal}$$
$$X \geqslant 16 \text{ in Binomial corresponds to } \quad X > 15.5 \text{ in Normal}$$
$$X > 16 \text{ in Binomial corresponds to } \quad X > 16.5 \text{ in Normal}$$
$$6 \leqslant X \leqslant 13 \text{ in Binomial corresponds to } 5.5 < X < 13.5 \text{ in Normal}$$
$$6 < X < 13 \text{ in Binomial corresponds to } 6.5 < X < 12.5 \text{ in Normal}$$

and so on.

This *continuity correction* is necessary because we are approximating a discrete distribution (the Binomial, which can only take integer values) by a continuous one (the Normal).

Conditions for using the Normal approximation

A Normal distribution will not always provide a good approximation to a Binomial distribution B(n, p).

A quick 'rule-of-thumb' is:

B(n, p) may be approximated by a Normal distribution, provided that np and nq are both greater than 5.

However, a better procedure is as follows:

A Normal distribution gives reasonable probabilities up to about three standard deviations each side of the mean. If this takes us outside the range of possible values of the Binomial distribution (i.e. 0 to n), then the Normal model predicts values where they cannot be, and so we would not expect it to give a good approximation. Thus

> B(n, p) may be approximated by the Normal distribution having
> mean $\mu = np$ and standard deviation $\sigma = \sqrt{npq}$.
> The approximation will be a good one if both the values $\mu \pm 3\sigma$ lie
> within the range 0 to n.

Example 1

Which of the following Binomial distributions could be satisfactorily approximated by a Normal distribution?

(i) B(20, 0.7)   (ii) B(40, 0.4)   (iii) B(50, 0.95)?

(i)   B(20, 0.7) has $\mu = 14$, $\sigma = 2.049$
      $\mu \pm 3\sigma = 14 \pm 6.15 = 7.85, 20.15$. These values are very nearly both within the range 0 to 20, so we would expect the Normal distribution to give a reasonable approximation (and we saw earlier that it did).

(ii) B(40, 0.4) has $\mu = 40 \times 0.4 = 16$, $\sigma = \sqrt{40 \times 0.4 \times 0.6} = 3.098$
      $\mu \pm 3\sigma = 16 \pm 9.30 = 6.70, 25.30$. These values are well within the range 0–40, so we would expect the Normal distribution to give a very good approximation.

(iii) B(50, 0.95) has $\mu = 50 \times 0.95 = 47.5$, $\sigma = \sqrt{50 \times 0.95 \times 0.05} = 1.541$
      $\mu \pm 3\sigma = 47.5 \pm 4.62 = 42.88, 52.12$.

The Normal distribution predicts values up to about 52, whereas the maximum possible value for the Binomial distribution is 50; so we would *not* expect the Normal distribution to give a good approximation (see Fig. 5.19).

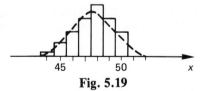

Fig. 5.19

Example 2

It is known from experience that 12% of the microchips produced by a machine are defective. Given that a sample of 150 is taken from the production line and tested, find the probability that more than 20 of them will be defective.

The number X of defectives in the sample has the Binomial distribution B(150, 0.12). This has mean $\mu = 150 \times 0.12 = 18$, and standard deviation

$$\sigma = \sqrt{150 \times 0.12 \times 0.88} = 3.980$$

Note that $\mu \pm 3\sigma = 6.06$, 29.94, which both lie well within the range 0 to 150; so this Binomial distribution can be approximated by a Normal distribution.

We require $P(X > 20)$. This includes $X = 21, 22, 23, \ldots$ but not $\ldots 18, 19, 20$, so the dividing line should be taken at 20.5.

$$P(X > 20 \text{ in Binomial}) \approx P(X > 20.5 \text{ in Normal}) = P\left(Z > \frac{20.5 - 18}{3.980}\right)$$

$$= P(Z > 0.628)$$

$$= 1 - \Phi(0.628) = 1 - 0.7350$$

$$= 0.2650$$

Example 3

In a certain city, 50 people take the driving test each day, and the probability that any one person passes the test may be taken to be $\frac{1}{3}$. Find the number of passes which is exceeded on at least 95% of days.

The number X of people who pass the test in a day has the Binomial distribution $B(50, \frac{1}{3})$, which we shall approximate by a Normal distribution having mean $\mu = 50 \times \frac{1}{3} = 16.667$, and standard deviation $\sigma = \sqrt{(50 \times \frac{1}{3} \times \frac{2}{3})} = 3.333$. We require the value of k such that $P(X > k)$ is at least 0.95.

Now $P(X > k \text{ in Binomial}) \approx P(X > k + 0.5 \text{ in Normal})$
so $k + 0.5$ corresponds to z where $\Phi(z) = 0.05$

$z = -1.645$, (Fig. 5.20).

Thus $\dfrac{(k + 0.5) - 16.667}{3.333} = -1.645$

$k = 10.68$

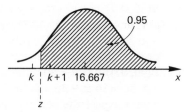

Fig. 5.20

However, k must be an integer. To make the probability *more* than 0.95 (rather than less) we must clearly decrease k to the next integer below; thus we may take $k = 10$.

More than 10 people should pass the driving test on at least 95% of days.

Exercise 5.3   The Normal approximation to the Binomial distribution

1   Which of the following Binomial distributions could be satisfactorily approximated by a Normal distribution?
(i)   B(15, 0.4)   (ii) B(40, 0.25)   (iii) B(55, 0.9)
(iv) B(100, 0.02)   (v) B(300, 0.08)

2   The random variable X has the Binomial distribution B(25, 0.2)
(a) Find the mean and standard deviation of X.

(b) Find the probability that $X = 2$
  (i) directly from the Binomial distribution
  (ii) using the Normal approximation.
(c) Use the Normal approximation to find the probability that $X \geqslant 7$.

3 Given that a coin is tossed 200 times, find the mean and standard deviation of the number of heads.
  Find the probability that the number of heads is between 90 and 110 inclusive.

4 A die is thown 120 times. Find the probability that there are less than 15 sixes
  (i) if the die is a fair one
  (ii) if the die is biased so that the probability of scoring a six is $\dfrac{1}{10}$.

5 A multiple-choice test consists of 45 questions, and for each question there is a choice of 4 answers. For a student who simply guesses the answer to each question, find the probability that
  (i)   he gets at least 16 answers right
  (ii) he gets less than 10 answers right
  (iii) he gets between 8 and 12 answers right (inclusive).

6 In a certain town, 46% of the population are under 30 years of age. If a random sample is taken, find the probability that more than half of the people in the sample are under 30
  (i) for a sample of 100 people
  (ii) for a sample of 225 people.

7 A secretarial agency has 80 clients, and the probability that a client will require a secretary on a given day is $\frac{1}{5}$.
  Find the smallest number of secretaries which should be available if the agency is to be able to satisfy all the requests on at least 99% of days.
  Find also the probability that less than half this number of secretaries will be required.

8 On average 2% of the bolts produced by a machine are defective. For a batch of 25 000 bolts, find two numbers between which the number of defective bolts will lie with probability approximately 0.95.

6
Paired Data and Correlation

6.1 Scatter diagrams and covariance

We now consider situations where two variables (x and y) are defined simultaneously. For example, we might have the height and weight of a person, or the examination marks of a candidate in two subjects, or the length and breadth of a leaf, and so on.

Example 1
A sample of 12 pupils took two tests, with the following results:

Mark in pure mathematics (x)	36	25	48	29	19	43	57	55	39	33	45	20
Mark in statistics (y)	39	32	35	23	13	44	35	42	30	31	25	25

One pupil scored 36 in pure mathematics and 39 in statistics; a second pupil scored 25 in pure mathematics and 32 in statistics, and so on.

Thus the data occurs in pairs, 36 and 39, 25 and 32, ... When we plot the pairs (x, y) on a graph, we obtain a *scatter diagram*. In Fig. 6.1 we plot the points (36, 39), (25, 32), ..., (20, 25).

The points are fairly well scattered, but as x increases then, generally speaking, so does y (i.e. pupils who scored higher marks in pure mathematics tend also to have scored higher marks in statistics). This underlying trend could reasonably be represented by a straight line.

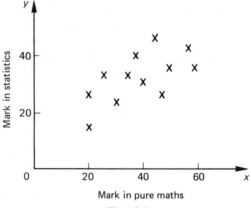

Fig. 6.1

In general, if the points on a scatter diagram are scattered at random and there is no discernible relationship between x and y, we say that the variables are *independent*.

If there *is* some connection between the variables, then this association may be.

linear (if the underlying trend can be represented by a straight line)
or *non-linear* (if the trend is better represented by a curve).

If *y* tends to increase as *x* increases, we say there is *positive correlation*.
If *y* tends to decrease as *x* increases, we say there is *negative correlation*.

Otherwise we say there is *no correlation* (this includes the case where *x* and *y* are independent; it also includes cases where *x* and *y* are related, but part of the scatter diagram shows positive correlation and a roughly equal part shows negative correlation).

The degree of correlation may be described as *good*, *fair* or *poor*, depending on how close the points are to a straight line.

In Example 1, the scatter diagram shows that there is fair positive correlation; the underlying relationship could be linear.

Examples of scatter diagrams

Figures 6.2–4 show some cases with varying degrees of correlation.

Fig. 6.2 Good negative correlation; the underlying trend is linear.

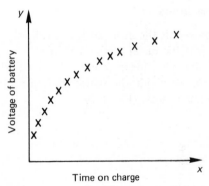

Fig. 6.3 Good positive correlation; but the relationship is non-linear.

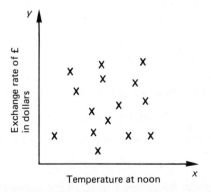

Fig. 6.4 No correlation; the variables appear to be independent.

The importance of scale

The appearance of a scatter diagram can be altered by changing the scales. The scatter diagram for Example 1 is redrawn in Fig. 6.5 with a different scale on the *y*-axis.

This diagram might suggest that the points lie very close to a straight line, which is misleading.

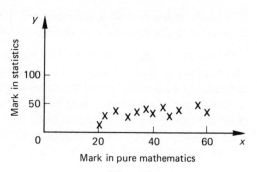

Fig. 6.5

The scales should be arranged so that the range of *x* and the range of *y* are represented by approximately equal lengths on the axes.

Covariance

Suppose $\bar{x}$ is the mean of the *x*-values and $\bar{y}$ is the mean of the *y*-values.

Using the 'mean point' $(\bar{x}, \bar{y})$ as the centre, we divide the scatter diagram into four quadrants, (Fig. 6.6).

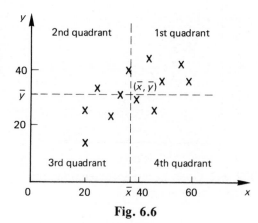

Fig. 6.6

For Example 1 we have

$$\bar{x} = \frac{\sum x}{n} = \frac{449}{12} = 37.42 \text{ and } \bar{y} = \frac{\sum y}{n} = \frac{374}{12} = 31.17$$

If there is positive correlation then the points lie mainly in the

1st and 3rd quadrants.

If there is negative correlation then the points lie mainly in the

2nd and 4th quadrants.

If there is no correlation then the points are scattered approximately

equally in all four quadrants.

Suppose (x, y) is a point in the scatter diagram and we now consider the sign of $(x-\bar{x})(y-\bar{y})$:

When (x, y) is	$(x-\bar{x})$ is	$(y-\bar{y})$ is	$(x-\bar{x})(y-\bar{y})$ is
In the 1st quadrant, i.e. $x > \bar{x}, y > \bar{y}$	+	+	+
In the 2nd quadrant, i.e. $x < \bar{x}, y > \bar{y}$	−	+	−
In the 3rd quadrant, i.e. $x < \bar{x}, y < \bar{y}$	−	−	+
In the 4th quadrant, i.e. $x > \bar{x}, y < \bar{y}$	+	−	−

Thus $(x-\bar{x})(y-\bar{y})$ is positive in the 1st and 3rd quadrants
and negative in the 2nd and 4th quadrants.
We define the *covariance*, cov (x, y), to be the average value of $(x-\bar{x})(y-\bar{y})$.
For n points $(x_1, y_1), (x_2, y_2),...,(x_n, y_n)$ we have

$$\text{cov } (x, y) = \frac{\sum_{i=1}^{n} (x_i-\bar{x})(y_i - \bar{y})}{n}$$

which we shall usually write simply as cov $(x, y) = \dfrac{\sum (x-\bar{x})(y-\bar{y})}{n}$

If there is positive correlation, then cov (x, y) is positive
If there is negative correlation, then cov (x, y) is negative
If there is no correlation, then cov (x, y) is close to zero.
Note (1) The units of covariance are (units of x) × (units of y)
(2) The name 'covariance' suggests a link with variance. In fact when the y-values are equal to the x-values, the covariance

$$\text{cov } (x, y) \text{ becomes cov } (x, x) = \frac{\sum (x-\bar{x})(x-\bar{x})}{n} = \frac{\sum (x-\bar{x})^2}{n}$$

which is the variance of the x-values.

Alternative formula

We have $\text{cov } (x, y) = \dfrac{\sum_{i=1}^{n} (x_i-\bar{x})(y_i - \bar{y})}{n} = \dfrac{\sum (x_i y_i - x_i \bar{y} - \bar{x} y_i + \bar{x}\bar{y})}{n}$

$$= \frac{\sum x_i y_i}{n} - \bar{y} \frac{\sum x_i}{n} - \bar{x} \frac{\sum y_i}{n} + \frac{n\bar{x}\bar{y}}{n}$$

$$= \frac{\sum x_i y_i}{n} - \bar{y}\bar{x} - \bar{x}\bar{y} + \bar{x}\bar{y}$$

$$= \frac{\sum x_i y_i}{n} - \bar{x}\bar{y}$$

Thus

$$\text{cov}(x, y) = \frac{\sum xy}{n} - \bar{x}\bar{y}$$

$$= \frac{\sum xy}{n} - \left(\frac{\sum x}{n}\right)\left(\frac{\sum y}{n}\right)$$

This formula is the one most often used when calculating the covariance for a set of data; it corresponds to the formula $\dfrac{\sum x^2}{n} - \bar{x}^2$ for the variance.

Example 2
Calculate the covariance for the marks in pure mathematics and statistics given in Example 1.

The table sets out the necessary working.

x	y	xy
36	39	1 404
25	32	800
48	35	1 680
29	23	667
19	13	247
43	44	1 892
57	35	1 995
55	42	2 310
39	30	1 170
33	31	1 023
45	25	1 125
20	25	500
449	374	14 813

$$\text{cov}(x, y) = \frac{\sum xy}{n} - \left(\frac{\sum x}{n}\right)\left(\frac{\sum y}{n}\right)$$

$$= \frac{14813}{12} - \left(\frac{449}{12}\right)\left(\frac{374}{12}\right)$$

$$= 68.26$$

Covariance of linear functions

Suppose that pairs (x, y) are related to pairs (u, v) by $x = au + b$, $y = cv + d$.
Then $\bar{x} = a\bar{u} + b$, and so $x - \bar{x} = au + b - (a\bar{u} + b) = a(u - \bar{u})$
Similarly $\bar{y} = c\bar{v} + d$, and so $y - \bar{y} = c(v - \bar{v})$

Thus $\text{cov}(x, y) = \dfrac{\sum(x - \bar{x})(y - \bar{y})}{n}$

$$= \frac{\sum a(u - \bar{u})\, c(v - \bar{v})}{n}$$

$$= \frac{ac \sum (u - \bar{u})(v - \bar{v})}{n}$$

$$= ac\, \text{cov}(u, v)$$

We use this result when we have 'coded' the x and y values by

$$u = \frac{x-b}{a}, \ v = \frac{y-d}{c}.$$

Example 3

15 pairs (x, y) were coded as $u = \dfrac{x-850}{25}$, $v = \dfrac{y-4.7}{0.1}$ and the covariance of u and v was calculated as cov $(u, v) = -7.84$.

Find the covariance of x and y.

Since $x = 25u + 850$ and $y = 0.1v + 4.7$, we have

$$\text{cov} \ (x, y) = 25 \times 0.1 \times \text{cov} \ (u, v)$$

$$= -19.6.$$

Exercise 6.1   Scatter diagrams and covariance

1   Comment on the relationships illustrated by the scatter diagram in Fig. 6.7(a–g). State the sign (positive or negative) and degree (good, fair or poor) of any correlation, and state whether the underlying trend (if any) is linear or non-linear.

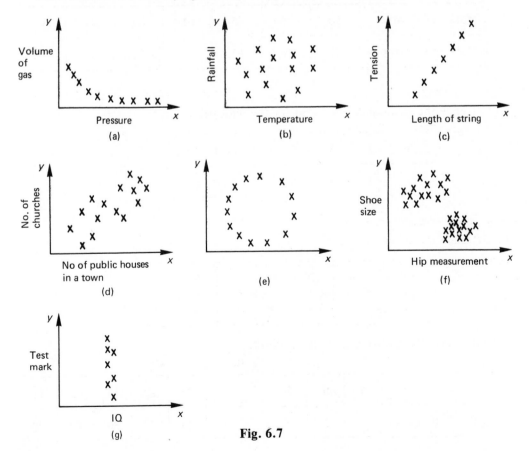

Fig. 6.7

2 Plot scatter diagrams for the three sets of data in the tables below. Choose scales so that the ranges of x and y occupy approximately equal lengths. Comment on the relationship between x and y.

(a)

x Price (pence)	69	91	80	70	110	101	92	129	114	118
y Volume of sales (1000s)	19.4	17.1	16.0	17.6	13.8	15.6	15.9	13.6	15.8	14.4

(b)

x Speed (km h^{-1})	34	40	54	76	90	108	114	140
y Fuel consumption (ℓ per 100 km)	12.0	10.1	8.8	8.2	7.1	7.4	9.6	10.9

(c)

x Mark in History	55	50	60	41	90	55	53
y Mark in Geography	60	55	55	70 ·	90	70	66
x Mark in History	50	46	45	61	49	44	37
y Mark in Geography	71	58	65	65	62	52	55

3 Calculate the covariance cov (x, y) for the data below.

x	2	3	5	4	1	3	2	8
y	1	5	6	4	0	1	2	10

4 Calculate the covariance for

u	12	10	3	5	1
v	3	7	8	6	9

and deduce the covariance for the following:

(i)

x	13	11	4	6	2
y	0	4	5	3	6

(ii)

s	22	20	13	15	11
t	30	70	80	60	90

(iii)

p	6	5	1.5	2.5	0.5
q	7.3	7.7	7.8	7.6	7.9

(iv)

w	14	10	-4	0	-8
z	-20	0	5	-5	10

6.2 The coefficient of correlation

The sign of the covariance $\text{cov}(x, y)$ indicates whether there is positive or negative correlation, but the numerical value (for example, $\text{cov}(x, y) = -19.6$ in Example 3 of Section 6.1) is difficult to interpret, as it depends on the units in which x and y are measured.

The *(product moment) coefficient of correlation*, r is defined as:

$$r = \frac{\text{cov}(x, y)}{s_x s_y}$$

where s_x and s_y are the standard deviations of x and of y.

The units of $\text{cov}(x, y)$ are (units of x) $\times$ (units of y) and the units of $s_x s_y$ are also (units of x) $\times$ (units of y), so the value of r does not depend on the units in which x and y are measured.

We shall later show that

$$-1 \leqslant r \leqslant 1$$

Example 1
Calculate the coefficient of correlation for the pairs

x	0	1	2	3	4
y	2	5	8	11	14

The work is set out in the table below, and Fig. 6.8 shows the scatter diagram.

x	y	x^2	y^2	xy
0	2	0	4	0
1	5	1	25	5
2	8	4	64	16
3	11	9	121	33
4	14	16	196	56
10	40	30	410	110

$$\text{cov}(x, y) = \frac{110}{5} - \left(\frac{10}{5}\right)\left(\frac{40}{5}\right) = 6$$

$$s_x = \sqrt{\frac{30}{5} - \left(\frac{10}{5}\right)^2} = 1.4142$$

$$s_y = \sqrt{\frac{410}{5} - \left(\frac{40}{5}\right)^2} = 4.2426$$

$$r = \frac{6}{1.4142 \times 4.2426} = 1$$

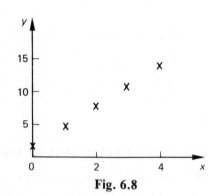

Fig. 6.8

Example 2

Calculate the coefficient of correlation for the following pairs of marks.

Mark in pure mathematics (x)	36	25	48	29	19	43	57	55	39	33	45	20
Mark in statistics (y)	39	32	35	23	13	44	35	42	30	31	25	25

The scatter diagram is shown in Fig. 6.9.

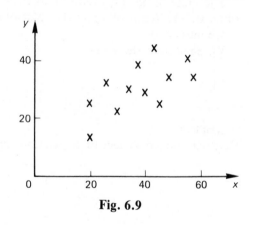

We have $\sum x = 449$, $\sum x^2 = 18\,585$

$\sum y = 374$, $\sum y^2 = 12\,504$

$\sum xy = 14\,813$

$\text{cov}(x, y) = 68.26$

$$s_x = \sqrt{\frac{18\,585}{12} - \left(\frac{449}{12}\right)^2} = 12.20$$

$$s_y = \sqrt{\frac{12\,504}{12} - \left(\frac{374}{12}\right)^2} = 8.405$$

Fig. 6.9

$$r = \frac{68.26}{12.20 \times 8.405}$$

$$= 0.666$$

Example 3

Find the coefficient of correlation for the following pairs:

x	17	12	5	21	24	8	26	10
y	4.0	3.2	4.3	4.7	3.4	5.0	4.4	4.0

The scatter diagram is shown in Fig. 6.10

We have $\sum x = 123,$ $\sum x^2 = 2\,315$

$\sum y = 33.0,$ $\sum y^2 = 138.74$

$\sum xy = 502.6$

cov $(x, y) = -0.5969$

$s_x = 7.279,$ $s_y = 0.5717$

$r = -0.143$

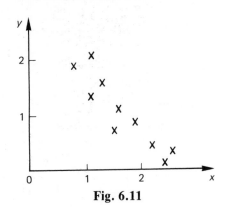

Fig. 6.10

Example 4

Find the coefficient of correlation for the following pairs:

x	1.1	1.5	1.9	1.3	1.6	2.2	0.8	2.6	1.1	2.5
y	1.3	0.7	0.8	1.5	1.1	0.4	1.8	0.3	2.0	0.1

The scatter diagram is shown in Fig. 6.11.

Using a calculator,

$\bar{x} = 1.66,$ $s_x = 0.5886$

$\bar{y} = 1.00,$ $s_y = 0.6148$

Also $\sum xy = 13.26,$

so cov $(x, y) = \dfrac{13.26}{10} - 1.66 \times 1.00$

$= -0.334$

$r = \dfrac{-0.334}{0.5886 \times 0.6148}$

$= -0.923$

Fig. 6.11

Example 5

Calculate the coefficient of correlation for the following pairs:

x	3.2	5.2	6.6	2.6	3.8	2.3	2.2	6.0	2.5	4.4	2.4
y	9	5	4	13	7	23	27	4	15	6	18

The scatter diagram is shown in Fig. 6.12

We have $\bar{x} = 3.7455$, $s_x = 1.5144$

$\bar{y} = 11.9091$, $s_y = 7.6212$

$\sum xy = 385.0$

$\text{cov}(x, y) = -9.6055$

$r = -0.832$

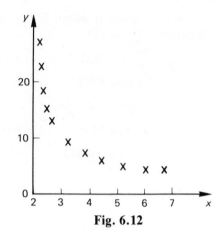

Fig. 6.12

Example 6

Calculate the coefficient of correlation for the following pairs:

x	1.3	2.5	2.0	0.2	0.8	1.0	0.1	0.4
y	0.71	0.81	0.75	0.79	0.73	0.72	0.80	0.77
x	2.3	2.1	1.5	1.8	0.5	0.7	2.2	
y	0.78	0.76	0.72	0.73	0.75	0.74	0.77	

The scatter diagram is shown in Fig. 6.13.

We have $\bar{x} = 1.2933$, $s_x = 0.7945$

$\bar{y} = 0.7553$, $s_y = 0.02986$

$\sum xy = 14.669$

$\text{cov}(x, y) = 0.0011$

$r = 0.047$

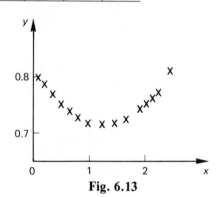

Fig. 6.13

Interpretation of *r*

If $r = +1$,   then all the points lie exactly on a straight line with positive gradient; this is perfect positive correlation.

If $r = -1$,   then all the points lie exactly on a straight line with negative gradient; this is perfect negative correlation.

Examples 1–6 show that the value of *r* indicates how good the correlation is; the closer *r* is to ± 1, the closer the points are to a straight line.

It is important to remember that r is only a measure of *linear* association.

In Examples 1–4 the underlying trend is linear.

In Example 5, there is almost an exact relationship between x and y; however the relationship is non-linear. The value $r = -0.832$ is a measure of how close the points are to a straight line, which is somewhat irrelevant in this situation.

In Example 6, r is approximately zero, yet there is a very strong (non-linear) relationship between x and y.

If r is close to ± 1, we can deduce that the points lie close to a straight line.

However, if $r \approx 0$, we can conclude only that there is no *linear* relationship. It is always advisable to draw a scatter diagram before deducing too much from the value of r.

Cause and effect

Even when there is good correlation between x and y, we cannot necessarily conclude that the changes in x are actually causing the changes in y. For example, if for each town, we let x be the number of churches and y be the number of violent crimes committed in the year 1983, we shall probably find that there is good positive correlation between x and y. This does not imply that a higher number of churches causes an increase in crime; it simply means that both x and y tend to increase with the size of the town.

Linear functions

If $x = au + b$ and $y = cv + d$ (where a and c are positive) then $\operatorname{cov}(x, y) = ac \operatorname{cov}(u, v)$.
Also $s_x = as_u$ and $s_y = cs_v$
Thus

$$\frac{\operatorname{cov}(x, y)}{s_x s_y} = \frac{ac \operatorname{cov}(u, v)}{as_u \, cs_v} = \frac{\operatorname{cov}(u, v)}{s_u s_v}$$

Hence the coefficient of correlation between x and y is the same as the coefficient of correlation between u and v.

If x and y are coded as $u = \dfrac{x - b}{a}$, $v = \dfrac{y - d}{c}$ and we calculate the coefficient of correlation between u and v, then this is also the coefficient of correlation between x and y.

Example 7
Calculate the coefficient of correlation, and the covariance, for the following data:

x	174.7	175.0	175.7	174.2	175.3	175.9
y	6740	6720	6700	6640	6800	6900

We first code the data.

$u = \dfrac{x - 175.0}{0.1}$	-3	0	7	-8	3	9
$v = \dfrac{y - 6700}{20}$	2	1	0	-3	5	10

Hence, $\sum u = 8$, $\sum v = 15$, $\sum u^2 = 212$, $\sum v^2 = 139$, $\sum uv = 123$

$$\text{cov } (u, v) = \frac{123}{6} - \left(\frac{8}{6}\right)\left(\frac{15}{6}\right) = 17.167$$

$$s_u = \sqrt{\frac{212}{6} - \left(\frac{8}{6}\right)^2} = 5.793, \quad s_v = \sqrt{\frac{139}{6} - \left(\frac{15}{6}\right)^2} = 4.113$$

$$r = \frac{17.167}{5.793 \times 4.113} = 0.721$$

$$\text{cov } (x, y) = 0.1 \times 20 \times 17.167 = 34.33$$

Exercise 6.2   Coefficient of correlation

1   What can you deduce if a calculation of the coefficient of correlation r yields
    (i) $r = 0.95$ (ii) $r = -0.8$ (iii) $r = 1$ (iv) $r = -0.02$ (v) $r = -1.2$?

2   Calculate the coefficient of correlation for the following

x No of girls in family	3	1	2	1	0	4
y No of boys in family	0	2	2	1	3	1

3   Calculate the coefficient of correlation for the following

x Mark in Paper 1	83	18	58	66	62	50	59	32	5	28
y Mark in Paper 2	73	46	50	86	76	63	51	61	36	36

4   Use the coding $u = \dfrac{x - 360}{5}$, $v = \dfrac{y - 18.70}{0.01}$ to calculate the coefficient of correlation for the following

x Temperature (°C)	335	350	355	360	370	380	400
y Breaking strain (kN)	18.78	18.76	18.75	18.72	18.73	18.72	18.70

State the value of the covariance cov (x, y)

5   Calculate the coefficient of correlation for the following

x Duration of contract (weeks)	5.4	2.5	4.7	0.7	0.9	1.7	0.2	4.8
y Percentage profit	9	23	17	6	11	21	1	14
x Duration of contract	5.7	0.5	2.1	5.8	4.3	1.1	3.4	
y Percentage profit	1	10	24	4	20	17	24	

given that $\sum x = 43.8$, $\sum x^2 = 186.82$, $\sum y = 202$, $\sum y^2 = 3652$, $\sum xy = 573.8$.
What do you deduce from this value? Plot a scatter diagram to check your conclusions.

6   A set of 50 pairs (x, y) has $\sum x = 100$, $\sum x^2 = 208$, $\sum y = 900$, $\sum y^2 = 16\,650$, $\sum xy = 1836$.
Calculate the coefficient of correlation.

7   For a set of n pairs (x, y) express $\sum x$, $\sum y$, $\sum x^2$, $\sum y^2$, $\sum xy$ in terms of n, $\bar{x}$, $\bar{y}$, s_x, s_y, and cov (x, y).
A set of 50 pairs (x, y) has $\bar{x} = 15$, $\bar{y} = 4$, $s_x = 2.6$, $s_y = 0.8$, and the coefficient of correlation is $r = 0.6$.
Find cov (x, y) and $\sum xy$.

8   A set of 20 pairs (x, y) has $\bar{x} = 12$, $\bar{y} = 22$, $s_x = 2.5$, $s_y = 4$, $r = 0.1$.
Find $\sum x$, $\sum y$, $\sum x^2$, $\sum y^2$ and $\sum xy$.
Another set of 10 pairs (x, y) has $\bar{x} = 30$, $\bar{y} = 54$, $s_x = 3$, $s_y = 2$, $r = -0.4$.
Taking all 30 pairs together, calculate the coefficient of correlation between x and y.
By sketching a scatter diagram, explain the result which you have obtained.

9   A set of 12 pairs (x, y) has $\bar{x} = 5.6$, $\bar{y} = 12$, $s_x = 1.2$, $s_y = 2$, $r = 0.94$.
Given that two extra pairs $(10, 2)$ and $(11.5, 3)$ are added, calculate the coefficient of correlation for all 14 pairs together.

6.3   Frequency tables for paired data

When there are many pairs (x, y), the data may be given in the form of a frequency table.

Example

Pairs of values (x, y) occur with frequencies given in the table. Find the coefficient of correlation between x and y.

x \ y	6	7	8	9
1	3	1	0	0
2	2	4	1	0
3	1	6	5	1
4	0	3	4	2
5	0	0	0	1

The figures inside this table are the frequencies, so for example,
the pair $(1, 6)$ occurs 3 times
the pair $(2, 7)$ occurs 4 times   and there are 34 pairs altogether.
The value $x = 1$ occurs 4 times, $x = 2$ occurs 7 times, and so on. By adding the frequencies in the rows, we obtain the frequency distribution of x, which we can use to find $\bar{x}$ and s_x.
    Similarly, by adding the frequencies in the columns we obtain the frequency distribution of y.
    The completed table is shown below

x \ y	6	7	8	9	f	xf	x^2f
1	3	1	0	0	4	4	4
2	2	4	1	0	7	14	28
3	1	6	5	1	13	39	117
4	0	3	4	2	9	36	144
5	0	0	0	1	1	5	25
f	6	14	10	4	34	98	318
yf	36	98	80	36	250		
y^2f	216	686	640	324	1866		

Now $s_x = \sqrt{\dfrac{\sum x^2f}{\sum f} - \left(\dfrac{\sum xf}{\sum f}\right)^2} = \sqrt{\dfrac{318}{34} - \left(\dfrac{98}{34}\right)^2} = 1.0222$

and $s_y = \sqrt{\dfrac{\sum y^2f}{\sum f} - \left(\dfrac{\sum yf}{\sum f}\right)^2} = \sqrt{\dfrac{1866}{34} - \left(\dfrac{250}{34}\right)^2} = 0.9037$

The covariance $\text{cov}(x, y) = \dfrac{\sum xyf}{\sum f} - \left(\dfrac{\sum xf}{\sum f}\right)\left(\dfrac{\sum yf}{\sum f}\right)$

To find $\sum xyf$, we compute xyf for each of the 20 compartments in the table, and add these values; in the first compartment, $x = 1$, $y = 6$, $f = 3$, so $xyf = 18$, and so on.

We obtain $\sum xyf = $
$$
\begin{aligned}
& 18 + 7 + 0 + 0 \\
& + 24 + 56 + 16 + 0 \\
& + 18 + 126 + 120 + 27 \\
& + 0 + 84 + 128 + 72 \\
& + 0 + 0 + 0 + 45 = 741
\end{aligned}
$$

Thus $\quad \text{cov}(x, y) = \dfrac{\sum xyf}{\sum f} - \left(\dfrac{\sum xf}{\sum f}\right)\left(\dfrac{\sum yf}{\sum f}\right)$

$$
= \frac{741}{34} - \left(\frac{98}{34}\right)\left(\frac{250}{34}\right)
$$

$$
= 0.6003
$$

and hence $\quad r = \dfrac{0.6003}{1.0222 \times 0.9037}$

$$
= 0.650
$$

Note that a calculator could be used to find $\bar{x}$ and s_x (by entering 1×4, 2×7, 3×13, 4×9, 5×1), and similarly to find $\bar{y}$ and s_y, in which case the xf, x^2f, yf and y^2f columns would not be needed.

The x and y values may be grouped; we would then use the mid-interval values. Also the values of x and y may be coded before carrying out the calculations.

Exercise 6.3 Frequency tables for paired data

1 96 applicants for a job were each given an initiative test and a general knowledge test. Copy and complete the following table

Score in general knowledge test

			0–19	20–39	40–59			
		y	9.5	29.5	49.5			
		v	-1	0	1			
x	u					f	uf	u^2f
6–15	10.5	-1	13	8	2	23	-23	23
16–25	20.5	0	12	10	6	28	0	0
26–35	30.5	1	4	15	15			
36–45	40.5	2	0	3	8			
		f	29			96	33	
		vf	-29					
		v^2f	29			60	$\sum uvf =$	

Score in initiative test

(i)   Explain how the x values (10.5, 20.5, ...) are calculated.

(ii)   The x values are coded as $u = \dfrac{x - 20.5}{10}$. Explain how the y-values are coded.

(iii)   Explain what the figure 13 in the table means.

(iv)   Calculate the coefficient of correlation between the scores.

(v)   State the covariance of the two scores, cov (x, y).

2   The following table shows the swimming times (for 200 m) and the shooting scores for 54 competitors in the National Triathlon Championships 1981. Calculate the coefficient of correlation between the swimming time and the shooting score.

Shooting score

		60–69	70–79	80–89	90–99
	130–140	1	2	3	0
	140–150	3	4	9	5
Swimming time (seconds)	150–160	4	1	4	5
	160–170	2	2	4	1
	170–180	1	0	1	0
	180–190	0	0	1	1

6.4   Rank correlation

Ranking

A set of numbers is ranked by assigning rank 1 to the highest value, rank 2 to the next highest value, and so on.

Example 1

Rank the values of x and the values of y in the following data.

x	83	99	108	69	77	58	106	90
y	98.6	101.3	98.9	100.1	97.3	97.1	101.7	99.5

We rank the x-values and the y-values separately, and the results are given in the table below.

Considering the values of x, 108 is the highest value, 106 is the second highest, 99 is the third highest, and so on.

Considering the values of y, 101.7 is the highest value, 101.3 is the second highest, and so on.

x	y	Ranks of x u	of y v
83	98.6	5	6
99	101.3	3	2
108	98.9	1	5
69	100.1	7	3
77	97.3	6	7
58	97.1	8	8
106	101.7	2	1
90	99.5	4	4

Spearman's coefficient of rank correlation

If we calculate the (product moment) coefficient of correlation (as defined in Section 6.2), using the ranks (u, v) instead of the original values (x, y), we obtain

Spearman's coefficient of rank correlation r_s

Example 2
Calculate Spearman's coefficient of rank correlation for the data given in Example 1.

From the ranks (u, v), we have

$$\sum u = 5 + 3 + \cdots + 4 = 36 \qquad \sum u^2 = 5^2 + 3^2 + \cdots + 4^2 = 204$$

$$\sum v = 6 + 2 + \cdots + 4 = 36 \qquad \sum v^2 = 6^2 + 2^2 + \cdots + 4^2 = 204$$

$$\sum uv = 5 \times 6 + 3 \times 2 + \cdots + 4 \times 4 = 186$$

Hence $\quad \text{cov}\,(u, v) = \dfrac{186}{8} - \left(\dfrac{36}{8}\right)\left(\dfrac{36}{8}\right) = 3$

$$s_u = \sqrt{\dfrac{204}{8} - \left(\dfrac{36}{8}\right)^2} = 2.291 \qquad s_v = 2.291$$

$$r_s = \dfrac{\text{cov}\,(u, v)}{s_u s_v}$$

$$= 0.57$$

In general, if there are n pairs of values (x, y), then the x ranks (u) will be the numbers $1, 2, \ldots, n$ in some order

and so $\quad \sum u = \dfrac{n(n + 1)}{2} \qquad$ and $\qquad \sum u^2 = \dfrac{n(n + 1)(2n + 1)}{6}$

Similarly, the *y*-ranks (v) will satisfy

$$\sum v = \frac{n(n+1)}{2} \qquad \text{and} \qquad \sum v^2 = \frac{n(n+1)(2n+1)}{6}$$

So $\quad r_s = \dfrac{\text{cov}(u,v)}{s_u s_v}$

$$= \frac{\dfrac{\sum uv}{n} - \left(\dfrac{\sum u}{n}\right)\left(\dfrac{\sum v}{n}\right)}{\sqrt{\dfrac{\sum u^2}{n} - \left(\dfrac{\sum u}{n}\right)^2}\sqrt{\dfrac{\sum v^2}{n} - \left(\dfrac{\sum v}{n}\right)^2}}$$

$$= \frac{\dfrac{\sum uv}{n} - [\tfrac{1}{2}(n+1)]^2}{\tfrac{1}{6}(n+1)(2n+1) - [\tfrac{1}{2}(n+1)]^2}$$

$$= \frac{12 \sum uv - 3n(n+1)^2}{2n(n+1)(2n+1) - 3n(n+1)^2}$$

This formula is simplified if we express $\sum uv$ in terms of $\sum (u-v)^2$

We have $\quad \sum (u-v)^2 = \sum (u^2 - 2uv + v^2)$

$$= \sum u^2 - 2\sum uv + \sum v^2$$

$$= \tfrac{1}{3}n(n+1)(2n+1) - 2\sum uv$$

thus

$$12 \sum uv = 2n(n+1)(2n+1) - 6 \sum (u-v)^2$$

and so

$$r_s = \frac{2n(n+1)(2n+1) - 6\sum (u-v)^2 - 3n(n+1)^2}{2n(n+1)(2n+1) - 3n(n+1)^2}$$

$$= 1 - \frac{6 \sum (u-v)^2}{n(n+1)(4n+2-3n-3)}$$

$$= 1 - \frac{6 \sum (u-v)^2}{n(n^2-1)}$$

Hence, if $d = u - v$ is the difference between the ranks of corresponding values,

$$\boxed{r_s = 1 - \frac{6 \sum d^2}{n(n^2-1)}}$$

Example 3
Calculate Spearman's coefficient of rank correlation for the data given in Example 1.

The table shows the ranks for x and y and the values of d^2

x	y	Ranks of x	of y	d^2
83	98.6	5	6	1
99	101.3	3	2	1
108	98.9	1	5	16
69	100.1	7	3	16
77	97.3	6	7	1
58	97.1	8	8	0
106	101.7	2	1	1
90	99.5	4	4	0
				$\sum d^2 = 36$

Now $n = 8$, and thus $r_s = 1 - \dfrac{6 \times 36}{8 \times 63}$

$$= 0.57 \text{ (as before)}$$

Example 4

The five finalists in a music competition were A, B, C, D, E and the two judges placed them as follows:

First judge: 1st B, 2nd E, 3rd D, 4th A, 5th C
Second judge: 1st C, 2nd B, 3rd A, 4th D, 5th E

Calculate Spearman's coefficient of rank correlation between the placings of the two judges.

We do not know the marks awarded by the judges, but we are given the rankings.

Finalist	Ranks First judge	Second judge	d^2
A	4	3	1
B	1	2	1
C	5	1	16
D	3	4	1
E	2	5	9
			$\sum d^2 = 28$

Now $n = 5$ and so $r_s = 1 - \dfrac{6 \times 28}{5 \times 24}$

$$= -0.40$$

The negative value of r_s indicates that high rankings by one judge tend to be associated with low rankings by the other judge; so these two judges do not agree at all well.

Tied ranks

When two or more x-values (or y-values) are equal, each is given the average rank for the tied places. For example,

if the 2nd and 3rd highest values are equal, each is assigned the rank 2.5;

if the 7th, 8th and 9th highest values are equal, each is assigned rank 8, and so on.

In the derivation of the formula

$$r_s = 1 - \frac{6 \sum d^2}{n(n^2 - 1)}$$

it was assumed that no ranks were tied. It is possible to modify the formula to allow for tied ranks (see Exercise 6.4, Question 7), but this is not usually done, since it is rarely necessary to know the *exact* value of r_s.

Example 5

Calculate Spearman's coefficient of rank correlation for the following test marks.

Mark in pure mathematics x	Mark in statistics y	Ranks		d^2
		Pure mathematics	Statistics	
36	39	7	3	16
25	32	10	6	16
48	35	3	4.5	2.25
29	23	9	11	4
19	13	12	12	0
43	44	5	1	16
57	35	1	4.5	12.25
55	42	2	2	0
39	30	6	8	4
33	31	8	7	1
45	25	4	9.5	30.25
20	25	11	9.5	2.25

$$\sum d^2 = 104$$

Now $n = 12$, and so $r_s \approx 1 - \dfrac{6 \times 104}{12 \times 143}$

$$= 0.64$$

Note that this value is quite close to the (product moment) coefficient of correlation between x and y, which is $r = 0.666$ (see Example 2, p. 102).

Use of rank correlation

Spearman's coefficient of rank correlation is easy to calculate, and it may be used to give a quick indication of the degree of correlation between two variables x and y. However, by using only the ranks, much information about x and y is lost; for example, the amount by which one value is greater than another, and the nature of the relationship (linear or

non-linear). It cannot be assumed in general that r_s will be very close to the product moment correlation coefficient r.

Rank correlation is particularly suitable when the actual numerical values of x and y are not very reliable (for example, if they are the results of subjective assessments), but they can reasonably be used to rank the values.

Exercise 6.4   Rank correlation

1   Calculate a coefficient of rank correlation between the following temperatures and rainfalls (recorded in 1981).

	Jan	Feb	Mar	Apr	May	Jun	Jul	Aug	Sep	Oct	Nov	Dec
Temperature (°C)	5.2	3.7	8.3	8.2	11.6	13.8	16.0	16.6	15.1	8.9	8.1	1.4
Rainfall (mm)	58	52	153	64	91	49	55	48	141	124	69	94

2   Calculate a coefficient of rank correlation for the following data.

Rate of inflation (%)	11.7	11.9	12.0	11.7	11.1	10.3	9.8	9.4	9.1	8.6
Percentage increase in average earnings	10.8	11.1	10.7	10.8	11.1	10.9	10.5	10.1	10.4	9.5

3   Calculate a coefficient of rank correlation between the following mock examination marks and the A level grades obtained as given below.

Mark in mock examination	76	41	78	59	14	29	61	86	32	64	51
Grade in A level	A	B	B	C	D	E	B	A	D	C	E

4   Calculate a coefficient of rank correlation between the values of x and y given in the following table

x	2	10	82	5	1
y	5	3	1	4	22

How do you interpret this value?

What can you deduce about the product moment coefficient of correlation between x and y? Check your assertions by calculating its value.

5 The six finalists in a talent competition were A, B, C, D, E, F, and three judges placed them as follows:

First judge:	1st E,	2nd C,	3rd A,	4th B,	5th F,	6th D
Second judge:	1st E,	2nd F,	3rd B,	4th C,	5th D,	6th A
Third judge:	1st A,	2nd E,	3rd B,	4th = C, F		6th D

Calculate coefficients of rank correlation between the placings of
(i)   the first judge and the second judge
(ii) the first judge and the third judge
(iii) the second judge and the third judge.
Which two judges
(iv) agree the most
(v) disagree the most?

6 A firm assesses its job applicants at interview by awarding marks (out of 10) for appearance, qualifications, experience, personality, and alertness. Eight successful applicants had been assessed as follows

Applicant:		P	Q	R	S	T	U	V	W
Mark for Appearance	(a)	8	6	2	9	5	7	0	3
Qualifications	(b)	4	10	8	5	2	1	9	9
Experience	(c)	0	3	7	10	0	8	6	4
Personality	(d)	9	5	3	2	10	4	1	8
Alertness	(e)	7	4	1	4	10	9	8	2

Calculate coefficients of rank correlation between the marks for
(i)   appearance and personality
(ii) qualifications and alertness.
(iii) After one year's employment, these eight people were awarded a mark (out of 10) for the job effectiveness, as follows:

	P	Q	R	S	T	U	V	W
Job effectiveness	4	5	3	6	6	8	9	0

By calculating coefficients of rank correlation, determine which of the five assessed qualities correlates best with job effectiveness.

(iv) In the past, the overall assessment at interview has been obtained by adding the five marks $(a + b + c + d + e)$. How well does this correlate with job effectiveness?

(v) It is now suggested that qualifications and personality should be ignored, and the overall assessment calculated as $a + 2c + 3e$.

Explain why this has been suggested.

Calculate a coefficient of rank correlation between this new assessment and job effectiveness.

7 For the following data

x	13	15	15	20	20	20	21	21	24
y	65	60	76	62	70	75	76	80	70

find Spearman's coefficient of rank correlation

(i) by calculating the product moment coefficient of the ranks (giving your answer to 6 decimal places)

(ii) by using the formula $r_s = 1 - \dfrac{6 \sum d^2}{n(n^2 - 1)}$

Why are these two values not the same?

(iii) The formula can be modified to allow for tied ranks, as follows:

If there are a_2 pairs of tied ranks, a_3 triples of tied ranks, and so on,

$$\text{let } C = 3a_2 + 12a_3 + 30a_4 + 60a_5 + \cdots$$

(in this example, $a_2 = 4$ and $a_3 = 1$, so $C = 24$)

Then $r_s \approx 1 - \dfrac{6 \sum d^2}{n(n^2 - 1) - C}$

For the given data, verify that this formula gives a value which is correct to 4 decimal places.

8 Use the modified formula

$$r_s = 1 - \frac{6 \sum d^2}{n(n^2 - 1) - C} \qquad \text{(see Question 7)}$$

to calculate Spearman's coefficient of rank correlation for the data given in Question 2.

Comment on your result.

7
Regression

7.1   The line of regression of *y* on *x*

There are many situations where we may wish to predict the value of one variable (say *y*) from knowledge of a related variable (say *x*). For example, we might want to predict the number of articles which will be sold at a given price, or the yield of a crop for a given amount of fertilizer, or the duration of unconsciousness induced by a given dose of anaesthetic, and so on.

Here we shall suppose that we have a sample of corresponding pairs of values (*x*, *y*), and that we wish to predict the value of *y* for a given value of *x*.

The scatter diagram is very useful in deciding how this might be done.

Example 1
The points in Fig. 7.1 lie very close to a straight line.
Clearly this line could be used to predict values of *y*.

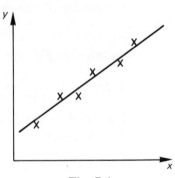

Fig. 7.1

Example 2
In Fig. 7.2 the points lie close to a curve. We can draw the curve and use it to predict values of *y*.

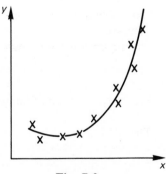

Fig. 7.2

Example 3

A sample of ten pairs (x, y) is given below:

x	0.3	0.5	1.0	1.1	1.3	1.4	1.7	2.1	2.3	2.6
y	1.0	0.5	1.8	0.6	0.9	1.7	2.2	1.9	1.3	2.2

Estimate the value of y when $x = 2.5$.

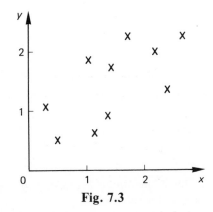

The scatter diagram (Fig. 7.3) shows fair positive correlation.

The relationship between x and y could be linear, but it is not at all obvious how we might draw the 'best' straight line through this pattern of points.

Fig. 7.3

Clearly any prediction we make in this situation will be subject to error, but ideally we should predict the average value of y for the given value of x.

We can draw an ellipse round the points in the scatter diagram (Fig. 7.4); we assume that if we had a much larger sample of pairs, then the points would fill this ellipse.

For a given value of x, the possible values of y are represented by a vertical chord of the ellipse; for the average value of y we take the mid-point of this chord.

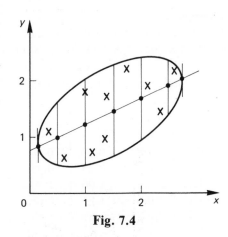

Fig. 7.4

We therefore draw the line joining the mid-points of the vertical chords. We can then use this line to predict values of y.

When $x = 2.5$, we estimate $y = 1.9$.

The method used in Example 3 involves drawing an accurate ellipse round the given points, and the results will vary depending on exactly how the ellipse is drawn.

An alternative method, which enables us to *calculate* a suitable line, is the *method of least squares*.

Consider points (x_1, y_1), (x_2, y_2), ..., (x_n, y_n) on a scatter diagram. For any straight line drawn on the diagram, let $e_1, e_2, ..., e_n$ be the vertical distances from the points to the line (Fig. 7.5).

By varying the straight line we can alter these 'errors' $e_1, e_2, ..., e_n$; it is desirable that these errors should be as small as possible.

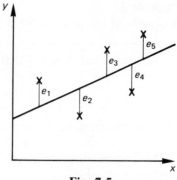

Fig. 7.5

We find the straight line for which the sum of the squares of these errors, $e_1^2 + e_2^2 + \cdots + e_n^2$, is a minimum. We shall show later that the equation of this line is:

$$y - \bar{y} = \frac{\mathrm{cov}\,(x, y)}{s_x^2}\,(x - \bar{x})$$

This is the *line of regression of y on x*, which may be used to predict the value of y for a given value of x.

Note that this line passes through the 'mean point' $(\bar{x}, \bar{y})$, and has gradient $\mathrm{cov}\,(x, y)/s_x^2$.
$\bar{x}$ and $\bar{y}$ (the means of x and y), s_x^2 (the variance of x), and the covariance $\mathrm{cov}\,(x, y)$, are calculated in the usual way from the sample of pairs (x, y).

Example 4

Calculate the equation of the line of regression of y on x for the sample of the ten pairs given in Example 3, and hence estimate the value of y when $x = 2.5$.

The table sets out the working.

x	y	x^2	xy
0.3	1.0	0.09	0.30
0.5	0.5	0.25	0.25
1.0	1.8	1.00	1.80
1.1	0.6	1.21	0.66
1.3	0.9	1.69	1.17
1.4	1.7	1.96	2.38
1.7	2.2	2.89	3.74
2.1	1.9	4.41	3.99
2.3	1.3	5.29	2.99
2.6	2.2	6.76	5.72
14.3	14.1	25.55	23.00

Note that there is no need to calculate $\sum y^2$

$$\bar{x} = \frac{\sum x}{n} = \frac{14.3}{10} = 1.43$$

$$\bar{y} = \frac{\sum y}{n} = \frac{14.1}{10} = 1.41$$

$$s_x^2 = \frac{\sum x^2}{n} - \left(\frac{\sum x}{n}\right)^2$$

$$= \frac{25.55}{10} - \left(\frac{14.3}{10}\right)^2 = 0.5101$$

$$\mathrm{cov}\,(x, y) = \frac{\sum xy}{n} - \left(\frac{\sum x}{n}\right)\left(\frac{\sum y}{n}\right)$$

$$= \frac{23.00}{10} - \left(\frac{14.3}{10}\right)\left(\frac{14.1}{10}\right)$$

$$= 0.2837$$

The line of regression of y on x is:

$$y - 1.41 = \frac{0.2837}{0.5101}(x - 1.43)$$

$$y = 0.5562x + 0.6147$$

When $x = 2.5$, we estimate $\qquad y = 0.5562 \times 2.5 + 0.6147$

$$\approx 2.01$$

We can easily plot the line of regression on the scatter diagram, (e.g. when $x = 0$, $y = 0.61$ and when $x = 2.5$, $y = 2.01$).

We see that this line is almost the same as the one obtained in Example 3.

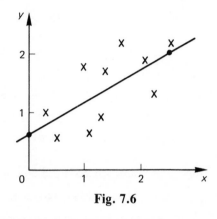

Fig. 7.6

Example 5

Find the equation of the line of regression of y on x for the following pairs of values, and hence estimate the value of y (i) when $x = 6.86$, (ii) when $x = 7.20$

x	6.72	6.79	6.80	6.81	6.84	6.88	6.89
y	1850	1650	1350	1500	1400	1350	1200

The scatter diagram (Fig. 7.7) shows that the relationship between x and y could be linear, so it is reasonable to use the line of regression to estimate values of y.

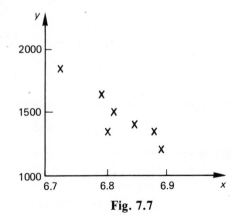

Fig. 7.7

We first code the data:

$u = \dfrac{x - 6.80}{0.01}$	-8	-1	0	1	4	8	9
$v = \dfrac{y - 1400}{50}$	9	5	-1	2	0	-1	-4

$$\sum u = 13, \quad \sum u^2 = 227, \quad \sum v = 10, \quad \sum uv = -119$$

$$\bar{u} = \frac{13}{7} = 1.8571, \qquad \bar{v} = \frac{10}{7} = 1.4286$$

$$s_u^2 = \frac{227}{7} - \left(\frac{13}{7}\right)^2 = 28.9796$$

$$\text{cov}\,(u, v) = \frac{-119}{7} - \left(\frac{13}{7}\right)\left(\frac{10}{7}\right) = -19.6531$$

Thus $\bar{x} = 0.01 \times 1.8571 + 6.80 = 6.81857$

$\bar{y} = 50 \times 1.4286 + 1400 = 1471.43$

$s_x^2 = 0.01^2 \times 28.9796 = 0.0028980$

$\text{cov}\,(x, y) = 0.01 \times 50 \times (-19.6531) = -9.8266$

The line of regression of y on x is

$$y - 1471.43 = \frac{-9.8266}{0.0028980}\,(x - 6.81857)$$

$$y = 24592 - 3390.8x$$

(i)   When $x = 6.86$, we estimate $y = 1331$
(ii) When $x = 7.20$, we estimate $y = 178$.
  Alternatively, we can find the line of regression of v on u, which is

$$v - \bar{v} = \frac{\text{cov}\,(u, v)}{s_u^2}\,(u - \bar{u})$$

$$v - 1.4286 = \frac{-19.6531}{28.9796}\,(u - 1.8571)$$

$$v = 2.688 - 0.6782u$$

(i)   When $x = 6.86$, $u = 6$; we estimate $v = 2.688 - 0.6782 \times 6 = -1.3812$

and so $y = 50 \times (-1.3812) + 1400 = 1331$

(ii) When $x = 7.20$, $u = 40$; we estimate $v = -24.44$

and so $y = 178$

We could find the line of regression of y on x by substituting into the equation for the

line of regression of v on u, $v = 2.688 - 0.6782u$

so

$$\frac{y - 1400}{50} = 2.688 - 0.6782\left(\frac{x - 6.80}{0.01}\right)$$

$$y - 1400 = 134.4 - 3391x + 23058.8$$

$$y = 24593 - 3391x$$

Example 6

An experiment was repeated 3 times at each of 4 different temperatures, with the following results

Temperature ($x°C$)	Time taken (y minutes)		
10	5.7	6.2	7.9
30	3.4	6.0	5.6
40	4.6	3.4	3.1
70	1.7	0.8	1.1

Find the line of regression of y on x.

Method 1

Considering the 12 pairs $(10, 5.7)$, $(10, 6.2)$, and so on, we have

$$\sum x = 450, \ \sum x^2 = 22\,500, \ \sum y = 49.5, \ \sum y^2 = 259.33, \ \sum xy = 1344$$

giving $\bar{x} = 37.5$, $s_x^2 = 468.75$, $\bar{y} = 4.125$, $s_y^2 = 4.595$, cov $(x, y) = -42.6875$

The line of regression of y on x is

$$y - 4.125 = \frac{-42.6875}{468.75}(x - 37.5)$$

$$y = 7.54 - 0.09107x$$

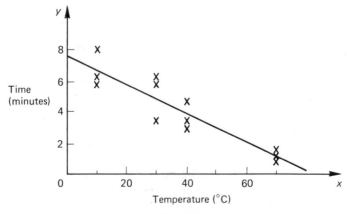

Fig. 7.8

Method 2

Let w be the average of the 3 times taken at each temperature. Considering just the 4 pairs shown below

x	10	30	40	70
w	6.6	5.0	3.7	1.2

we have $\sum x = 150$, $\sum x^2 = 7500$, $\sum w = 16.5$, $\sum w^2 = 83.69$, $\sum xw = 448$ giving $\bar{x} = 37.5$, $s_x^2 = 468.75$, $\bar{w} = 4.125$, $s_w^2 = 3.907$, cov $(x, w) = -42.6875$.

Note that $\bar{x}$, s_x^2, $\bar{w}$ and cov (x, w) are the same as the corresponding quantities for the 12 pairs (x, y); hence the line of regression of w on x is

$$w = 7.54 - 0.09107x$$

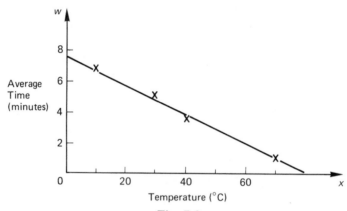

Fig. 7.9

Hence in a situation like this we can find the line of regression of y on x by considering just the average value of y for each value of x (but note that this will only work if we have the same number of y-values for each value of x).

However, the coefficient of correlation between x and y is *not* the same as the coefficient of correlation between x and w (since $s_y \neq s_w$; this is also clear from the scatter diagrams).

Reliability of estimates

Given a sample of pairs (x, y), it is always possible to calculate the equation of the line of regression of y on x, and then to use this to estimate values of y. However, it should not be assumed that this always gives satisfactory results.

(1) When there is a good linear relationship (as revealed by a scatter diagram, or by the coefficient of correlation $r \approx \pm 1$), the value of y predicted by the line of regression should be a reliable estimate, provided that x lies within the region occupied by the given points.

We should always be cautious when estimating a value of y corresponding to a value of x which is far away from the given points. We cannot be sure that the linear relationship will continue to hold (for example, a material may reach its elastic limit if it is stretched too far, or a liquid may boil if it is heated above a certain temperature).

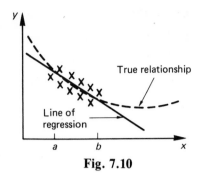

Fig. 7.10

Also, if the true relationship is given by a slight curve (as in Fig. 7.10), the line of regression will give good results when x is between a and b, but not otherwise.

The process of estimating values beyond the limits of the given points is called *extrapolation*, and it is often unreliable.

In Example 5 above, the estimate of y when $x = 6.86$ should be reasonably reliable, since there is a good linear relationship and $x = 6.86$ lies within the region of the given points ($6.7 < x < 6.9$).

However, $x = 7.20$ lies well outside this region and the corresponding estimate of y should not be regarded as a reliable one.

(2) When the underlying trend is linear, but the correlation is only fair (as in Examples 3 and 4 above), then no estimate of y could be totally reliable, since there is always a range of possible values of y corresponding to a given value of x. However the line of regression does give the average value of y corresponding to each value of x, and so it is still an appropriate method for estimating values of y; it is the best we can do in the circumstances.

(3) When there is a clear non-linear relationship between x and y (Fig. 7.11), the line of regression is quite useless for estimating values of y since it no longer gives the average value of y corresponding to each value of x. Here we could obviously do better by finding the equation of the curve. We shall discuss this problem further in Section 7.3.

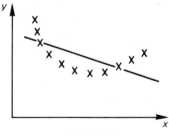

Fig. 7.11

(4) When there is no correlation between x and y (Fig. 7.12), then cov $(x, y) \approx 0$, and the line of regression of y on x becomes $y - \bar{y} = 0$, or $y = \bar{y}$. We simply estimate y as its mean value $\bar{y}$, whatever the value of x. As expected, knowledge of the value of x does not help us to predict the value of y.

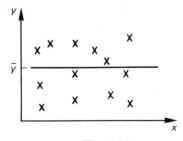

Fig. 7.12

Exercise 7.1   The line of regression of *y* on *x*

1   Find the line of regression of *y* on *x* for the following data:

x	2	4	5	3	7	5
y	1	3	2	2	5	4

2   Find the line of regression of *v* on *u* for:

u	− 3	− 1	0	2	5
v	4	0	− 1	− 2	− 6

and hence find the line of regression of *y* on *x* for:

x	16.4	16.8	17.0	17.4	18.0
y	775	675	650	625	525

$$\left(\text{Note that } u = \frac{x-17.0}{0.2}, \ v = \frac{y-675}{25} \right)$$

3   Find the line of regression of *y* on *x* for the following:

Girth of tree trunk (*x* m)	2.4	4.2	4.7	2.5	1.7
Height of tree (*y* m)	4.5	12.5	12.0	10.0	9.0
Girth of tree trunk (*x* m)	5.7	6.4	5.1	1.0	3.7
Height of tree (*y* m)	11.5	18.0	17.5	4.5	6.0

and hence estimate the height of a tree whose girth is 3.2 m

4   The following table shows the retail price index (*y*) at 12-monthly intervals (*x* is the time in months since 1 July 1977).

Date − 1st July	1977	1978	1979	1980	1981
x months	0	12	24	36	48
y	182	197	223	264	295

Draw a scatter diagram, and hence show that linear regression could be suitable for

these variables. Find the line of regression of y on x, and hence estimate the retail price index
(i)   on 1 January 1981 (i.e. $x = 42$)
(ii) on 1 July 1976
(iii) on 1 July 1990
How reliable do you think your estimates are?

5   The following table refers to 12 electric motors.

Power rating (x kW)	0.5	2.6	3.7	1.6	0.9	3.2
Efficiency (y %)	88	20	44	24	46	26
Power rating (x kW)	0.8	2.9	3.5	1.3	1.9	0.7
Efficiency (y %)	60	22	38	34	20	74

Given that $\sum x = 23.6$, $\sum x^2 = 61.4$, $\sum y = 496$, $\sum xy = 800.6$, find the line of regression of y on x, and hence estimate the efficiency of a motor with power rating 1.4 kW.
Draw a scatter diagram, and comment on your use of the line of regression.

6   Using the data in the following table, calculate the line of regression of son's height (y cm) on father's height (x cm), and hence estimate the height of a son whose father is 182 cm tall.

Son's height (cm)

x \ y	150–160	160–170	170–180	180–190
150–160	3	2	1	0
160–170	5	13	8	6
170–180	9	16	24	10
180–190	2	4	10	12

Father's height (cm)

7   Eight students spent some time (x hours) studying some complicated legal documents, and were then tested on the contents.

Hours spent studying x	0.6	2.3	1.4	3.0	0.8	3.2	1.1	1.9
Percentage of correct answers y	15	40	28	44	24	50	22	35

Find the line of regression of y on x for these eight students. A ninth student spent 5.0 hours studying, and scored 20% of correct answers. Find the line of regression of y on x for all nine students.
Draw a scatter diagram showing all nine points, and plot the line of regression for the first eight students, and the line of regression for all nine. Comment on the effect of the ninth student.

8 The following table shows the yield obtained (*y* kg) when a certain chemical process was repeated for differing amounts (*x* g) of a catalyst.

Amount of catalyst *x* (g)	Yield *y* (kg)
21	4.6, 5.4, 4.3, 4.5, 4.2
22	5.5, 4.9, 4.5, 5.3, 5.9
23	4.7, 6.0, 5.8, 5.1, 5.7
24	6.2, 5.9, 5.8, 5.9, 6.1
25	6.3, 5.8, 6.7, 6.7, 6.9
26	6.1, 7.4, 7.2, 6.6, 6.7

By first finding the average yield for each value of *x*, calculate the equation of the line of regression of *y* on *x*.

Estimate the yield when 23.7 grams of catalyst is used.

7.2 The two regression lines

Suppose we have a sample of pairs (x, y), and we now wish to predict a value of x from a given value of y.

Example 1

A sample of ten pairs (x, y) is given below (as in Section 7.1, Examples 3 and 4).

x	0.3	0.5	1.0	1.1	1.3	1.4	1.7	2.1	2.3	2.6
y	1.0	0.5	1.8	0.6	0.9	1.7	2.2	1.9	1.3	2.2

Estimate the value of x when $y = 2.0$.

We can draw an ellipse round the points in the scatter diagram (Fig. 7.13), as before. For a given value of y, the possible values of x are represented by a horizontal chord of the ellipse; for the average value of x we take the mid-point of this chord.

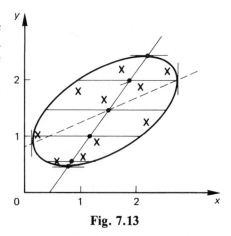

Fig. 7.13

We therefore draw the line joining the mid-points of the horizontal chords. We can use this line to predict values of x.

Note that this is a different line from that used to predict values of y (shown dotted on the diagram).

When $y = 2.0$, we estimate $x = 1.85$.

When we apply the method of least squares, we find the straight line which minimises

$$d_1^2 + d_2^2 + \cdots + d_n^2, \text{ where } d_1, d_2, \ldots, d_n$$

are the *horizontal* distances from the points to the line (Fig. 7.14).

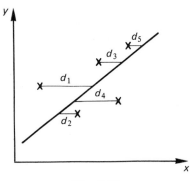

Fig. 7.14

The roles of x and y have been interchanged, and so the equation of this line is

$$x - \bar{x} = \frac{\text{cov }(y, x)}{s_y^2}(y - \bar{y})$$

and since cov $(y, x) =$ cov (x, y), this is

$$x - \bar{x} = \frac{\text{cov }(x, y)}{s_y^2}(y - \bar{y})$$

This is the *line of regression of x on y*, which may be used in suitable circumstances (i.e. when the underlying trend is linear) to predict the value of x for a given value of y.

Example 2
Calculate the equation of the line of regression of x on y for the sample of ten pairs given in Example 1, and hence estimate the value of x when $y = 2.0$.

We have $\sum x = 14.3$, $\sum y = 14.1$, $\sum y^2 = 23.53$, $\sum xy = 23.00$

Thus $\bar{x} = \dfrac{14.3}{10} = 1.43$, $\bar{y} = \dfrac{14.1}{10} = 1.41$

$$s_y^2 = \frac{23.53}{10} - \left(\frac{14.1}{10}\right)^2 = 0.3649$$

$$\text{cov }(x, y) = \frac{23.00}{10} - \left(\frac{14.3}{10}\right)\left(\frac{14.1}{10}\right) = 0.2837$$

The line of regression of x on y is

$$x - 1.43 = \frac{0.2837}{0.3649}(y - 1.41)$$

$$x = 0.7775y + 0.3338$$

When $y = 2.0$, we estimate

$$x = 0.7775 \times 2.0 + 0.3338$$

$$\approx 1.89$$

We can plot this line of regression on the scatter diagram (Fig. 7.15) (e.g. when $y = 0$, $x = 0.33$ and when $y = 2.0$, $x = 1.89$), and compare it with the line of regression of y on x.

Note that the two lines meet at (1.43, 1.41), the 'mean point'.

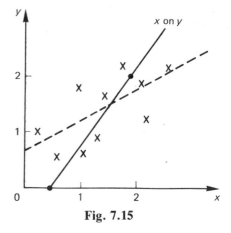

Fig. 7.15

The two lines of regression are always different (unless the points lie exactly on a straight line).

When there is negative correlation, the relative positions of the two lines are as shown on the diagram.

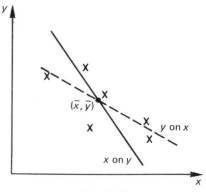

Fig. 7.16

We must always take care to use the correct line.

To estimate y from a given value of x, we use the line of regression of y on x:

$$y - \bar{y} = \frac{\text{cov}\,(x, y)}{s_x^2}\,(x - \bar{x})$$

To estimate x from a given value of y, we use the line of regression of x on y:

$$x - \bar{x} = \frac{\text{cov}\,(x, y)}{s_y^2}\,(y - \bar{y})$$

Example 3

120 readings of temperature ($x°$C) and humidity ($y\%$) were taken. The results were coded as

$$u = \frac{x - 20}{0.5} \qquad v = \frac{y - 50}{10}$$

and it was found that

$$\sum u = 723, \quad \sum u^2 = 6252, \quad \sum v = 96, \quad \sum v^2 = 150, \quad \sum uv = 222.$$

(i) Estimate the temperature when the humidity is 53%
(ii) Estimate the humidity when the temperature is 12.5°C.
Comment on the reliability of these estimates.

We have $\bar{u} = \dfrac{723}{120} = 6.025,$ $s_u^2 = \dfrac{6252}{120} - \left(\dfrac{723}{120}\right)^2 = 15.7994$

$\bar{v} = \dfrac{96}{120} = 0.8,$ $s_v^2 = \dfrac{150}{120} - \left(\dfrac{96}{120}\right)^2 = 0.61$

$\text{cov}\,(u, v) = \dfrac{222}{120} - \left(\dfrac{723}{120}\right)\left(\dfrac{96}{120}\right) = -2.97$

Thus $\bar{x} = 0.5 \times 6.025 + 20 = 23.0125$

$\bar{y} = 10 \times 0.8 + 50 \quad = 58$

$s_x^2 = 0.5^2 \times 15.7994 \ = 3.9498$

$s_y^2 = 10^2 \times 0.61 \qquad = 61$

$\text{cov}\,(x, y) = 0.5 \times 10 \times (-2.97) = -14.85$

(i) We are given $y = 53$ and we wish to estimate x; so we use the line of regression of x on y, which is

$$x - 23.0125 = \frac{-14.85}{61}\,(y - 58)$$

$$x = 37.132 - 0.2434y$$

When $y = 53$, we estimate the temperature as

$$x = 37.132 - 0.2434 \times 53$$

$$= 24.2°\text{C}$$

(ii) We are given $x = 12.5$ and we wish to estimate y; so we use the line of regression of y on x, which is

$$y - 58 = \frac{-14.85}{3.9498}\,(x - 23.0125)$$

$$y = 144.52 - 3.760x$$

When $x = 12.5$, we estimate the humidity as

$$y = 144.52 - 3.760 \times 12.5$$

$$= 97.5\%$$

To check the reliability of these estimates, we cannot draw a scatter diagram (because we are not given the individual readings), but we can calculate the coefficient of correlation.

$$r = \frac{\text{cov}\,(x,\,y)}{s_x s_y} = \frac{-14.85}{1.987 \times 7.810}$$

$$= -0.957$$

This indicates very good correlation, so the lines of regression should give reliable estimates provided that we remain in the region of the given points.

We have $\bar{x} = 23.0125$, $s_x = 1.987$; it is reasonable to assume that the given values of x lie within three standard deviations of the mean, i.e. between about 17 and 29°C.

Similarly, since $\bar{y} = 58$, $s_y = 7.810$, the given values of y lie between about 35 and 81%.

(i) Since $y = 53$ lies within the range 35–81%, we would expect our estimate of the temperature to be reliable.

(ii) Since $x = 12.5$ is well outside the range 17–29°C, we would not expect our estimate of the humidity to be a reliable one.

It is possible to work 'backwards' from the equations of the lines of regression to obtain information about the original pairs of values (x, y).

Both lines of regression pass through the 'mean point' $(\bar{x}, \bar{y})$, and if the equations are written as, for y on x: $y = m_1 x + c_1$

for x on y: $x = m_2 y + c_2$

then $m_1 = \dfrac{\text{cov}\,(x,\,y)}{s_x^2}$ and $m_2 = \dfrac{\text{cov}\,(x,\,y)}{s_y^2}$

Note that $m_1 m_2 = \dfrac{\{\text{cov}\,(x,\,y)\}^2}{s_x^2 s_y^2} = \left[\dfrac{\text{cov}\,(x,\,y)}{s_x s_y}\right]^2$

i.e. $\boxed{m_1 m_2 = r^2}$

Example 4

For a sample of second-hand cars of a certain type, the age (x years) and the price (y pounds) were noted. The line of regression of price on age was found to be $y = 4800 - 600x$ and the line of regression of age on price was found to be $x = 6.6 - 0.001y$.

Find

(i) the mean age and the mean price of the cars in the sample

(ii) the coefficient of correlation between age and price

(iii) the ratio $\dfrac{\text{standard deviation of price}}{\text{standard deviation of age}}$.

(i) Both the lines of regression pass through the mean point $(\bar{x}, \bar{y})$, so we can find $(\bar{x}, \bar{y})$ as the point of intersection of the two lines.

Solving $y = 4800 - 600x$
          $x = 6.6 - 0.001y$

as simultaneous equations, we have
$$x = 6.6 - 0.001(4800 - 600x)$$

$$x = \frac{1.8}{0.4} = 4.5$$

and $y = 4800 - 600 \times 4.5 = 2100$

The lines intersect at the point $(4.5, 2100)$; hence $\bar{x} = 4.5$, $\bar{y} = 2100$. The mean age is 4.5 years and the mean price is £2100

(ii) Using the result $m_1 m_2 = r^2$, we have $m_1 = -600$ and $m_2 = -0.001$

and so $r^2 = (-600) \times (-0.001) = 0.6$

$$r = \pm 0.775$$

Since m_1 and m_2 are negative, there is negative correlation;

thus $r = -0.775$

(iii) We have $\dfrac{\text{cov}(x, y)}{s_x^2} = -600$ and $\dfrac{\text{cov}(x, y)}{s_y^2} = -0.001$

Dividing, $\dfrac{s_y^2}{s_x^2} = \dfrac{-600}{-0.001} = 600\,000$

Hence $\dfrac{s_y}{s_x} = 744.6$

Example 5
Two values of x (denoted by a and b) are missing from the following data

x	4	3	a	9	b
y	5	6	8	9	12

but it is known that the line of regression of x on y (calculated from the complete table) is $x = 1.7y - 5.2$.
Calculate the missing values a and b.

Using the five values of y, we have $\bar{y} = 8$, $s_y^2 = 6$
Since the line of regression passes through the mean point $(\bar{x}, \bar{y})$, we have

$$\bar{x} = 1.7\bar{y} - 5.2 = 1.7 \times 8 - 5.2 = 8.4$$

and thus $\sum x = n\bar{x} = 5 \times 8.4 = 42$

Hence $4 + 3 + a + 9 + b = 42$

$$a + b = 26$$

Also, $\dfrac{\text{cov}(x, y)}{s_y^2} = 1.7$,   so $\text{cov}(x, y) = 1.7 \times 6 = 10.2$

Now $\text{cov}(x, y) = \dfrac{\sum xy}{n} - \bar{x}\bar{y},$ thus $\sum xy = n(\text{cov}(x, y) + \bar{x}\bar{y})$

$$= 5(10.2 + 8.4 \times 8)$$

$$= 387$$

Hence $20 + 18 + 8a + 81 + 12b = 387$

$$2a + 3b = 67$$

Solving $\left.\begin{array}{l} a + b = 26 \\ 2a + 3b = 67 \end{array}\right\}$

gives $a = 11$ and $b = 15$

Exercise 7.2   The two regression lines

1   The following table shows the weekly rainfall (x cm) and the number of tourists (y thousand) visiting a certain beauty spot, for 9 successive weeks.

Rainfall (x cm)	4.5	3.0	5.2	5.0	2.1	0	0	1.2	3.2
No of tourists (y thousand)	5.0	8.0	0.8	4.2	4.8	7.4	9.4	8.6	2.6

Find the line of regression of y on x, and also the line of regression of x on y.
(i)   In the next week the rainfall was 1.6 cm. Estimate the number of tourists.
(ii)   In the week after that there were 3200 tourists. Estimate the rainfall.
(iii)   Draw a scatter diagram, and plot both lines of regression, marking clearly which is which.

2   The temperature ($x°$C) and volume of a liquid (y ml) were measured. 80 pairs of values (x, y) were obtained, and

$$\sum x = 5200, \ \sum x^2 = 356\,000, \ \sum y = 16\,000, \ \sum y^2 = 3\,205\,120, \ \sum xy = 1\,049\,420.$$

Calculate the coefficient of correlation; what does this value tell you about the use of lines of regression for estimating values?
   Use appropriate lines of regression to estimate
(i)   the volume when the temperature is $78°$C
(ii)   the temperature when the volume is 260 ml.
   Do you consider your estimates to be reliable?

3   A set of pupils take examinations in mathematics and physics. The mathematics marks (x) have mean 48 and standard deviation 18; the physics marks (y) have mean 58 and standard deviation 12; and the coefficient of correlation between the mathematics and the physics marks is 0.6.
   Find the line of regression of y on x, and the line of regression of x on y. Hence

estimate
(i)   the mathematics mark for a candidate who missed the examination but scored 71 in the physics examination
(ii) the physics mark for a candidate who missed the examination but scored 25 in the mathematics examination.

4   At a seaside resort, the wind speed has mean 15 km h^{-1} and standard deviation 8 km h^{-1}. The line of regression of temperature $(y°\text{C})$ on wind speed $(x \text{ km h}^{-1})$ is $y = 21.6875 - 0.3125x$ and the coefficient of correlation between wind speed and temperature is -0.5.
(i)   Find the mean temperature.
(ii) Find the covariance cov (x, y).
(iii) Find the standard deviation of the temperature.
(iv) Find the line of regression of x on y.
(v)   Estimate the wind speed on a day when the temperature is $21°\text{C}$.

5   For a certain group of people, the line of regression of weight $(y \text{ kg})$ on height $(x \text{ cm})$ is $y = 0.56x - 28.32$, and the line of regression of x on y is $x = 0.88y + 112.16$.
(i)   Estimate the height of someone weighing 78 kg.
(ii) Estimate the weight of someone whose height is 180.8 cm.
(iii) Find the mean height and the mean weight.
(iv) Find the coefficient of correlation between the heights and the weights.
(v)   Find $\dfrac{s_y}{s_x}$ (where s_x is the standard deviation of x, etc.).

6   The following data (when completed) gives the line of regression of y on x as $y = 2x - 2$

x	1	6	3	2	3
y	0		6	2	

(i)   Find $\bar{x}$ and s_x^2
(ii) Use the line of regression to find $\bar{y}$ and cov (x, y)
      Deduce the values of $\sum y$ and $\sum xy$
(iii) Calculate the missing values of y.

7   The line of regression of y on x is $y = 13.6 - 0.8x$ and the original data is published as

x	1	2	3	4	5	6	7	8
y	14.3	9.5	11.2	11.0	10.4	9.8	8.1	6.9

but one of the values of y has been misprinted.
      Use the values of x and the line of regression to calculate $\sum y$, and $\sum xy$.
      Say which value of y is wrong, and correct it.

7.3   Non-linear regression

When the scatter diagram shows that there is a definite non-linear relationship between x and y, we should like to find the equation of a curve which closely 'fits' the given points. To do this we shall need to know the general form that the equation might take.

Sometimes we can define two new variables, say w and z, which are related linearly, i.e. $w = mz + c$; we then consider the line of regression of w on z.

For example, if the relationship is of the form $y = ax^2 + b$, we define $z = x^2$, then $y = az + b$ and by considering the line of regression of y on z, we can estimate a and b.

Similarly,

for $y = a \sin x + b$, we define $z = \sin x$, then $y = az + b$

for $y^2 = \dfrac{a}{x} + b$, we define $w = y^2$, $z = \dfrac{1}{x}$, then $w = az + b$

for $y = \dfrac{1}{ax + b}$, i.e. $\dfrac{1}{y} = ax + b$, define $w = \dfrac{1}{y}$, then $w = ax + b$

for $y = ax^2 + bx$, i.e. $\dfrac{y}{x} = ax + b$, define $w = \dfrac{y}{x}$, then $w = ax + b$,

and so on.

Example 1
The following values were obtained from an experiment.

x	0.10	0.15	0.22	0.37	0.76	1.53
y	5.6	4.4	3.9	3.6	3.5	3.5

Show that the relationship could be of the form $y = \dfrac{a}{x^2} + b$ and estimate the values of a and b.

Hence predict the value of y when $x = 0.30$.

The scatter diagram (Fig. 7.17) shows that the points lie on a curve.

If $y = \dfrac{a}{x^2} + b$, we define $z = \dfrac{1}{x^2}$

and then $y = az + b$. We shall consider the line of regression of y on z.

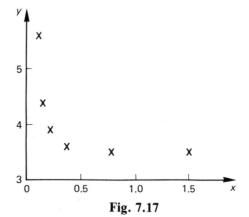

Fig. 7.17

$z = \dfrac{1}{x^2}$	100	44.4444	20.6612	7.3046	1.7313	0.4272
y	5.6	4.4	3.9	3.6	3.5	3.5

We have $\bar{z} = 29.0948,$ $s_z = 35.0706$

$\bar{y} = 4.0833,$ $s_y = 0.7470$

$\text{cov}(z, y) = 26.1939$

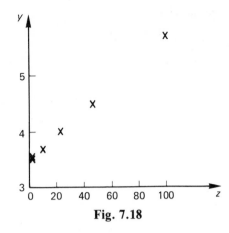

Fig. 7.18

If we plot y against z (Fig. 7.18), we find that the points are very close to a straight line. This confirms that the given relationship $\left(y = \dfrac{a}{x^2} + b \right)$ is a suitable one.

(Alternatively, we could calculate the coefficient of correlation between z and y. This gives

$$r = \frac{26.1939}{35.0706 \times 0.7470}$$

$$= 0.9998$$

which shows that z and y are related linearly.)

The line of regression of y on z is

$$y - 4.0833 = \frac{26.1939}{35.0706^2} (z - 29.0948)$$

$$y = 0.0213z + 3.464$$

$$y = \frac{0.0213}{x^2} + 3.464 \qquad \text{(i.e. } a = 0.0213, \ b = 3.464)$$

When $x = 0.30$, we predict $\qquad y = \dfrac{0.0213}{0.30^2} + 3.464$

$$= 3.70$$

We now consider two important relationships which can be reduced to linear form by taking logarithms. It does not matter what base we use for the logarithms (so long as all the logarithms are to the same base). In practice we shall use either

base 10 (common logarithms $\log_{10} x$)

or base e (natural logarithms $\log_e x = \ln x$, where e ≈ 2.71828)

The relationship $y = ax^b$

Taking logarithms, we have $\log y = \log (ax^b) = \log a + \log (x^b)$

$$= \log a + b \log x$$

Define $w = \log y$, $z = \log x$, then $w = bz + \log a$, so we find the line of regression of $\log y$ on $\log x$.

Example 2

Fit a curve of the form $y = ax^b$ to the following data:

x	2	5	10	20	50	120	500
y	5.8	8.7	12.6	18.1	28.4	43.5	89.2

If $y = ax^b$, then $\log y = b \log x + \log a$.

Let $w = \log y$, $z = \log x$; then $w = bz + \log a$

$z = \log_{10} x$	0.3010	0.6990	1.0000	1.3010	1.6990	2.0792	2.6990
$w = \log_{10} y$	0.7634	0.9395	1.1004	1.2577	1.4533	1.6385	1.9504

We have $\bar{z} = 1.3969$, $s_z = 0.7660$

$\bar{w} = 1.3005$, $s_w = 0.3818$

$\text{cov}(z, w) = 0.2923$

Note that the coefficient of correlation between z and w is

$$r = \frac{0.2923}{0.7660 \times 0.3818} = 0.9996$$

confirming that the given relationship is suitable.

The line of regression of w on z is

$$w - 1.3005 = \frac{0.2923}{0.7660^2}(z - 1.3969)$$

i.e. $w = 0.498z + 0.6046$

Comparing this equation with $w = bz + \log a$

we have $b = 0.498$, $\log a = 0.6046$

$$a = 10^{0.6046}$$

$$= 4.02$$

Hence the relationship is $y = 4.02x^{0.498}$

The relationship $y = ab^x$

Taking logarithms, we have $\log y = \log(ab^x) = \log a + \log(b^x)$

$$= \log a + x \log b$$

Define $w = \log y$, then $w = (\log b)x + \log a$, so we find the line of regression of $\log y$ on x.

Example 3

Fit a curve of the form $y = 2 + ab^x$ to the following data:

x	-1.5	0	1.5	3.0	4.5
y	47.5	11.2	3.0	2.4	2.1

If $y = 2 + ab^x$, then $y - 2 = ab^x$

$$\log(y - 2) = x \log b + \log a$$

Let $w = \log(y - 2)$, then $w = (\log b)x + \log a$

x	-1.5	0	1.5	3.0	4.5
$w = \log_{10}(y - 2)$	1.6580	0.9638	0	-0.3979	-1

We have $\bar{x} = 1.5,$ $s_x^2 = 4.5$

$\bar{w} = 0.2448,$ $\text{cov}(x, w) = -2.0033$

The line of regression of w on x is

$$w - 0.2448 = \frac{-2.0033}{4.5}(x - 1.5)$$

i.e. $w = -0.4452x + 0.9126$

Comparing this equation with $w = (\log b)x + \log a$

we have $\log b = -0.4452$ and $\log a = 0.9126$
$$b = 0.359 \qquad\qquad a = 8.18$$

Hence the relationship is $y = 2 + 8.18(0.359)^x$.

Exercise 7.3   Non-linear regression

1   For each of the ollowing relationships, state two variables which are related linearly;
i.e. $w = mz + c$, and give m and c in terms of a and b
(i) $y = ax^3 + b$ (ii) $y^2 = ax + b$ (iii) $y = ax^{-b}$
(iv) $y = a(x - 2)^b$ (v) $y = ab^{-x^2}$ (vi) $y = 4 + ax^b$
(vii) $y = ax + bx^3$
(viii) $y = \dfrac{a + bx}{x^3}$

2   Find the coefficient of correlation between x and y for the following:

x	-2.3	-0.9	0.2	1.8	4.0
y	8.6	6.4	6.1	7.6	13.8

Find also the coefficient of correlation between x^2 and y, and comment on its value.
Find the line of regression of y on x^2, and hence estimate the value of y when $x = 3.2$.

3 The following values of x and y are available,

x	0.3	0.5	0.7	0.9
y	915	145	45	18

and it is thought that they satisfy a relationship of the form $y = ax^b$.

Find the coefficient of correlation between $\log x$ and $\log y$, and explain why this confirms the suggested relationship.

By finding the line of regression of $\log y$ on $\log x$, estimate the values of a and b.

4 An athlete's best times for various distances are

Distance (x metres)	100	200	400	800	1500	10 000
Best time (y seconds)	11.2	21.8	51.5	110.3	220.3	1775

Assuming that x and y satisfy approximately a relationship of the form $y = ax^b$, find values of a and b by considering the line of regression of $\log y$ on $\log x$.

Estimate the athlete's time for
(i) 5000 m   (ii) the Marathon (42 200 m)
How reliable do you consider these estimates to be?

5 The population of a certain country is given in the following table

Date (x years after 1960)	0	2	4	6	8	10
Population (y millions)	30.5	29.5	24.0	13.0	14.0	13.5
Date (x years after 1960)	12	14	16	18	20	22
Population (y millions)	12.0	7.5	9.0	8.5	7.5	5.0

Plot these points on a graph.

Theoretical considerations suggest that a relationship of the form $y = ab^{-x}$ might be appropriate. Find the line of regression of $\log y$ on x, and hence estimate a and b.

Calculate the predicted values of y for $x = 0, 2, 4, ..., 22$ and plot the fitted curve $y = ab^{-x}$ on your graph.

Would you be prepared to predict the population of the country in the year 2000?

6 By considering the line of regression of $\dfrac{y-1}{x}$ on x^2 fit a relationship of the form

$y = 1 + ax + bx^3$ to the following data

x	-1	1	2	3	4	5
y	-3.0	5.0	8.1	10.5	10.6	8.3

7.4 Regression from first principles

In this section we shall look at the method of least squares in more detail. We shall prove that the coefficient of correlation r always lies between -1 and $+1$, and we shall derive the equation

$$y - \bar{y} = \frac{\text{cov}(x, y)}{s_x^2}(x - \bar{x})$$

for the line of regression of y on x.

Suppose that we have n pairs of values (x_1, y_1), (x_2, y_2), ..., (x_n, y_n) and we are considering the use of a straight line $y = mx + c$ to predict values of y from given values of x.

When $x = x_1$,

  the value of y predicted by the line $y = mx + c$ is $(mx_1 + c)$
  whereas the true value of y is y_1,
  and hence the 'error' is $e_1 = y_1 - mx_1 - c$

e_1 is the vertical distance of the point (x_1, y_1) from the line $y = mx + c$.

Similarly we define $e_2 = y_2 - mx_2 - c$

$e_3 = y_3 - mx_3 - c$

$\vdots$

$e_n = y_n - mx_n - c$

These errors may be

positive (if the point is above the line)
negative (if the point is below the line)
or zero (if the point is actually on the line).

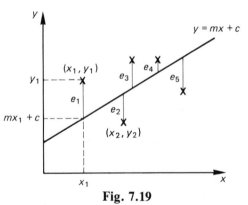

Fig. 7.19

The sum of the squares of these errors is

$$E = e_1^2 + e_2^2 + \cdots + e_n^2 = (y_1 - mx_1 - c)^2 + (y_2 - mx_2 - c)^2 + \cdots + (y_n - mx_n - c)^2$$

$$= \sum_{i=1}^{n}(y_i - mx_i - c)^2$$

It is clearly desirable that e_1, e_2, ..., e_n should be (numerically) as small as possible; so we should like E to be small.

If we use a different straight line to predict the values of y, we shall change the values of m and c, and hence we shall change the value of E.

In the *method of least squares*, we find the values of m and c which make E a minimum; then the corresponding straight line $y = mx + c$ is the line of regression of y on x.

The errors $e_1, e_2, ..., e_n$ are called the *residuals*,

and $E = e_1^2 + e_2^2 + \cdots + e_n^2$ is the *residual sum of squares* (RSS)

Example 1
Find, from first principles, the equation of the line of regression of y on x for the 4 pairs of values

x	1	2	4	7
y	2	5	11	9

and find the corresponding residual sum of squares.

If the line $y = mx + c$ is used to predict the values of y, the sum of the squares of the errors is

$$E = (y_1 - mx_1 - c)^2 + \cdots + (y_n - mx_n - c)^2$$
$$= (2 - m - c)^2 + (5 - 2m - c)^2 + (11 - 4m - c)^2 + (9 - 7m - c)^2$$
$$= 70m^2 + 28mc + 4c^2 - 238m - 54c + 231$$

E is a function of two variables m and c.
First, regarding E as a function of m (with c constant), for a minimum value, we require
$\dfrac{\partial E}{\partial m} = 0$, i.e. $140m + 28c - 238 = 0$
Now, regarding E as a function of c (with m constant), for a minimum value, we require
$\dfrac{\partial E}{\partial c} = 0$, i.e. $28m + 8c - 54 = 0$

The minimum value of E occurs when $140m + 28c - 238 = 0$
and $28m + 8c - 54 = 0$

Solving these as simultaneous equations gives $m = \dfrac{7}{6}$ and $c = \dfrac{8}{3}$

Hence the line of regression of y on x is $y = \dfrac{7}{6}x + \dfrac{8}{3}$
The residual sum of squares is the corresponding minimum value of E

i.e. $E = 70 \times \left(\dfrac{7}{6}\right)^2 + 28 \times \dfrac{7}{6} \times \dfrac{8}{3} + 4 \times \left(\dfrac{8}{3}\right)^2 - 238 \times \dfrac{7}{6} - 54 \times \dfrac{8}{3} + 231$

$= \dfrac{121}{6}$

≈ 20.167

Note   (1) The individual residuals are $e_1 = y_1 - \left(\dfrac{7}{6}x_1 + \dfrac{8}{3}\right)$

$= 2 - 3.833 = -1.833$

$e_2 = 5 - 5 = 0$

$$e_3 = 11 - 7.333 = 3.667$$

$$e_4 = 9 - 10.833 = -1.833$$

We observe that $\sum e_i = 0$

and $\sum e_i^2 = 20.167$ (as before)

(2) We can verify that the line of regression is the same as that given by

$$y - \bar{y} = \frac{\text{cov}(x, y)}{s_x^2}(x - \bar{x})$$

We have $\bar{x} = 3.5$, $s_x^2 = 5.25$, $\bar{y} = 6.75$, cov $(x, y) = 6.125$
Hence the line of regression of y on x is

$$y - 6.75 = \frac{6.125}{5.25}(x - 3.5) = \frac{7}{6}(x - 3.5)$$

$$y = \frac{7}{6}x + \frac{8}{3}$$ (as before)

Starting with n general points $(x_1, y_1), (x_2, y_2), ..., (x_n, y_n)$, we could follow the process used in the above Example to derive the equation of the line of regression. However, this would be a tedious algebraic exercise, and it is easier to work with standardised variables.

Suppose that the x-values, $x_1, x_2, ..., x_n$ have mean $\bar{x}$ and standard deviation s_x.

Let $u_i = \dfrac{x_i - \bar{x}}{s_x}$. Then $u_1, u_2, ..., u_n$ are the standardised values of x, giving each value as the number of standard deviations from the mean.

We have $\bar{u} = 0$ and $s_u = 1$

It follows that $\displaystyle\sum_{i=1}^{n} u_i = n\bar{u} = 0$

and $\displaystyle\sum_{i=1}^{n} u_i^2 = n(\bar{u}^2 + s_u^2) = n$

Similarly let $v_i = \dfrac{y_i - \bar{y}}{s_y}$ be the standardised values of y.

Then we have $\bar{v} = 0$ and $s_v = 1$

and $\displaystyle\sum_{i=1}^{n} v_i = 0$, $\displaystyle\sum_{i=1}^{n} v_i^2 = n$

The coefficient of correlation between x and y is

$$r = \frac{\text{cov}(x, y)}{s_x s_y} = \frac{\dfrac{1}{n}\displaystyle\sum_{i=1}^{n}(x_i - \bar{x})(y_i - \bar{y})}{s_x s_y}$$

$$= \frac{1}{n}\sum_{i=1}^{n}\left(\frac{x_i - \bar{x}}{s_x}\right)\left(\frac{y_i - \bar{y}}{s_y}\right)$$

$$= \frac{1}{n}\sum_{i=1}^{n} u_i v_i$$

and thus
$$nr = \sum_{i=1}^{n} u_i v_i$$

If we use the line $v = mu + c$ to predict values of v from given values of u, the sum of the squares of the errors is

$$E = \sum_{i=1}^{n} (v_i - mu_i - c)^2 = \sum_{i=1}^{n} (v_i^2 + m^2 u_i^2 + c^2 - 2mu_i v_i - 2cv_i + 2mcu_i)$$

$$= \sum_{i=1}^{n} v_i^2 + m^2 \sum_{i=1}^{n} u_i^2 + nc^2 - 2m \sum_{i=1}^{n} u_i v_i - 2c \sum_{i=1}^{n} v_i + 2mc \sum_{i=1}^{n} u_i$$

$$= n + m^2 n + nc^2 - 2mnr - 0 + 0$$

$$= n(m^2 - 2mr) + n + nc^2$$

$$= n(m^2 - 2mr + r^2) - nr^2 + n + nc^2$$

$$= n(m - r)^2 + nc^2 + n(1 - r^2)$$

The values of m and c which make E a minimum are $m = r$ and $c = 0$; thus the line of regression of v on u is $v = ru$; and the minimum value of E (i.e. the residual sum of squares) is $E = n(1 - r^2)$.

Now E is a sum of squares, which cannot be negative, and so

$$n(1 - r^2) \geqslant 0$$

$$\text{hence} \quad r^2 \leqslant 1$$

$$\text{i.e.} \quad -1 \leqslant r \leqslant 1$$

If $r = \pm 1$, then $r^2 = 1$, so the minimum value is $E = 0$; this means that all the residuals are zero, and hence all the points lie exactly on the straight line.

If r^2 is small, then the minimum value $E = n(1 - r^2)$ is large, so the points do not lie close to the straight line.

We have now proved the properties of r which were discovered in Chapter 6.

We have also proved that the line of regression of v on u is $v = ru$; similarly, the line of regression of u on v is $u = rv$.

Figures 7.20 and 7.21 show the lines of regression in the cases
(1) when there is positive correlation (Fig 7.20)

$$0 < r < 1$$

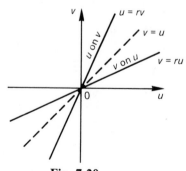

Fig. 7.20

(2) when there is negative correlation (Fig. 7.21)

$$-1 < r < 0$$

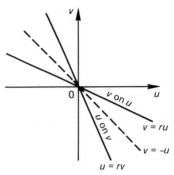

Fig. 7.21

Finally we shall establish the general equation for the line of regression of y on x.

We have
$$u = \frac{x - \bar{x}}{s_x}, \qquad v = \frac{y - \bar{y}}{s_y}, \qquad r = \frac{\mathrm{cov}\ (x,\ y)}{s_x s_y};$$

from the line of regression of v on u. i.e. $v = ru$ we obtain the line of regression of y on x as

$$\frac{y - \bar{y}}{s_y} = \frac{\mathrm{cov}\ (x,\ y)}{s_x s_y} \left(\frac{x - \bar{x}}{s_x} \right)$$

that is,

$$y - \bar{y} = \frac{\mathrm{cov}\ (x,\ y)}{s_x^2} (x - \bar{x})$$

Exercise 7.4 Regression from first principles

1 Points (x_i, y_i) are plotted on a scatter diagram, and $y = mx + c$ is any straight line. Interpret the lengths $e_i = y_i - mx_i - c$ on a diagram and explain why minimising the quantity $E = \sum e_i^2$ leads to the line of regression of y on x.

2 For the two points $(1, 3)$ and $(4, 7)$, show that
$$E = 17m^2 + 10mc + 2c^2 - 62m - 20c + 58.$$

Show that E is minimised when $34m + 10c - 62 = 0$ and $10m + 4c - 20 = 0$, and find the corresponding values of m and c. Find the minimum value of E, and interpret this value.

Write down the equation of the line of regression of y on x, and verify that it is the line joining the two given points.

3 For the three points $(0, 4)$, $(1, 2)$ and $(4, 8)$, express E in terms of m and c. Find the values of m and c which minimise E, and hence find the line of regression of y on x.

4 For the four points $(-4, 10)$, $(11, 0)$, $(2, 6)$, $(5, 4)$, find the values of m and c which minimise E, and calculate the minimum value of E. Explain this value, and write down the coefficient of correlation between x and y.

5 For the three points $(-5, 1)$, $(2, 4)$ and $(3, 7)$ show that

$$E = 38\left(m - \frac{12}{19}\right)^2 + 3(c - 4)^2 + \frac{54}{19}.$$

Hence write down the equation of the line of regression of y on x, and state the residual sum of squares (i.e. the corresponding value of E).

6 For the following data

Rainfall (x cm)	4.5	3.0	5.2	5.0	2.1	0	0	1.2	3.2
No of tourists (y thousand)	5.0	8.0	0.8	4.2	4.8	7.4	9.4	8.6	2.6

calculate $\bar{x}$, s_x, $\bar{y}$, s_y, and find the standardised values

$$u_i = \frac{x_i - \bar{x}}{s_x}, \qquad v_i = \frac{y_i - \bar{y}}{s_y}.$$

Use the formula $r = \sum u_i v_i / n$ to find the coefficient of correlation.

Plot the points (u_i, v_i) on a scatter diagram, and plot the line of regression of v on u $(v = ru)$ and the line of regression of u on v $(u = rv)$.

By first standardising, estimate
(i)   the value of y when $x = 1.6$
(ii) the value of x when $y = 3.2$
(See Exercise 7.2 Question 1).

8
Probability

Introduction

Probability is a measure of likelihood. For example, to say that the probability that a day is rainy is $\frac{1}{4}$ means that in the long run we would expect $\frac{1}{4}$ of the days to be rainy; thus out of 600 days, about 150 should be rainy.

Experimentally, the probability of an event A can be estimated as

$$P(A) \approx \frac{\text{number of trials for which event A occurred}}{\text{total number of trials}}$$

For example, if we surveyed 245 married couples and found that 38 of these had more than two children, we would estimate the probability that a couple has more than two children to be $\frac{38}{245} \approx 0.1551$.

This is only an approximate value; if we survey another group of married couples, we may well obtain a slightly different answer.

The probability of an event is a number between 0 and 1.

Sometimes it is possible to find the probability of an event theoretically.

If we can list all the possible outcomes *in such a way that they are all equally likely*, then the probability of an event A is

$$P(A) = \frac{\text{number of outcomes for which the event A occurs}}{\text{total number of possible outcomes}}$$

Example 1

Two dice are thrown.

We can list the 36 possible outcomes as:

$$
\begin{array}{cccccc}
(1,1) & (1,2) & (1,3) & (1,4) & (1,5) & (1,6) \\
(2,1) & (2,2) & (2,3) & (2,4) & (2,5) & (2,6) \\
(3,1) & (3,2) & (3,3) & (3,4) & (3,5) & (3,6) \\
(4,1) & (4,2) & (4,3) & (4,4) & (4,5) & (4,6) \\
(5,1) & (5,2) & (5,3) & (5,4) & (5,5) & (5,6) \\
(6,1) & (6,2) & (6,3) & (6,4) & (6,5) & (6,6)
\end{array}
$$

Hence $P(\text{double 3}) = \frac{1}{36}$

$P(\text{a 3 and a 4}) = \frac{2}{36}$

$$P(\text{at least one } 6) = \frac{11}{36}$$

$$P(\text{total is } 10) = \frac{3}{36}$$

$$P(\text{double}) = \frac{6}{36}$$

$$P(\text{numbers differ by } 2) = \frac{8}{36}$$

and so on.

Example 2

A bag contains 3 green, 5 blue and 2 red discs.

One disc is removed at random.

The possible outcomes can be listed as: green, blue or red, but these are not equally likely.

Each disc is equally likely to be removed, so if we list the outcomes as

$$G_1, G_2, G_3, B_1, B_2, B_3, B_4, B_5, R_1, R_2$$

then these are equally likely.

$$\text{Hence } P(\text{a green disc is removed}) = \frac{3}{10}$$

$$P(\text{disc removed is not red}) = \frac{8}{10}$$

and so on.

Probabilities may also be obtained from a theoretical distribution (such as a Normal distribution).

8.1   Multiplication and addition of probabilities

To find the probability of a complicated event, we can usually express it in terms of simpler events, for which the probabilities are either known or can be found as simple theoretical probabilities.

Roughly speaking, for 'and' we multiply the probabilities, and for 'or' we add them; but the indiscriminate use of these 'rules' can lead to error. We shall now state the appropriate rules precisely.

Multiplying probabilities

If A and B are any two events, the probability that A and B both occur is

$$P(A \text{ and } B) = P(A) \times P(B \mid A)$$

P(B│A), read as 'the probability of B, given A', is the probability that B occurs *if it is known that A has occurred.*

This is a *conditional probability.*

We can extend this result to three (or more) events.

For P(A and B and C), we consider 'A and B' as the first event and C as the second, so that

$$P\{(A \text{ and } B) \text{ and } C\} = P(A \text{ and } B) \times P(C \mid A \text{ and } B)$$

Then using P(A and B) = P(A) × P(B│A), we obtain

$$P(A \text{ and } B \text{ and } C) = P(A) \times P(B \mid A) \times P(C \mid A \text{ and } B)$$

and so on.

When multiplying probabilities in this way, each probability must be *conditional* on all the preceding events having occurred.

Example 1

In a small show-jumping competition, there are 8 competitors: 5 men and 3 ladies. The jumping order is decided by drawing names at random out of a hat. Find the probability that

(i)   The first rider is a man
(ii) The first two riders are both men
(iii) The second rider is a man
(iv) The last rider is a lady
(v)   The first rider is a lady and the last rider is a man
(vi) The ladies ride first, seventh and eighth.

(i)   The first rider is equally likely to be any of the 8 competitors, so P(1st is a man) $= \dfrac{5}{8}$

(ii) The probability that the first rider is a man is $\dfrac{5}{8}$.

Given that the first rider is a man, the second rider is equally likely to be any of the remaining 7 competitors (of which 4 are men).

Thus

P(1st is a man and 2nd is a man) = P(1st is a man) × P(2nd is a man │ 1st is a man)

$$= \frac{5}{8} \times \frac{4}{7}$$

$$= \frac{5}{14}$$

(iii) As we have no information about the first (or any other) rider, the second rider is equally likely to be any one of the 8 competitors, so P(2nd is a man) $= \dfrac{5}{8}$

(iv) The last rider is equally likely to be any of the 8 competitors, so P(last is a lady) $= \dfrac{3}{8}$

(v)   P(1st is a lady and last is a man) = P(1st is a lady) × P(last is a man | 1st is a lady)

$$= \frac{3}{8} \times \frac{5}{7}$$

$$= \frac{15}{56}$$

(vi) P(1st is a lady and 7th is a lady and 8th is a lady)
= P(1st is a lady) × P(7th is a lady | 1st is a lady) × P(8th is a lady | 1st and 7th are ladies)

$$= \frac{3}{8} \times \frac{2}{7} \times \frac{1}{6}$$

$$= \frac{1}{56}$$

Example 2
One bag contains 5 black and 4 white discs, and a second bag contains 3 black and 7 white discs.

Given that one disc is taken from each bag, find the probability that both these discs are black.

P(1st is black and 2nd is black) = P(1st is black) × P(2nd is black | 1st is black)

$$= \frac{5}{9} \times \frac{3}{10}$$

$$= \frac{1}{6}$$

Note that, in this Example, the probability that the second disc is black is not affected by the knowledge that the first disc is black. The two events 'the first disc is black' and 'the second disc is black' are said to be *independent*.

Adding probabilities

Suppose that A and B are two events which do not overlap (i.e. A and B cannot both occur. We may say that A and B are *exclusive* events, since if one of them occurs this excludes the possibility that the other one occurs).

Then the probability that either A occurs or B occurs is

$$P(A \text{ or } B) = P(A) + P(B)$$

This extends to three (or more) events.
If A, B, C, ..., are non-overlapping events (i.e. no two events can occur simultaneously, we may say that the events are *mutually exclusive*), then

$$P(A \text{ or } B \text{ or } C \text{ or } ...) = P(A) + P(B) + P(C) + \cdots$$

We can often split an event into several different (i.e. non-overlapping) ways in which it can occur. We then find the probability for each of these and add them up.

When using this method, it is important to ensure that the various cases considered do not overlap.

Example 3

Two discs are removed (without replacement) from a bag containing 3 green, 5 blue and 2 red discs. Find the probability that
(i) one is red and one is green
(ii) the two discs are the same colour.

(i) We may consider the two discs to be removed one at a time.
The required event (one red and one green) can occur in two different ways:
first is red and second is green
or first is green and second is red.

$$P(\text{1st is red and 2nd is green}) = \frac{2}{10} \times \frac{3}{9} = \frac{6}{90}$$

$$P(\text{1st is green and 2nd is red}) = \frac{3}{10} \times \frac{2}{9} = \frac{6}{90}$$

Hence $P(\text{one red and one green}) = \frac{6}{90} + \frac{6}{90} = \frac{12}{90} = \frac{2}{15}$

(ii) The required event (both discs are the same colour) can occur in three different ways: both green, both blue, or both red.

$$P(\text{both green}) = \frac{3}{10} \times \frac{2}{9} = \frac{6}{90}$$

$$P(\text{both blue}) = \frac{5}{10} \times \frac{4}{9} = \frac{20}{90}$$

$$P(\text{both red}) = \frac{2}{10} \times \frac{1}{9} = \frac{2}{90}$$

Hence $P(\text{same colour}) = \frac{6}{90} + \frac{20}{90} + \frac{2}{90} = \frac{28}{90} = \frac{14}{45}$

If A is any event, then 'not A' is the event that A does not occur. Clearly A and 'not A' cannot occur simultaneously, so we have

$$P(A \text{ or not } A) = P(A) + P(\text{not } A)$$

Also, either A or not A must occur, so $P(A \text{ or not } A) = 1$

Thus $P(A) + P(\text{not } A) = 1$, giving

$$P(\text{not } A) = 1 - P(A)$$

Sometimes it is easier to find the probability that an event does *not* occur. We can then subtract this probability from one to obtain the probability that the event does occur.

Example 4

A marksman fires a sequence of three shots at a target.
The probability that a shot hits the target is
0.4 if he has not already hit the target,
0.7 is he has previously hit the target once, or
0.9 if he has previously hit the target twice.
Find the probability that
(i)   his third shot hits the target
(ii)   exactly one shot hits the target.

This situation can conveniently be represented by a 'probability tree' (Fig. 8.1) (where H denotes a hit, and M a miss; in each case the probability of a miss is one minus the probability of a hit).

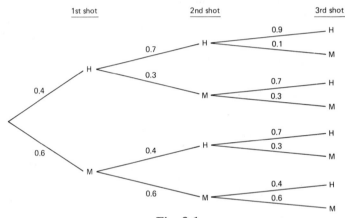

Fig. 8.1

Note that the probabilities on the tree are conditional on having reached the previous fork (for example, 0.9 is the probability that the third shot is a hit, given that the first and second shots were both hits). So the probability of any sequence may be obtained by multiplying the probabilities along the branches; for example, the probability that all three shots hit the target is $P(HHH) = 0.4 \times 0.7 \times 0.9$.

(i)   There are 4 sequences for which the third shot is a hit

$P(HHH) = 0.4 \times 0.7 \times 0.9 = 0.252$
$P(HMH) = 0.4 \times 0.3 \times 0.7 = 0.084$
$P(MHH) = 0.6 \times 0.4 \times 0.7 = 0.168$
$P(MMH) = 0.6 \times 0.6 \times 0.4 = \underline{0.144}$

$\qquad\qquad\qquad\qquad\qquad 0.648$

Hence   P(3rd shot is a hit) = 0.648.

(ii)   There are three sequences with exactly one hit.

$P(HMM) = 0.4 \times 0.3 \times 0.3 = 0.036$
$P(MHM) = 0.6 \times 0.4 \times 0.3 = 0.072$
$P(MMH) = 0.6 \times 0.6 \times 0.4 = \underline{0.144}$

$\qquad\qquad\qquad\qquad\qquad 0.252$

Hence   P(exactly one hit) = 0.252.

Example 5

In a table of random numbers, each digit is equally likely to be 0, 1, 2, 3, 4, 5, 6, 7, 8, or 9, and the digits are selected independently.

Given that a sequence of three digits is taken from such a table, find the probability that
(i)   the sequence is 555
(ii) there is at least one 7
(iii) all the digits are less than 8
(iv) the three digits are different.

(i)   P(1st is 5 and 2nd is 5 and 3rd is 5) $= \dfrac{1}{10} \times \dfrac{1}{10} \times \dfrac{1}{10} = 0.001$

(ii) If there is *not* at least one 7, then there are no 7s.

P(no 7s) = P(1st is not a 7 and 2nd is not a 7 and 3rd is not a 7)

$$= \dfrac{9}{10} \times \dfrac{9}{10} \times \dfrac{9}{10} = 0.729.$$

So P(at least one 7) = 1 − P(no 7s)

$$= 1 - 0.729$$

$$= 0.271$$

(iii) P(1st is < 8 and 2nd is < 8 and 3rd is < 8) $= \dfrac{8}{10} \times \dfrac{8}{10} \times \dfrac{8}{10} = 0.512.$

(iv) The probability that the 2nd digit is different from the 1st is $\dfrac{9}{10}$;

given that the first two digits are different, the probability that the 3rd is different from the first two is $\dfrac{8}{10}$.

Hence P(3 different digits) $= \dfrac{9}{10} \times \dfrac{8}{10} = 0.72.$

Example 6

Two cards are taken (without replacement) from a well-shuffled pack of 52 playing cards. Find the probability that at least one ace is taken and at least one spade is taken.

It might appear that one card must be an ace and the other a spade; however if one of the cards is the ace of spades, then both conditions are satisfied and it does not matter what the other card is.

We may assume that the cards are taken one at a time, and we consider four different cases, depending on the first card taken.

P(1st card is ace of spades and 2nd card is anything) $= \dfrac{1}{52} \times \dfrac{51}{51}$

P(1st card is an ace (not of spades) and 2nd card is a spade) $= \dfrac{3}{52} \times \dfrac{13}{51}$

P(1st card is a spade (not the ace) and 2nd card is an ace) $= \dfrac{12}{52} \times \dfrac{4}{51}$

P(1st card is not an ace nor a spade and 2nd card is ace of spades) $= \dfrac{36}{52} \times \dfrac{1}{51}$

Hence   P(at least one ace and at least one spade) $= \dfrac{51 + 39 + 48 + 36}{52 \times 51} = \dfrac{174}{2652}$

$$= \dfrac{29}{442}$$

Exercise 8.1   Multiplication and addition rules

1   A bag contains 7 red discs and 3 blue ones. Discs are removed one at a time without replacement. Find the probability that
   (i)   the first disc removed is red
   (ii)   the second disc removed is red
   (iii)   the sixth disc removed is blue
   (iv)   the first two discs are both red
   (v)   the second disc is red and the fourth disc is blue
   (vi)   the first two discs are the same colour
   (vii)   the first two discs are different colours
   (viii) of the first three discs removed, exactly two are blue
   (ix)   of the last three discs removed, at least one is blue
   (x)   the first blue disc is removed at the fourth draw
   (xi)   the last blue disc is removed at the eighth draw.

2   The probability that it will rain on a particular day is 0.4 if it rained on the previous day, but 0.1 if it did not rain on the previous day. Given that it rained on Monday, find the probability that
   (i)   it will rain on Tuesday and Wednesday
   (ii)   it will not rain on Tuesday and it will rain on Wednesday
   (iii) it will rain on Wednesday
   (iv)   it will rain on exactly two out of the next three days (i.e. Tuesday, Wednesday, Thursday).

3   In a simple learning experiment, a child attempts a simple task 4 times in succession. Each attempt results in either success (S) or failure (F). The probability of success at the first attempt is 0.3 and the probability of success increases at each attempt; it increases by 0.2 following a success or by 0.1 following a failure.
   Find the probability of each of the following sequences:
   (i) SSSS   (ii) SSSF   (iii) SSFS   (iv) SFSS   (v) FSSS.
   Deduce the probability that there are
   (vi)   exactly 3 successes
   (vii) at least 3 successes.

4   If a car is chosen at random, the probability that it is red is 0.2 and the probability that it is blue is 0.3. Find the probability that
   (i)   it is not red
   (ii)   it is not blue

(iii) it is not red and it is not blue.
   If two cars are chosen at random, find the probability that
(iv) they are both red
(v) one is red and the other is blue.

5 A die is thrown 4 times. Find the probability that
(i)   there are no sixes
(ii) there is at least one six
(iii) all the scores are different
(iv) the scores are alternately odd and even (but it does not matter which comes first).

6 Two cards are selected from a normal pack of 52 cards. Find the probability that
(i)   they are both aces
(ii) at least one of them is a spade
(iii) they are of different suits
(iv) one of them is the ace of spades.

7 The probability that a person has blue eyes is 0.4 and the probability that he is left-handed is 0.2; these may be assumed to be independent.
   Find the probability that a person
(i)   is not left-handed
(ii) has blue eyes and is left-handed
(iii) has blue eyes and is not left-handed.
   Given that two people are chosen at random, find the probability that one of them has blue eyes and is left-handed, and the other has blue eyes and is not left-handed.

8 At a party there are 20 guests, of whom 12 are lawyers, 6 are teachers and 2 are doctors. The hostess speaks to 2 guests at random. Find the probability that
(i)   neither of them is a lawyer
(ii) at least one of them is a teacher
(iii) they both belong to the same profession.

9 Four digits (between 0 and 9 inclusive) are taken from a random number table. Find the probability that
(i)   all the digits are odd
(ii) all the digits are different.

10 An application for a bank loan is dealt with by either the manager, his secretary, or the assistant manager, with probabilities 0.2, 0.5 and 0.3 respectively. The probability that the manager will grant a loan is 0.7, and the probabilities that the secretary and the assistant manager will grant a loan are 0.9 and 0.4 respectively.
   Find the probability that a loan will be granted.

11 Bag P contains 6 green and 3 yellow discs.
   Bag Q contains 2 green and 5 yellow discs.
      One disc is removed from each bag. Find the probability that these two discs are of the same colour.

12 Bag P contains 6 green and 3 yellow discs.
  Bag Q contains 2 green and 5 yellow discs.
  One disc is transferred from Bag P to Bag Q, and then one disc is removed from each bag. Find the probability that these last two discs are the same colour.

13 Bag P contains 6 green and 3 yellow discs.
  Bag Q contains 2 green and 5 yellow discs.
  One of the bags (either P or Q) is chosen at random, and one disc is transferred from the chosen bag to the other one. Then one disc is removed from each bag. Find the probability that these last two discs are the same colour.

8.2 Applications

We can apply the ideas of the previous section in situations where the probabilities are obtained from Normal distributions, Binomial distributions and so on.

Example 1
The lifetime of a spy satellite has a Normal distribution with mean 7 years and standard deviation 2 years. The lifetimes of different satellites may be assumed to be independent.
(i)   Given that two satellites are launched, find the probability that one of them lasts for less than 5 years and the other lasts for more than 10 years.
(ii) Given that six satellites are launched, find the probability that exactly two of them will still be working after 10 years.

(i)   If one satellite has lifetime T years then

$$P(T < 5) = P\left(Z < \frac{5-7}{2}\right) = P(Z < -1)$$

$$= \Phi(-1)$$

$$= 1 - 0.8413$$

$$= 0.1587$$

$$P(T > 10) = P\left(Z > \frac{10-7}{2}\right) = P(Z > 1.5)$$

$$= 1 - \Phi(1.5)$$

$$= 1 - 0.9332$$

$$= 0.0668$$

If two satellites have lifetimes T_1 and T_2, then the probability that one lasts less than 5 years and the other more than 10 years is

$$P(T_1 < 5 \text{ and } T_2 > 10) + P(T_1 > 10 \text{ and } T_2 < 5)$$

$$= 0.1587 \times 0.0668 + 0.0668 \times 0.1587$$

$$= 0.0212$$

(ii) If six satellites are launched, the number still working after 10 years will have a Binomial distribution B(6, 0.0668)

$$P(\text{exactly two are still working}) = \binom{6}{2} (0.0668)^2 (0.9332)^4$$

$$= 0.0508$$

Example 2

A man sits near the centre of a long row of chairs. He throws a coin; if it lands heads, he moves one place to the right, and if it lands tails, he moves one place to the left.

Given that he throws the coin 6 times, find the probability that he finishes two places from his starting position.

In 6 throws, the number of heads has a Binomial distribution B(6, $\frac{1}{2}$).
If there are no heads (and hence 6 tails), he will finish 6 places to the left.
If there is one head (and hence 5 tails), he will move 1 place to the right and 5 places to the left (in some order); so he will finish 4 places to the left, and so on.

Number of heads	0	1	2	3	4	5	6
Finishing position	6 places to left	4 places to left	2 places to left	start	2 places to right	4 places to right	6 places to right

The probability that he finishes 2 places from his starting position is

$$P(\text{2 heads}) + P(\text{4 heads}) = \binom{6}{2}\left(\frac{1}{2}\right)^2\left(\frac{1}{2}\right)^4 + \binom{6}{4}\left(\frac{1}{2}\right)^4\left(\frac{1}{2}\right)^2$$

$$= \frac{15}{64} + \frac{15}{64}$$

$$= \frac{15}{32}$$

Example 3

It is known that 70% of the microchips produced by a machine are faulty. The chips are individually tested, and good ones are packed in boxes of four.

Find the probability that, in order to produce a box of four good chips,
(i) exactly 20 chips need to be tested
(ii) more than 20 chips need to be tested.

(i) If 20 chips need to be tested, then 3 of the first 19 must be good, and the 20th chip must be good.
For the first 19 chips, the number of good ones has the Binomial distribution B(19, 0.3), so

$$P(\text{3 good ones}) = \binom{19}{3}(0.3)^3 (0.7)^{16}$$

$$P(\text{20th chip is good}) = 0.3$$

Hence $P(20 \text{ chips needed}) = 969 \times (0.3)^3 \times (0.7)^{16} \times (0.3)$

$$= 0.0261$$

(ii) We could find, as in part (i), the probabilities that 4, 5, 6, ..., 19, 20 chips are needed; add these up, and subtract from one. However the following method is quicker.

More than 20 chips will need to be tested if there are less than 4 good ones in the first 20 chips tested.

For the first 20 chips, the number of good ones has the Binomial distribution B(20, 0.3); hence

$$P(\text{more than 20 chips needed}) = P(\text{less than 4 good ones in first 20})$$

$$= P(\text{no good ones}) + P(1 \text{ good one})$$

$$+ P(2 \text{ good ones}) + P(3 \text{ good ones})$$

$$= (0.7)^{20} + \binom{20}{1}(0.3)(0.7)^{19}$$

$$+ \binom{20}{2}(0.3)^2(0.7)^{18} + \binom{20}{3}(0.3)^3(0.7)^{17}$$

$$= 0.00080 + 0.00684 + 0.02785 + 0.07160$$

$$= 0.1071$$

Example 4

Two players, A and B, take it in turns to throw a die, with A starting. Find the probability that A is the first to throw a six.

The probability that A throws a six on his first throw is $\frac{1}{6}$.

The probability that A and B do not throw a six on the first throw, then A throws a six on his second throw is $\frac{5}{6} \times \frac{5}{6} \times \frac{1}{6}$

The probability that A, B, A, B do not throw a six, then A throws a six on his third throw is $\frac{5}{6} \times \frac{5}{6} \times \frac{5}{6} \times \frac{5}{6} \times \frac{1}{6}$ and so on.

Hence the probability that A is the first to throw a six is

$$\frac{1}{6} + \left(\frac{5}{6} \times \frac{5}{6} \times \frac{1}{6}\right) + \left(\frac{5}{6} \times \frac{5}{6} \times \frac{5}{6} \times \frac{5}{6} \times \frac{1}{6}\right) + \cdots$$

This is an infinite geometric series, with the first term $\frac{1}{6}$ and common ratio $\frac{5}{6} \times \frac{5}{6} = \frac{25}{36}$; hence

$$P(\text{A is first to throw a six}) = \frac{\frac{1}{6}}{1 - \frac{25}{36}} = \frac{6}{11}$$

Exercise 8.2 Applications

1 The heights of men are Normally distributed with mean 175 cm and standard deviation 10 cm. Given that one man is chosen at random, find the probability that he is
   (i) shorter than 188 cm
   (ii) taller than 180 cm
   (iii) shorter than 188 cm and taller than 180 cm.

2 The volumes of ball-bearings produced by a machine are Normally distributed with mean 250 mm^3 and standard deviation 15 mm^3. Given that two ball-bearings are selected at random, find the probability that
   (i) both of them have volume greater than 240 mm^3
   (ii) one of them has volume greater than 240 mm^3 and the other has volume less than 240 mm^3.

3 The time taken by students to complete a test is distributed Normally with mean 38 minutes and standard deviation 12 minutes.
   Given that 10 students take the test, find
   (i) the probability that they will all finish the test when a time of 45 minutes is allowed.
   (ii) the time which should be allowed if the probability that all 10 students will finish is to be 0.5.

4 A coin is tossed until 3 heads have been obtained. Find the probability that
   (i) exactly 3 tosses are necessary
   (ii) exactly 8 tosses are necessary
   (iii) more than 15 tosses are necessary.

5 In a multiple choice test, each question offers a choice of 5 answers (only one of which is correct). The probability that a student knows the correct answer is $\frac{5}{8}$; if he does not know which answer is correct he selects one of the 5 answers at random.
   (i) Find the probability that he selects the correct answer to a question.
   (ii) Given that the test consists of 10 questions, find the probability that he gets at least 9 right.

6 A chain is made up of n links. Under a certain load, the probability that an individual link will break is 0.01, and the links behave independently.
   (i) Find the probability that the chain will break (i.e. at least one of the links breaks).
   (ii) Find the maximum number of links if the probability that the chain will break is to be less than 0.4.

7 For a family with 3 children, list the probabilities that there are 0, 1, 2, 3 boys.
   Given that two families, each having 3 children, are selected, find the probability that they have the same number of boys.

8 The number of cherries in a tin of 'Tropical' fruit salad is a random variable X having the distribution shown below

X	2	3	4	5	6
Probability	0.1	0.2	0.3	0.3	0.1

The number of cherries in a tin of 'Exotic' fruit salad is a random variable Y having the distribution below

Y	3	4	5	6	7	8
Probability	0.1	0.1	0.2	0.3	0.2	0.1

Given that one tin of each brand is chosen, find the probability that the 'Exotic' tin contains more cherries than the 'Tropical' tin.

9 The distance of a shot from the centre of a target is a random variable X cm having probability density function

$$f(x) = \begin{cases} \dfrac{k}{(x+1)^2} & \text{for} \quad 0 \leqslant x \leqslant 5 \\ 0 & \text{otherwise.} \end{cases}$$

(i) Find the probability that a shot lands between 2 and 3 cm from the centre of the target.
(ii) Given that four shots are fired, find the probability that exactly two of them land between 2 and 3 cm from the centre of the target.

10 A girl plays the following game: she throws a die; if the number showing is a 5 or 6 she moves 1 m forwards, otherwise she moves 1 m backwards. She repeats this for 5 throws of the die.

List her possible finishing positions and find the probability for each position. Hence find the probability that
(i) she finishes 1 m from her starting position
(ii) she finishes in front of her starting position.

11 Three players A, B and C take turns (in that order, with A starting) to throw a pair of dice. The winner is the first player to obtain a total score of exactly 9 on the two dice. For each player, find the probability that he wins the game.

12 The weights of apples are Normally distributed with mean 120 g and standard deviation 16 g. If two apples are selected, find the probability that at least one of them is heavier than 100 g and at least one of them is lighter than 130 g.

(Let the weights be X_1 and X_2, and consider the cases $X_1 \leqslant 100$, $100 < X_1 < 130$ and $X_1 \geqslant 130$.)

8.3 Venn diagrams and conditional probability

For a given situation, let S be the set of all possible outcomes. This is called the *sample space*.

An event is represented by the set of outcomes for which the event occurs. For example, when a die is thrown, the set of possible outcomes is

$$S = \{1, 2, 3, 4, 5, 6\}$$

The event 'the number showing is odd' corresponds to the set A = {1, 3, 5}, the event 'the number showing is divisible by 3' corresponds to the set B = {3, 6}, and so on.

Now suppose that A and B are the sets corresponding to any two events. We can represent the sets on a Venn diagram (Fig. 8.2).

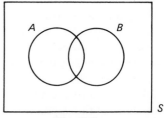

Fig. 8.2

The intersection A∩B contains the outcomes which are in both sets A and B, i.e. for which both events occur (Fig. 8.3).

A∩B corresponds to the event 'A and B'.

In the example above, A∩B = {3}, the only outcome for which the number showing is odd and divisible by 3.

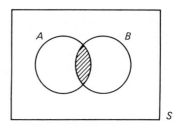

Fig. 8.3

The union A∪B contains the outcomes which are either in A or in B (or in both) (Fig. 8.4).

A∪B corresponds to the event 'A or B'

(note that this includes the possibility that both events occur).

In the example above, A∪B = {1, 3, 5, 6}, the set of outcomes for which the number showing is odd or divisible by 3.

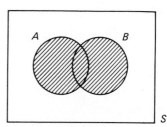

Fig. 8.4

The complement A′ contains the outcomes which are not in A (Fig. 8.5).

A′ corresponds to the event (not A).

In the example above, A′ = {2, 4, 6}, the set of outcomes for which the number showing is not odd.

Also B′ = {1, 2, 4, 5}, the set of outcomes for which the number showing is not divisible by 3.

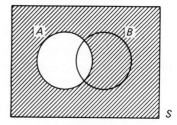

Fig. 8.5

Probabilities may be written in the regions of a Venn diagram. This approach is particularly useful when dealing with events which are not independent, and when the information is given in a complicated way.

Example 1
Paintings are offered for sale to three art galleries, the Alpha, the Beta and the Gamma. For any particular painting, the probability that:
Alpha wants to buy it is 0.59
Beta wants to buy it is 0.49
all three galleries want to buy it is 0.25
just one of the galleries wants to buy it is 0.34
Beta and Gamma both want to buy it is 0.28
Alpha and Beta both want to buy it, and Gamma does not, is 0.12
Gamma wants to buy it and the other two do not, is 0.07
Find the probability that
(i)   the Gamma gallery wants to buy it
(ii) none of the galleries wants to buy it
(iii) just two of the galleries want to buy it
(iv) neither the Alpha nor the Beta wants to buy it
(v)   the Alpha wants to buy it and the others do not.

Writing A for the event 'the Alpha gallery wants to buy it', and so on, we draw a Venn diagram (Fig. 8.6).

Let a, b, c, d, e, f, g, h be the probabilities corresponding to the eight regions of the Venn diagram.

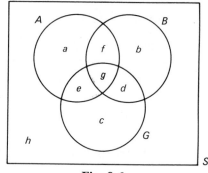

Fig. 8.6

We are given

$P(A) = 0.59,$	so	$a + e + f + g = 0.59$
$P(B) = 0.49,$	so	$b + d + f + g = 0.49$
$P(A \text{ and } B \text{ and } G) = 0.25,$	so	$g = 0.25$
$P(\text{just one}) = 0.34,$	so	$a + b + c = 0.34$
$P(B \text{ and } G) = 0.28,$	so	$d + g = 0.28$
$P(A \text{ and } B \text{ and not } G) = 0.12,$	so	$f = 0.12$
$P(G \text{ and not } A \text{ and not } B) = 0.07,$	so	$c = 0.07$
The probabilities must add up to one,	so	$a + b + c + d + e + f + g + h = 1.00$

Solving these equations, we obtain

$$g = 0.25, \ d = 0.03, \ f = 0.12, \ b = 0.09, \ c = 0.07, \ a = 0.18, \ e = 0.04, \ h = 0.22.$$

Figure 8.7 shows these values, and hence we obtain

(i)   P(G) = 0.07 + 0.04 + 0.25 + 0.03 = 0.39
(ii) P(none) = 0.22
(iii) P(just two) = 0.12 + 0.03 + 0.04 = 0.19
(iv) P(not A and not B) = 0.22 + 0.07 = 0.29
(v)   P(A and not B and not G) = 0.18

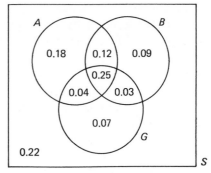

Fig. 8.7

Conditional probability

Suppose that an event A is known to have occurred; then the set A becomes effectively the sample space.

Another event B can now only occur in the intersection (A and B) (Fig. 8.8). The conditional probability of B, given that A has occurred, is the probability of this intersection compared with the probability of the sample space A, i.e.

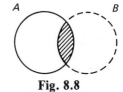

Fig. 8.8

$$P(B \mid A) = \frac{P(A \text{ and } B)}{P(A)}$$

When using this formula to calculate a conditional probability P(B | A), the probabilities P(A and B) and P(A) might be found from a Venn diagram, from a probability tree, from a theoretical distribution, or by any other method.

Note that, on multiplying by P(A), we obtain

$$P(A \text{ and } B) = P(A) \times P(B \mid A)$$

the multiplication law which we used earlier.

Independence

Two events A and B are said to be *independent* if the probability that B occurs is not affected by the knowledge that A has occurred; i.e. if

$$P(B \mid A) = P(B)$$

$$\text{Then} \quad \frac{P(A \text{ and } B)}{P(A)} = P(B)$$

$$\text{and so} \quad P(A \text{ and } B) = P(A) \times P(B)$$

> A and B are independent if P(B | A) = P(B)
> This is equivalent to P(A and B) = P(A) × P(B)

More generally, several events A_1, A_2, A_3, ..., are independent if the probability of each event is not affected by any knowledge about the others (for example, $P(A_3 | A_1)$, $P(A_3 | A_2)$ and $P(A_3 | A_1$ and $A_2)$ are all equal to $P(A_3)$).

Example 2
In a survey, 35% said they watched football but not cricket; 10% said that they watched cricket but not football, and 40% said that they did not watch either game.
   One person is selected at random from those in the survey.
(i)   Find the probability that he watches football, given that he watches cricket.
(ii) Find the probability that he watches football, given that he does not watch cricket.
(iii) Are the events 'he watches football' and 'he watches cricket' independent?

   Let F be the event 'he watches football' and C be the event 'he watches cricket'.
   We draw a Venn diagram, and write in the probability for each region;
$P(F$ and $C) = 0.15$ because the probabilities must add up to one.

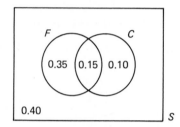

Fig. 8.9

(i) $P(F | C) = \dfrac{P(C \text{ and } F)}{P(C)} = \dfrac{0.15}{0.25} = 0.6$

(ii) $P(F | \text{not } C) = \dfrac{P(\text{not } C \text{ and } F)}{P(\text{not } C)} = \dfrac{0.35}{0.75} = \dfrac{7}{15} \approx 0.4667$

(iii) The events F and C are *not* independent; this can be seen in several ways.
   (a) $P(F | C) = 0.6$, but $P(F) = 0.5$
       Since $P(F | C) \neq P(F)$, F and C are not independent.
   (b) $P(F | C)$ and $P(F | \text{not } C)$ are not the same, so the events F and C cannot be independent.
       (If F and C were independent, then $P(F | C)$ and $P(F | \text{not } C)$ would both be equal to $P(F)$)
   (c) We have $P(F$ and $C) = 0.15$
       but $P(F) \times P(C) = 0.5 \times 0.25 = 0.125$
       Since $P(F$ and $C) \neq P(F) \times P(C)$, F and C are not independent.

Example 3
A certain disease is present in 1 in 2000 of the population. In a mass screening programme, a quick test for the disease is used. This test is not totally reliable; for someone who does have the disease the test gives a positive result (suggesting the presence of the disease) in 90% of cases; but for someone who does not have the disease, the test gives a positive result in 2% of cases.
(i)   Given that the test gives a positive result, find the probability that the person has the disease.
(ii) People for whom the test proves positive are recalled and retested. Find the probability that the person has the disease if this second test also proves positive.

(i)   The probability tree is shown in Fig. 8.10.

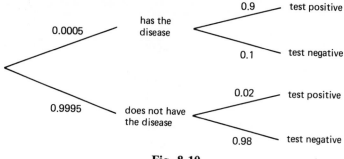

Fig. 8.10

We require the probability that someone has the disease, given that the test is positive. That is

$$P(\text{disease} \mid \text{test positive}) = \frac{P(\text{test positive and has disease})}{P(\text{test positive})}$$

$$= \frac{0.0005 \times 0.9}{0.0005 \times 0.9 + 0.9995 \times 0.02}$$

$$= \frac{0.00045}{0.02044}$$

$$= 0.0220$$

$\left(\text{This might appear to be a low probability; but the positive test result has increased}\right.$
the probability of having the disease from $\dfrac{1}{2000}$ to about $\left.\dfrac{1}{45}.\right)$

Note that we were given the conditional probability $P(\text{test positive} \mid \text{disease}) = 0.9$ and we have calculated the conditional probability the other way round: $P(\text{disease} \mid \text{test positive})$.

(ii)   We now extend the probability tree to include the second test.

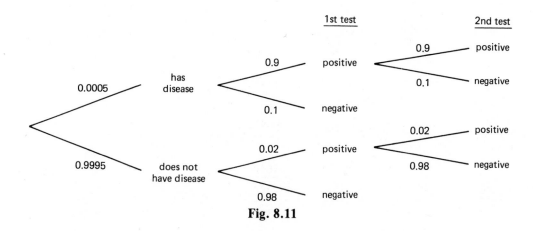

Fig. 8.11

$$P(\text{disease} \mid \text{both tests positive}) = \frac{P(\text{both tests positive and has disease})}{P(\text{both tests positive})}$$

$$= \frac{0.0005 \times 0.9 \times 0.9}{0.0005 \times 0.9 \times 0.9 + 0.9995 \times 0.02 \times 0.02}$$

$$= \frac{0.000405}{0.0008048}$$

$$= 0.5032$$

Example 4

The bolts produced by a machine have diameters which are Normally distributed with mean 10.4 mm and standard deviation 0.8 mm; but only those with diameters between 9 mm and 11 mm can be used.

Given that a bolt is usable, find the probability that its diameter is greater than 10 mm.

If a bolt has diameter X mm, we require the probability that $X > 10$, given that the bolt is usable (i.e. $9 < X < 11$)

$$P(X > 10 \mid 9 < X < 11) = \frac{P(9 < X < 11 \text{ and } X > 10)}{P(9 < X < 11)}$$

$$= \frac{P(10 < X < 11)}{P(9 < X < 11)}$$

Now $P(10 < X < 11) = P\left(\frac{10 - 10.4}{0.8} < Z < \frac{11 - 10.4}{0.8}\right)$

$$= P(-0.5 < Z < 0.75)$$

$$= 0.7734 - (1 - 0.6915)$$

$$= 0.4649$$

and $P(9 < X < 11) = P(-1.75 < Z < 0.75)$

$$= 0.7734 - (1 - 0.9599)$$

$$= 0.7333$$

so $P(X > 10 \mid \text{usable}) = \dfrac{P(10 < X < 11)}{P(9 < X < 11)}$

$$= \frac{0.4649}{0.7333}$$

$$= 0.6340$$

The general addition law

Suppose that A and B are any two events. If we add P(A) to P(B), the probability of the intersection (A and B) is counted twice (Fig. 8.12).

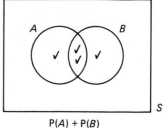

P(A) + P(B)

Fig. 8.12

Subtracting P(A and B), we obtain the probability of the union (A or B) (Fig. 8.13).

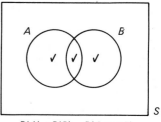

P(A) + P(B) − P(A and B)

Fig. 8.13

Thus

$$P(A \text{ or } B) = P(A) + P(B) - P(A \text{ and } B)$$

This is the 'general addition law'. It may be used to find one of P(A), P(B), P(A and B), P(A or B) when the other three are known.

Note that, if A and B are non-overlapping events, then P(A and B) = 0, so this rule reduces to

$$P(A \text{ or } B) = P(A) + P(B)$$

the result we used earlier.

Example 5

Three cards are taken (without replacement) from an ordinary pack of 52 cards. Find the probability that, in these three cards, there are
(i)   no spades
(ii) no aces
(iii) no spades and there are no aces
(iv) no spades or there are no aces (or both)
(v) no spades, given that there are no aces.

(i)   There are 39 cards in the pack which are not spades,

so $P(\text{no spades}) = \frac{39}{52} \times \frac{38}{51} \times \frac{37}{50} = \frac{54\,834}{132\,600} = \frac{703}{1700}$

(ii) There are 48 cards which are not aces,

so $P(\text{no aces}) = \frac{48}{52} \times \frac{47}{51} \times \frac{46}{50} = \frac{103\,776}{132\,600} = \frac{4324}{5525}$

(iii) There are 36 cards (2 to King of the suits hearts, diamonds and clubs) which are neither spades nor aces,

so P(no spades and no aces) $= \frac{36}{52} \times \frac{35}{51} \times \frac{34}{50} = \frac{42\,840}{132\,600} = \frac{21}{65}$

(iv) P(no spades or no aces) = P(no spades) + P(no aces) − P(no spades and no aces)

$$= \frac{54\,834}{132\,600} + \frac{103\,776}{132\,600} - \frac{42\,840}{132\,600}$$

$$= \frac{115\,770}{132\,600} = \frac{227}{260}$$

(v) P(no spades | no aces) $= \dfrac{\text{P(no aces and no spades)}}{\text{P(no aces)}}$

$$= \frac{42\,840}{132\,600} \Big/ \frac{103\,776}{132\,600}$$

$$= \frac{42\,840}{103\,776}$$

$$= \frac{1785}{4324}$$

Exercise 8.3 Venn diagrams and conditional probability

1 For a member of a certain book club, the probability that he has read book A is 0.4, the probability that he has read book B is 0.52 and the probability that he has read book C is 0.63. The probability that he has read both books A and B is 0.2, the probability that he has read both books B and C is 0.4, and the probability that he has read both books C and A is 0.3.

The probability that he has read all three books A, B and C is 0.15. Draw a Venn diagram and mark the probability corresponding to each region. Hence find the probability that a member
(i)   has read none of the books A, B or C
(ii) has read book B but not book A
(iii) has read at least two of the books A, B, C
(iv) has read book A or book C (or both).

2 In a group of 100 sixth-form students
        60 are studying A level mathematics
        48 are studying A level physics
        42 are studying A level chemistry
        13 are studying mathematics and physics but not chemistry
        22 are studying physics and chemistry but not mathematics
         2 are studying chemistry and mathematics but not physics
         3 are studying none of these subjects.
Draw a Venn diagram and find the number of students corresponding to each region.
    Given that one student is selected at random, find the probability that
(i)   he is studying mathematics but not physics
(ii) he is studying just one of the three subjects.
    Given that two students are selected from the group find the probability that at least one of them is taking all 3 subjects.

3 For any day, let A be the event 'the sun has shone', and let B be the event 'it has rain-ed'. The probability that the sun has shone is 0.5, the probability that it has rained is 0.3, and the probability that both events have occurred is 0.1.

(i)   Draw a Venn diagram showing the sets A and B, and mark the probability of each region.

(ii)   Find the probability that the sun has shone, given that it has rained.

(iii)   Find the probability that it has rained, given that the sun has shone.

(iv)   Find the probability that the sun has shone and it has not rained.

(v)   Find the probability that the sun has shone, given that it has not rained.

(vi)   Explain why the events A and B are not independent.

4 A car is examined for the MOT test by two garages 'Smiths' and 'Taylors'. The pro-bability that Smiths passes the car is 0.6, and the probability that Taylors passes the car is 0.8. Given that Smiths passes the car, the probability that Taylors passes it is 0.9.

Let S be the event 'Smiths passes the car' and T be 'Taylors passes the car'.

(i)   Are the events S and T independent?

(ii)   Find the probability that both garages pass the car.

(iii)   Draw a Venn diagram showing the sets S and T and the probability of each region.

(iv)   Find the probability that the car passes the test at one of the two garages or at both of them.

(v)   Given that Taylors passes the car, find the probability that Smiths passes it.

5 Two dice are thrown. Find the probability that

(i)   there is at least one 6

(ii)   the total score is 8

(iii)   there is at least one 6 *and* the total score is 8

(iv)   there is at least one 6 *or* the total score is 8 (or both)

(v)   there is at least one 6 *given* that the total score is 8.

6 A bag contains 3 red discs and 5 green discs. Two discs are removed. Find the prob-ability that

(i)   these two discs are of different colours

(ii)   the first disc is green and the second one red

(iii)   the first disc is green, given that the two discs are of different colours.

7 A box contains 12 eggs and the probability that an egg is bad is 0.1. Given that at least one of the eggs is bad, find the conditional probability that exactly two of the eggs are bad.

8 A machine produces bags of sugar whose weights are Normally distributed with mean 1012 g and standard deviation 20 g. Bags weighing less than 990 g are rejected.

For a bag which has been accepted, find the probability that it weighs less than 1020 g.

9 30% of vehicles brought for testing are in a dangerous condition. The probability that

a vehicle will pass the test is 0.6 if it is not in a dangerous condition, or 0.1 if it is in a dangerous condition. Find the probability
(i)   that a vehicle will pass the test
(ii) that a vehicle is in a dangerous condition given that it has passed the test
(iii) that a vehicle is in a dangerous condition given that it has not passed the test.

10 The adult populations of three towns P, Q and R are in the ratio $4:7:9$, and the proportions of Labour voters in these towns are 65%, 10% and 40% respectively.

A person is chosen at random from all those living in the three towns, and it is found that he votes Labour. What is the probability that he lives in town P?

8.4 Arrangements

When we have a set of equally likely outcomes, the probability of an event may be calculated as

$$\frac{\text{number of outcomes for which the event occurs}}{\text{total number of possible outcomes}}$$

If the number of outcomes is large, it may be impractical to list them all. However, all we need to know is how many outcomes there are, and for how many of these the required event occurs.

We now consider some methods for finding the number of outcomes.

Consider three objects A, B, C. These can be arranged in the following ways:

$$\text{ABC, ACB, BAC, BCA, CAB, CBA}$$

that is, in 6 different ways.
We may also see this as follows
there are 3 choices for the first object (A, B, or C);
when we have chosen the first object there are 2 choices for the second object;
when we have chosen the first two objects there is only one choice for the third object.
So the number of possible arrangements is $3 \times 2 \times 1 = 6$

Now consider the arrangements of four objects A, B, C, D.
There are 4 choices for the first object, then 3 for the second, 2 for the third and 1 for the fourth; so the number of arrangements is

$$4 \times 3 \times 2 \times 1 = 4! = 24$$

Without needing to list them all, we can say that there are 24 possible arrangements of the four objects A, B, C, D.

In general, if there are n objects, there will be n choices for the first, $(n-1)$ choices for the second, and so on, and the number of arrangements is

$$n(n-1)(n-2) \cdots \times 2 \times 1 = n!$$

---

n different objects can be arranged in a line in $n!$ ways

Example 1

Nine people A, B, C, D, E, F, G, H, I sit down, in random order, on a bench.
Find the probability that A is sitting next to B.

The number of possible arrangements of the 9 people on the bench is 9! We are told
that they sit in random order, which means that all of these possible arrangements are
equally likely.

We now need to find the number of arrangements in which A and B are sitting together.
If we consider 'AB' as a single unit, then the 8 items AB, C, D, E, F, G, H, I can be
arranged in 8! ways.

Similarly, using 'BA' as a single unit, we obtain another 8! arrangements with A and
B together. The number of arrangements with A and B together is thus $8! \times 2$.

Hence the probability that A and B are sitting together is $\dfrac{8! \times 2}{9!} = \dfrac{2}{9}$.

We now consider situations where some of the objects are indistinguishable. For
example, consider the arrangements of the letters of D E P L E T E. If we label the three
Es, i.e. $D E_1 P L E_2 T E_3$ we now have 7 different letters, so there are 7! arrangements.

However, any arrangement of the original letters, such as L P E T D E E,

will occur as $L P E_1 T D E_2 E_3$, $L P E_1 T D E_3 E_2$, $L P E_2 T D E_1 E_3$, and so on;

in fact in 3! ways corresponding to the arrangements of the three Es amongst themselves.

Each arrangement of D E P L E T E occurs 3! times among the 7! arrangements of
$D E_1 P L E_2 T E_3$;

hence the number of arrangements of DEPLETE is $\dfrac{7!}{3!} = 840$

Now consider arrangements of the letters of D E F I N I T I V E. The 10 letters
$D E_1 F I_1 N I_2 T I_3 V E_2$ can be arranged in 10! ways. However I_1, I_2, I_3 can be arranged
amongst themselves in 3! ways, and E_1, E_2 in 2! ways, so each arrangement of
D E F I N I T I V E occurs $3! \times 2!$ times among the 10! arrangements of
$D E_1 F I_1 N I_2 T I_3 V E_2$

Hence the number of arrangements of D E F I N I T I V E is $\dfrac{10!}{3!2!} = 302400$

In general,

> *n* objects, of which *r* are of one type (indistinguishable from each other)
> *s* are of a second type
> *t* are of a third type, and so on,
>
> can be arranged in a line in $\dfrac{n!}{r!\ s!\ t!\ \cdots}$ ways.

This result is often useful.

Example 2

In how many ways can the letters of INDEPENDENT be arranged?

Given that one of these arrangements is chosen at random find the probability that it contains the sequence ETE.

The word INDEPENDENT has 11 letters, including 3 Es, 3 Ns, and 2 Ds; so the number of arrangements is

$$\frac{11!}{3! \ 3! \ 2!} = 554\ 400$$

The arrangements containing ETE can be considered as the arrangements of the 9 items ETE, E, N, N, N, D, D, I, P; hence the number of arrangements is

$$\frac{9!}{3! \ 2!} = 30\ 240$$

The probability that an arrangement contains ETE is thus

$$\frac{30\ 240}{554\ 400} = \frac{3}{55}$$

Example 3
Five dice are thrown.
Find the probability that there are two 6s, two 4s and a 2 showing.

The probability of obtaining any particular arrangement, such as 6, 4, 4, 2, 6, is

$\frac{1}{6} \times \frac{1}{6} \times \frac{1}{6} \times \frac{1}{6} \times \frac{1}{6} = \left(\frac{1}{6}\right)^5$, and the number of arrangements is $\frac{5!}{2! \ 2!}$.

Hence P(two 6s, two 4s, one 2) $= \left(\frac{1}{6}\right)^5 \times \frac{5!}{2! \ 2!} = \frac{5}{1296}$.

When considering arrangements which do not use all the available objects, it is usually best to work from first principles, as shown in the following example.

Example 4
Find the numbers of arrangements of 3 letters which can be made
(i)   using letters from the word EQUATION
(ii) using letters form the word IMPLICIT.

(i)   EQUATION has 8 different letters. For an arrangement of 3 letters, there are 8 choices for the first letter, 7 choices for the second, and 6 choices for the third. Hence the number of arrangements is $8 \times 7 \times 6 = 336$
(ii) IMPLICIT has 6 different letters, I, M, P, L, C, T, and two extra I's. The number of arrangements containing 3 different letters is $6 \times 5 \times 4 = 120$
There are also some arrangements containing more than one I.
An arrangement containing just two Is must be II–, I–I, or –II, where the dash is one of the 5 other letters;
so the number of arrangements containing just two Is is $5 \times 3 = 15$
There is 1 arrangement, III, containing three Is
Hence the total number of arrangements is $120 + 15 + 1 = 136$

Circular arrangements
When objects are arranged in a circle (such as people sitting round a circular table), it is customary for arrangements which are simple rotations of each other to be regarded as the same.

For example, the two arrangements in Fig. 8.14 are regarded as the same.

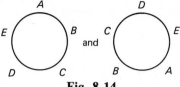

Fig. 8.14

Suppose there are *n* objects. We consider one of the objects, say A, to be fixed in a certain position (since any arrangement may be rotated until A occupies that position). Then the remaining $(n-1)$ objects can be arranged in $(n-1)!$ ways.

> *n* different objects can be arranged in a circle
> in $(n-1)!$ ways.

Example 5

Four men A, B, C, D and four women E, F, G, H sit round a circular table.
(i)   How many arrangements are possible if men and women alternate round the table?
(ii) Given that the 8 people sit down at random, find the probability that men and women will alternate round the table.

(i)   We consider A to be fixed.
      The other 3 men B, C, D must occupy the places marked
      ×; they can do so in 3! ways.
      The 4 women E, F, G, H must occupy the places marked
      ∘; they can do this in 4! ways.
      Hence the number of arrangements is $3! \times 4! = 144$

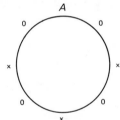

Fig. 8.15

(ii) The 8 people may sit in 7! arrangements, which are equally likely.
      Hence the probability that men and women will alternate is

$$\frac{3! \times 4!}{7!} = \frac{1}{35}$$

Note   If we have, for example, charms on a bracelet, then arrangements which are reflections of each other, such as those shown in Fig. 8.16, are also regarded as the same (since one can be obtained from the other by turning the bracelet over).

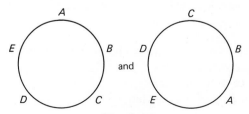

Fig. 8.16

The number of arrangements of n objects round a bracelet is then

$$\frac{(n-1)!}{2}$$

Exercise 8.4   Arrangements

1   7 people, Alison, Brian, Charles, Diana, Emma, Frances and George, sit on a bench. In how many ways can they be arranged?
How many arrangements are possible such that
(i)   Alison is sitting in the middle
(ii) Brian is sitting at an end
(iii) Diana and George are sitting at the ends
(iv) Alison is sitting next to Brian
(v)   Emma is sitting between Brian and George.
If the 7 people sit down at random, deduce the probabilities of each of the five events (i), (ii), (iii), (iv) and (v).

2   Three men, Paul, Quentin and Roger, and three women, Susan, Tina and Una sit round a circular table. How many different arrangements are there?
How many arrangements are there if
(i)   Paul is sitting next to Una
(ii) Quentin is sitting opposite a woman
(iii) Men and women alternate round the table?

3   In how many ways can all the letters of COMPUTER be arranged?
(i)   How many arrangements contain the word 'PET'?
(ii) In how many arrangements do the three vowels come together?

4   How many arrangements of three letters can be made using letters from the word FORTUNE?
How many of these three-letter arrangements
(i)   contain the letter R
(ii) contain both E and T?

5   How many arrangements can be made using all the letters of the following words:
(i) VARIANCE   (ii) CALCULUS   (iii) MATHEMATICIAN?

6   How many arrangements can be made using all the letters of STATISTICS?
(i)   How many of these arrangements begin with T?
(ii) In how many of the arrangements are the three Ss together?
(iii) Given that one arrangement is chosen at random, find the probability that the A is next to the C.

7   Consider arrangements of three letters chosen from SWEETHEART.
How many arrangements are there
(i)   containing three different letters
(ii) containing two Es and another letter

(iii) containing two Ts and another letter
(iv) containing three Es
(v) altogether?
Given that one of these three letter arrangements is chosen at random, find the probability that it
(a) contains at least two Es
(b) does not contain a T
(c) contains at least one T.

8 Three dice are thrown.
(i) How many arrangements are possible showing a 4, a 5, and a 6? What is the probability of obtaining each arrangement? Deduce the probability of obtaining a 4, a 5 and a 6.
(ii) Write down the number of arrangements of two 6s and a 3, and deduce the probability of obtaining two 6s and a 3.

9 Four dice are thrown.
(i) Write down the number of arrangements of two 6s and two 5s and deduce the probability of obtaining two 6s and two 5s
(ii) Find the probability of obtaining two 6s, a 5 and a 4.

10 Five random digits (between 0 and 9 inclusive) are chosen. Find the probability that there are
(i) three 7s, a 5 and a 2
(ii) two 9s, two 6s and a 0.

8.5 Selections

Suppose that 8 people, A, B, C, D, E, F, G, H, wish to go to a pop concert, but there are only 3 tickets available to them. We must choose 3 people from the original 8; how many different selections are possible?

The number of arrangements of 3 people from these 8 is $8 \times 7 \times 6$. However, the order in which the people are selected does not matter; for example, AHD, DAH, HDA, and so on, are different arrangements, but they represent the same selection of 3 people namely A, D and H).

Each selection (such as A, D and H) will occur in 3! arrangements amongst the $8 \times 7 \times 6$ arrangements of 3 people.

Hence the number of different selections of 3 people chosen from 8 is

$$\frac{8 \times 7 \times 6}{3!} = 56$$

Note that $8 \times 7 \times 6 = \dfrac{8!}{5!}$, so the number of selections may be written as $\dfrac{8!}{5! \, 3!}$

In general, the number of arrangements of r objects chosen from n different ones is

$$n(n-1)(n-2) \cdots (n-r+1) = \frac{n!}{(n-r)!}$$

Each selection of r objects will occur in $r!$ arrangements among the total of $\dfrac{n!}{(n-r)!}$ arrangements.

Hence the number of selections of r objects chosen from n is $\dfrac{n!}{(n-r)!r!}$

Note that this is the Binomial coefficient $\dbinom{n}{r}$

> The number of ways of selecting r objects from n different ones (no regard being paid to the order of selection) is
>
> $$\binom{n}{r} = \frac{n!}{(n-r)!\ r!}$$

Example 1

The 3 available tickets for a pop concert are allocated at random among the 8 people A, B, C, D, E, F, G, H wishing to go.
Find the probability that
(i)   none of A, B, C goes to the concert.
(ii) A goes to the concert and B does not go.

The number of ways of selecting three people from eight is $\binom{8}{3} = 56$; and all these selections are equally likely.

(i)   If A, B, and C do not go, then the 3 people must be chosen from the other 5 (D, E, F, G, H); so the number of ways of selecting them is $\binom{5}{3} = 10$.

Hence

$$\text{P(none of A, B, C goes)} = \tfrac{10}{56} = \tfrac{5}{28}$$

(ii) If A goes and B does not, then one of the people selected must be A; the other 2 may be selected from 6 (C, D, E, F, G, H); so the number of ways of selecting them is $\binom{6}{2} = 15$.

Hence

$$\text{P(A goes and B does not)} = \tfrac{15}{56}.$$

Example 2

Four children are selected at random from a class containing 5 boys and 7 girls. Find the probability that
(i)   2 boys and 2 girls are selected
(ii) at least 2 girls are selected.

The number of ways of selecting 4 children from 12 is $\binom{12}{4} = 495$; and these selections are all equally likely.

(i)   2 boys may be selected in $\binom{5}{2}$ ways

and 2 girls may be selected in $\binom{7}{2}$ ways, so the number of ways of selecting 2 boys

and 2 girls is $\binom{5}{2} \times \binom{7}{2} = 10 \times 21 = 210$.

Hence

$$P(2 \text{ boys and } 2 \text{ girls}) = \tfrac{210}{495} = \tfrac{14}{33}.$$

(ii) If at least 2 girls are selected, then we must have *either* 2 boys and 2 girls *or* 1 boy
and 3 girls *or* 0 boys and 4 girls.

$$P(2 \text{ boys and } 2 \text{ girls}) = \tfrac{210}{495}$$

Similarly

$$P(1 \text{ boy and } 3 \text{ girls}) = \frac{\binom{5}{1} \times \binom{7}{3}}{495} = \frac{5 \times 35}{495} = \frac{175}{495}$$

$$P(0 \text{ boys and } 4 \text{ girls}) = \frac{\binom{7}{4}}{495} = \frac{35}{495}$$

Hence

$$P(\text{at least } 2 \text{ girls}) = \tfrac{210}{495} + \tfrac{175}{495} + \tfrac{35}{495} = \tfrac{420}{495} = \tfrac{28}{33}$$

Example 3

A hand of 13 cards is dealt from a well-shuffled ordinary pack of 52 playing cards. Find
the probability that this hand contains
(i)   exactly 4 spades
(ii) exactly 3 diamonds and exactly 2 clubs
(iii) 4 spades, 4 hearts, 3 diamonds and 2 clubs.
  The number of different hands of 13 cards chosen from 52 is

$$\binom{52}{13} = \frac{52!}{39! \; 13!} \approx 6.3501 \times 10^{11}$$

Since the pack is well shuffled, all these hands are equally likely.
(i)   In the pack there are 13 spades and 39 cards which are not spades. The number of
hands containing exactly 4 spades (and hence 9 cards which are not spades) is

$$\binom{13}{4} \times \binom{39}{9}$$

Hence

$$P(\text{exactly } 4 \text{ spades}) = \frac{\binom{13}{4} \times \binom{39}{9}}{\binom{52}{13}} \approx 0.2386$$

(ii) In the pack there are 13 diamonds, 13 clubs, and 26 others.
The number of hands containing 3 diamonds, 2 clubs and 8 others is

$$\binom{13}{3} \times \binom{13}{2} \times \binom{26}{8}$$

Hence

$$P(\text{exactly 3 diamonds and exactly 2 clubs}) = \frac{\binom{13}{3} \times \binom{13}{2} \times \binom{26}{8}}{\binom{52}{13}}$$

$$\approx 0.0549$$

(iii) In the pack there are 13 spades, 13 hearts, 13 diamonds and 13 clubs.
The number of hands containing 4 spades, 4 hearts, 3 diamonds and 2 clubs is

$$\binom{13}{4} \times \binom{13}{4} \times \binom{13}{3} \times \binom{13}{2}$$

Hence

$$P(\text{4 spades, 4 hearts, 3 diamonds, 2 clubs}) = \frac{\binom{13}{4} \times \binom{13}{4} \times \binom{13}{3} \times \binom{13}{2}}{\binom{52}{13}}$$

$$\approx 0.0180$$

Example 4
A group of 15 people are going out for the evening. There are 6 tickets for the theatre, 5 tickets for a dance and 4 tickets for a concert.
(i)   In how many ways can the group be divided up between the theatre, the dance and the concert?
(ii) In how many ways can the group be divided up if two particular people (say A and B) refuse to go to the theatre, and another person (say C) insists on going to the dance?
(iii) If in fact the tickets are allocated at random among the 15 people find the probability that A, B and C will be satisfied.

(i)   The 6 people for the theatre may be chosen in $\binom{15}{6}$ ways; when these have been chosen, the 5 people for the dance must be chosen from the remaining 9 people, i.e. in $\binom{9}{5}$ possible ways; then 4 people are left for the concert.

$$\text{The number of ways is } \binom{15}{6} \times \binom{9}{5} \times \binom{4}{4}$$

$$= 5005 \times 126 \times 1$$

$$= 630\,630$$

(ii) The 6 people for the theatre must now be chosen from 12 (i.e. excluding A, B, and C), so there are $\binom{12}{6}$ ways;

for the dance, C must be chosen, and the other 4 may be chosen from 8 (i.e. excluding C and the 6 chosen for the theatre), so there are $\binom{8}{4}$ ways; then there are 4 people left for the concert.

$$\text{The number of ways is } \binom{12}{6} \times \binom{8}{4} \times \binom{4}{4}$$

$$= 924 \times 70 \times 1$$

$$= 64\,680$$

(iii) If the tickets are allocated at random, then the 630 630 possible ways are equally likely, so the probability that A, B and C will be satisfied is

$$\frac{64\,680}{630\,630} = \frac{4}{39}$$

Exercise 8.5 Selections

1 How many ways are there of choosing
   (i)   3 prizewinners from a class of 8 children
   (ii)   2 different digits from {0, 1, 2, ..., 9}
   (iii) a football team of 11 players from a squad of 15
   (iv)   4 different letters from the word COMBINATORIAL
   (v)   5 cards from a normal pack of 52
   (vi)   7 houses to visit in a street containing 20 houses
   (vii) 5 questions to answer from an examination paper of 9 questions
   (viii) 6 people to make redundant from a staff of 80?

2 A committee of 5 people is to be selected from the following 10 people: Mr Quixotte, Mr Renoir, Mr Smith, Mr Thatcher, Mr Urquhart, Mr Varsity, Mr Wilson, Miss Xavier, Miss Young and Miss Zapper.
   In how many ways can the committee be chosen?
   What is the number of ways if
   (i)   Mr Wilson must be on the committee
   (ii)   all three women must be on the committee
   (iii) the committee must contain four men and one woman
   (iv)   the committee must contain three men and two women
   (v)   Mr Smith is on the committee and Miss Young is not
   (vi)   the committee contains Mr Quixotte, two other men, and two women
   (vii) the committee must contain at least one woman
   (viii) the committee must contain either Mr Renoir or Mr Smith (or both of them)
   (ix)   Miss Xavier refuses to serve on the committee if Mr Quixotte is on it
   (x)   Mr Thatcher will only serve on the committee if Miss Zapper is also on it?

3 A touring cricket squad of 17 players has 8 batsmen, 2 wicket keepers and 7 bowlers. The manager decides that a team must consist of *either* 6 batsmen, 1 wicket keeper and 4 bowlers *or* 5 batsmen, 1 wicket keeper and 5 bowlers. How many possible teams are there?

    If the 17 names are placed in a hat, and 11 are selected at random, find the probability that these 11 players form a possible team.

4 5 cards are selected at random from a normal pack of 52 cards.
In how many ways can this be done?
Find the probabilities that
(i)   exactly two Queens are selected
(ii) two Queens and three Kings are selected.

5 A hand of 13 cards is dealt at random from a normal pack of 52 cards.
Find the probabilities that the hand contains
(i) no aces   (ii) exactly one ace   (iii) exactly two aces   (iv) exactly three aces
(v) all four aces.

6 A hand of 13 cards is dealt at random from a normal pack of 52 cards. Find the probabilities that the hand contains
(i)   exactly 5 spades   (ii) at least 3 spades
(iii) 5 spades, 3 hearts, 3 diamonds and 2 clubs.

7 10 mathematicians are to be split into two groups of 5: one group to be taught by Mr Einstein and the other group to be taught by Mr Newton. In how many ways can this be done?

8 A party of 12 tourists is to be split into a group of 5 to visit the British Museum, a group of 4 to visit the Science Museum and a group of 3 to visit the Houses of Parliament. In how many ways can this be done? If Jackie and Ken are two members of the party, and the splitting up is done at random, find the probability that Jackie and Ken are in the same group.

9 A certain kind of sweet is available in 8 different flavours. They are sold in packets of 10 sweets which always contain 3 of one flavour, 2 of a second flavour, 2 of a third flavour, 1 of a fourth flavour, 1 of a fifth flavour and 1 of a sixth flavour. How many different packets are possible?

10 An examination paper is split into two sections. Section A has 8 questions and Section B has 6 questions. The rubric says 'Answer six questions choosing at least two questions from each Section'. In how many ways can a candidate choose his questions?

    A careless candidate answers six questions chosen at random from the whole paper, regardless of section. Find the probability that he satisfies the rubric.

8.6 Harder examples

The results about arrangements and selections may be used together with the multiplication and addition rules.

Example 1

Four dice are thrown. Find the probability that
(i) one number appears three times and another number appears once
(ii) one number appears twice, and two other numbers appear once each.

(i) We first consider a particular case: three 6s and one 5.

$$6 \; 6 \; 6 \; 5 \text{ can be arranged in } \frac{4!}{3!} \text{ ways,}$$

and the probability of obtaining any particular arrangement is $\left(\frac{1}{6}\right)^4$

$$\text{so P(three 6s, one 5)} = \left(\frac{1}{6}\right)^4 \times \frac{4!}{3!}$$

There are 6 choices for the number which appears three times; then 5 choices for the single number. The probability of any such combination (such as three 4s and one 6) will be the same as P(three 6s, one 5).

Hence

$$\text{P(three of one number, one of another)} = \left(\frac{1}{6}\right)^4 \times \frac{4!}{3!} \times 6 \times 5$$

$$= \frac{5}{54}$$

(ii) $6 \; 6 \; 5 \; 4$ can be arranged in $\frac{4!}{2!}$ ways, so

$$\text{P(two 6s, one 5, one 4)} = \left(\frac{1}{6}\right)^4 \times \frac{4!}{2!}$$

There are 6 choices for the number which appears twice; then $\binom{5}{2}$ choices for the two single numbers (*not* 5×4 choices, since, for example, one 3 and one 2 is the same as one 2 and one 3). Hence

$$\text{P(two of one number, one of another, one of a third)} = \left(\frac{1}{6}\right)^4 \times \frac{4!}{2!} \times 6 \times \binom{5}{2}$$

$$= \frac{5}{9}$$

Example 2

A hand of 13 cards is dealt from a well-shuffled ordinary pack of 52 playing cards. Find the probability that the hand contains four cards of one suit, four cards of another suit, three cards of a third suit and two cards of the fourth suit.

$$\text{We have P(4 spades, 4 hearts, 3 diamonds, 2 clubs)} = \frac{\binom{13}{4} \times \binom{13}{4} \times \binom{13}{3} \times \binom{13}{2}}{\binom{52}{13}}$$

The two suits having four cards can be chosen in $\binom{4}{2}$ ways; then there are two choices for the suit having three cards; and then 1 choice for the suit having two cards. Hence

$$P(4-4-3-2 \text{ distribution}) = \frac{\binom{13}{4} \times \binom{13}{4} \times \binom{13}{3} \times \binom{13}{2}}{\binom{52}{13}} \times \binom{4}{2} \times 2 \times 1$$

$$\approx 0.2155$$

Exercise 8.6   Harder questions

1   Five dice are thrown. Find the probability that
   (i)   one number appears three times, and another twice (a 'full house').
   (ii) two numbers appear twice each, and a third number appears once.

2   A hand of 13 cards is dealt from a normal pack of 52 cards.
   Find the probability that there are five of some suit, three of another suit, three of a third suit and two of the fourth suit.

3   A sequence of 6 digits is read from a random number table. Find the probability that the sequence consists of three different numbers each occurring twice.

4   A bag contains 5 green, 3 red and 2 blue discs. Given that 4 discs are removed (without replacement) find the probability that at least one disc of each colour is removed.

5   When asked a certain question, 50% of people reply 'Yes', 40% reply 'No' and the remaining 10% reply 'Don't know'.
   10 people, selected at random, are each asked the question.
   (i)   How many arrangements are there of 5 'Yes's, 3 'No's and 2 'Don't know's?
   (ii) Deduce the probability that 5 people answer 'Yes', 3 answer 'No', and 2 answer 'Don't know'.
   (iii) Find the probability that 6 people answer 'Yes', 2 answer 'No', and 2 answer 'Don't know'.

8.7   Miscellaneous probability

There are sometimes several methods for answering a particular probability question. Of course, any valid method will give the correct answer, but knowing which method will give the simplest solution is largely a matter of experience.

Example 1
Seven people, A, B, C, D, E, F, G sit down, in random order, on a bench. Find the probability that A and B are sitting next to each other.

Method 1   The total number of arrangements of the 7 people is 7! and the number of

arrangements with A and B together is 6! × 2; so

$$P(\text{A and B together}) = \frac{6! \times 2}{7!} = \frac{2}{7}$$

Method 2   The probability that A is sitting at an end is $\frac{2}{7}$; in this case only one of the remaining 6 seats is next to A, so

$$P(\text{A at an end, and B next to A}) = \frac{2}{7} \times \frac{1}{6} = \frac{2}{42}$$

The probability that A is not at an end is $\frac{5}{7}$; then two of the remaining 6 seats are next to A, so

$$P(\text{A not at an end, and B next to A}) = \frac{5}{7} \times \frac{2}{6} = \frac{10}{42}$$

Hence

$$P(\text{A and B together}) = \frac{2}{42} + \frac{10}{42} = \frac{12}{42} = \frac{2}{7}$$

Method 3   The 2 places occupied by A and B can be chosen in $\binom{7}{2} = 21$ ways and there are 6 pairs of adjacent places, so

$$P(\text{A and B together}) = \frac{6}{21} = \frac{2}{7}$$

Example 2

Three discs are taken (without replacement) from a bag containing 6 red discs, 3 green discs and 1 blue disc.

Find the probability that 2 red discs and 1 green disc are taken.

Method 1 $P(2R, 1G) = P(RRG) + P(RGR) + P(GRR)$

$$= \tfrac{6}{10} \times \tfrac{5}{9} \times \tfrac{3}{8} + \tfrac{6}{10} \times \tfrac{3}{9} \times \tfrac{5}{8} + \tfrac{3}{10} \times \tfrac{6}{9} \times \tfrac{5}{8}$$

$$= \tfrac{3}{8}$$

Method 2   The number of ways of selecting 3 discs from the bag is $\binom{10}{3}$, and the number of ways of selecting 2 red discs and 1 green disc is $\binom{6}{2} \times \binom{3}{1}$

$$\text{So } P(2R, 1G) = \frac{\binom{6}{2} \times \binom{3}{1}}{\binom{10}{3}} = \frac{15 \times 3}{120} = \frac{3}{8}$$

Example 3

Four dice are thrown. Find the probability that there are four different numbers showing.

Method 1 $P(6\ 5\ 4\ 3, \text{ in any order}) = \left(\frac{1}{6}\right)^4 \times 4!$

The four numbers can be chosen in $\binom{6}{4}$ ways, so

$$P(\text{four different numbers}) = \left(\frac{1}{6}\right)^4 \times 4! \times \binom{6}{4} = \frac{5}{18}$$

Method 2   If the four numbers are different, then the 2nd must be different from the 1st, the 3rd different from the first two, and the 4th different from the first three; so

$$P(\text{four different numbers}) = \tfrac{5}{6} \times \tfrac{4}{6} \times \tfrac{3}{6} = \tfrac{5}{18}$$

Glossary of terms

The following terms are important ones used in formal probability theory.

The *sample space* S is the set of all possible outcomes.

An *event* A is a subset of S
$A \cap B$ is the event (A and B)
$A \cup B$ is the event (A or B)
  A' is the event (not A)

The *conditional probability* of B, given A, is defined as

$$P(B \mid A) = \frac{P(A \cap B)}{P(A)}$$

Two events A,B are *mutually exclusive* if $A \cap B = \emptyset$
                     exhaustive   if $A \cup B = S$
                     independent   if $P(A \cap B) = P(A) \times P(B)$

Events $A_1, A_2, ..., A_n$ are

      mutually exclusive if $A_i \cap A_j = \emptyset$ for all $i \neq j$
      exhaustive   if $A_1 \cup A_2 \cup \cdots \cup A_n = S$
      independent   if $P(A_i \cap A_j) = P(A_i) \times P(A_j)$,
                  $P(A_i \cap A_j \cap A_k) = P(A_i) \times P(A_j) \times P(A_k)$,
               $P(A_i \cap A_j \cap A_k \cap A_l) = P(A_i) \times P(A_j) \times P(A_k) \times P(A_l)$,
      and so on, for all distinct $i, j, k, l, ...$

Multiplication Law: $P(A \cap B) = P(A) \times P(B \mid A)$
Addition Law: $P(A \cup B) = P(A) + P(B) - P(A \cap B)$

The number of *permutations* (arrangements) of r objects chosen from n is

$$_nP_r = n(n-1)(n-2) \cdots (n-r+1) = \frac{n!}{(n-r)!}$$

The number of *combinations* (selections) of r objects chosen from n is

$$_nC_r = \binom{n}{r} = \frac{n!}{(n-r)!r!}$$

Exercise 8.7 Miscellaneous probability questions

1   A die is biased so that the faces showing 1 and 6 are twice as likely to land uppermost as those showing 2, 3, 4, and 5.
     This die is thrown twice. Find the probability that the total score is at least 10.

2   The probability that it will rain on a particular day is 0.4 if it rained on the previous day, but 0.1 if it did not rain on the previous day. It rained on Monday (see Exercise 8.1, Question 2.)
     (i)   Find the probability that it will rain on Tuesday.
     (ii) Find the probability that it will rain on Wednesday.
     (iii) By considering the two cases 'it rains on Wednesday' and 'it does not rain on Wednesday', find the probability that it will rain on Thursday.
     (iv) Similarly, find the probability that it will rain on Friday.

3   Five digits (between 0 and 9 inclusive) are taken from a random number table. Find the probability that
     (i)   all the digits are less than or equal to 8
     (ii) all the digits are less than or equal to 7
     (iii) the highest digit chosen is an 8.

4   A bridge hand of 13 cards is dealt from a normal pack of 52 cards. Find the probability that the hand contains no Jacks, Queens, Kings, or Aces.

5   A certain fencer always cheats. In any competition, the probability that his cheating will be detected is 0.3; on the first occasion he is discovered cheating, he is warned; on the second occasion he is disqualified from taking part in any further competitions.
     Find the probability that
     (i)   he is warned at his 5th competition
     (ii) he is disqualified at his 7th competition
     (iii) he takes part in 10 competitions without being disqualified.

6   A train departs at either 2.35, 2.36 or 2.37 with probabilities 0.6, 0.3, 0.1 respectively, and the journey time is Normally distributed with mean 25 minutes and standard deviation 2 minutes.
     Find the probability that the train arrives at its destination before 3 o'clock.

7   Urn A contains 6 red discs, 3 yellow and 1 black.
     Urn B contains 3 red discs, 2 yellow, 1 green and 1 black.
     One disc is transferred from Urn A to Urn B; then one disc is transferred from Urn B to Urn A.
     Two discs are now taken from Urn A. Find the probability that they are both red.

8   5 children are selected at random from a class containing 10 boys and 20 girls. Find the probability that exactly 2 boys are selected.

9   The probability that a marriage will end in divorce is $\frac{1}{3}$. Given that five married couples are selected at random, find the probability that exactly two of them will become divorced.

10 A telephone enquiry is answered by one of three people, Fred, George, or Hilda, with probability 0.7, 0.1, 0.2 respectively. The probabilities that these people can answer an enquiry satisfactorily are respectively 0.8, 1 and 0.4.
  (i)   Find the probability that an enquiry is answered satisfactorily.
  (ii) Given that an enquiry was answered satisfactorily, find the probability that it was answered by Hilda.

If I have two enquiries to make, find the probability that both are answered satisfactorily when

  (iii) I ring up on two separate occasions.
  (iv) I ring up once and put both enquiries to the person who answers.

11 Two players A and B contest a snooker match, which consists of a series of frames. The probabilities that A and B win any particular frame are $\frac{2}{3}$ and $\frac{1}{3}$ respectively, and the frames may be assumed to be independent. The match ends when one of the players has won 8 frames; then that player becomes the champion.
  (i)   Find the probability that the match ends after exactly 12 frames.
  (ii) Given that the match ends after exactly 12 frames, find the probability that
      (a) A becomes the champion
      (b) B becomes the champion.

12 The letters of the word PARALLEL are placed in a hat, and drawn out one at a time (without replacement). Find the probability that
  (i)   the first two letters are both Ls
  (ii) the E is drawn before the P
  (iii) the R is drawn before the first L
  (iv) the two As are drawn in succession
  (v)   the letters are drawn in alphabetical order.

13 A dinner party of 14 people sit down at random at two tables, one laid for 8 and the other laid for 6. Find the probability that two specified people are sitting at the same table.

14 A school-leaver applies to two firms, 'Corlands' and 'Marks', for a job. The probability that he is offered a job by Corlands is 0.5, and the probability that he is offered a job by Marks is 0.6. The probability that he is offered a job by at least one of these firms is 0.7.
Find the probability that
  (i)   he is offered a job by both firms
  (ii) he is offered a job by Corlands, but not by Marks
  (iii) he is offered a job by Corlands, given that he is not offered a job by Marks.

In Questions 15–25 most of the 'answers' given are not valid. For each question, say whether the 'answer' given is correct. If not, explain where the argument breaks down, and correct it.

15 If a die is thrown 3 times, find the probability that there is at least one six.

Answer   P(at least one six) = P(1st throw is a six) + P(2nd throw is a six)
                                                     + P(3rd throw is a six)

$$= \frac{1}{6} + \frac{1}{6} + \frac{1}{6}$$

$$= \frac{1}{2}$$

16 Given that three cards are taken from a normal pack of 52 cards, find the probability that one of them is the Queen of Hearts.

Answer   P(QH) = P(1st is QH) + P(2nd is QH) + P(3rd is QH)

$$= \frac{1}{52} + \frac{1}{52} + \frac{1}{52}$$

$$= \frac{3}{52}$$

17 A coin is tossed twice. Find the probability that there is one head and one tail.

Answer   There are 3 possible outcomes: '2 heads', '2 tails' or '1 of each'.

Hence P(1 of each) $= \dfrac{1}{3}$

18 In a sequence of 4 random digits, find the probability that the highest number is a 6.

Answer   This can occur as 6*** or *6** or **6* or ***6 where * represents a digit between 0 and 6 inclusive.

P(highest number is 6) $= \dfrac{1}{10} \times \dfrac{7}{10} \times \dfrac{7}{10} \times \dfrac{7}{10} \times 4 = 0.1372$

19 The probability that a pebble weighs less than 60 gram is 0.84, and the probability that it weighs less than 20 gram is 0.28. Find the probability that a pebble weighs less than 60 gram and more than 20 gram.

Answer   P(less than 60 and more than 20) = P(less than 60) × P(more than 20)

$$= 0.84 \times 0.72 = 0.6048$$

20 If two dice are thrown, find the probability that
   (i) the total score is 8 *and* the numbers showing differ by 2
   (ii) the total score is 8 *or* the numbers showing differ by 2

Answer   P(total is 8) $= \dfrac{5}{36}$, P(differ by 2) $= \dfrac{8}{36}$

   (i)   P(total is 8 and differ by 2) $= \dfrac{5}{36} \times \dfrac{8}{36} = \dfrac{5}{162}$

   (ii) P(total is 8 or differ by 2) $= \dfrac{5}{36} + \dfrac{8}{36} = \dfrac{13}{36}$

21 In a drawer there are 4 different pairs of socks. Given that 2 socks are taken at random, find the probability that they form a pair.

Answer   Number of possible selections $= \binom{8}{2} = 28$

There are 4 pairs, hence $P(\text{pair}) = \dfrac{4}{28} = \dfrac{1}{7}$

22 A family has 3 children. Given that at least one child is a boy, find the probability that all three children are boys.

Answer   As one child is known to be a boy, we require

$$P(\text{all boys} \mid \text{at least one is a boy}) = P(\text{other two are boys}) = \left(\frac{1}{2}\right)^2 = \frac{1}{4}$$

23 Three dice are thrown. Given that there are no sixes, find the probability that there are no fives.

Answer   As there are no sixes, each die must show $1, 2, 3, 4$ or 5

$$P(\text{no fives} \mid \text{no sixes}) = \left(\frac{4}{5}\right)^3 = \frac{64}{125}$$

24 An organiser needs 4 people for a certain task. The probability that a person will agree to help when asked is $\frac{1}{4}$. Find the probability that the 4 helpers will be obtained after asking 10 people.

Answer   When 10 people are asked, the number who agree to help is $B(10, \frac{1}{4})$

$$P(\text{4 people agree}) = \binom{10}{4} \times \left(\frac{1}{4}\right)^4 \times \left(\frac{3}{4}\right)^6 \approx 0.146.$$

25 Given that four cards are taken from a normal pack, find the probability that there are two of one suit and two of another suit.

Answer $P(\text{2 spades and 2 hearts}) = \dfrac{\binom{13}{2} \times \binom{13}{2}}{\binom{52}{4}} = \dfrac{468}{20\,825}$

$$P(\text{2 of one suit, 2 of another}) = \frac{468}{20\,825} \times 4 \times 3 \approx 0.2697.$$

9
Sums and Differences of Random Variables

9.1 Joint distributions of two discrete random variables

Joint probability distributions

Suppose that X and Y are two random variables defined at the same time. In this section we shall assume that the random variables are discrete, say X has possible values $x_1, x_2, \ldots, x_m$ and Y has possible values $y_1, y_2, \ldots, y_n$.

The *joint distribution* of X and Y gives the probabilities that each possible pair of values, $X = x_i$ and $Y = y_j$, occurs.

These probabilities may be conveniently arranged in an $(m \times n)$ table as shown in the following example.

Example 1

Three coins (one 50p, one 20p, and one 10p) are tossed and random variables X and Y are defined as follows:

X is the number of heads showing on the 50p and 20p coins

(so that $X = 0$, 1, or 2)

Y is the number of heads showing on all three coins

(so that $Y = 0$, 1, 2 or 3).

(i) Find the joint distribution of X and Y
(ii) Find the mean and variance of X and of Y.

(i) We consider the 8 equally likely outcomes and note the values of X and Y for each outcome.

50p	20p	10p	X	Y
H	H	H	2	3
H	H	T	2	2
H	T	H	1	2
H	T	T	1	1
T	H	H	1	2
T	H	T	1	1
T	T	H	0	1
T	T	T	0	0

Hence P($X = 2$ and $Y = 2$) $= \frac{1}{8}$, P($X = 1$ and $Y = 2$) $= \frac{2}{8}$

P($X = 0$ and $Y = 2$) $= 0$, and so on.

We write these probabilities in a table.

X \ Y	0	1	2	3
0	$\frac{1}{8}$	$\frac{1}{8}$	0	0
1	0	$\frac{2}{8}$	$\frac{2}{8}$	0
2	0	0	$\frac{1}{8}$	$\frac{1}{8}$

This is the joint distribution of X and Y. Note that all the probabilities in the table add up to one.

(ii) The distribution of X can be found from the joint distribution.

P($X = 0$) = P($X = 0$ and $Y = 0$) + P($X = 0$ and $Y = 1$) + P($X = 0$ and $Y = 2$)

$$+ \text{P}(X = 0 \text{ and } Y = 3).$$

$$= \frac{1}{8} \quad + \quad \frac{1}{8} \quad + \quad 0 \quad + \quad 0 \quad = \frac{1}{4}$$

This is the sum of the probabilities in the first row of the table. Similarly, adding the probabilities in the second and third rows,

$$\text{P}(X = 1) = 0 + \tfrac{2}{8} + \tfrac{2}{8} + 0 = \tfrac{2}{4}$$
$$\text{P}(X = 2) = 0 + 0 + \tfrac{1}{8} + \tfrac{1}{8} = \tfrac{1}{4}$$

Hence the distribution of X is

X	0	1	2
probability	$\frac{1}{4}$	$\frac{2}{4}$	$\frac{1}{4}$

In the usual way we can now calculate

$$\text{E}[X] = 0 \times \frac{1}{4} + 1 \times \frac{2}{4} + 2 \times \frac{1}{4} = 1$$

$$\text{E}[X^2] = 0 \times \frac{1}{4} + 1 \times \frac{2}{4} + 4 \times \frac{1}{4} = \frac{3}{2}$$

$$\text{var}(X) = \frac{3}{2} - 1^2 = \frac{1}{2}$$

The row totals are usually written at the right-hand side of the table; this is sometimes called the 'marginal distribution' of X.

The joint distribution table is completed as

X \ Y	0	1	2	3	
0	$\frac{1}{8}$	$\frac{1}{8}$	0	0	$\frac{1}{4}$
1	0	$\frac{2}{8}$	$\frac{2}{8}$	0	$\frac{2}{4}$
2	0	0	$\frac{1}{8}$	$\frac{1}{8}$	$\frac{1}{4}$
	$\frac{1}{8}$	$\frac{3}{8}$	$\frac{3}{8}$	$\frac{1}{8}$	

Similarly, by adding the probabilities in the columns as above, we obtain the distribution of Y:

Y	0	1	2	3
probability	$\dfrac{1}{8}$	$\dfrac{3}{8}$	$\dfrac{3}{8}$	$\dfrac{1}{8}$

Hence

$$E[Y] = \frac{3}{2}$$

$$E[Y^2] = 3,$$

$$\operatorname{var}(Y) = 3 - \left(\frac{3}{2}\right)^2 = \frac{3}{4}.$$

Covariance and correlation

The table of probabilities is analogous to the frequency table for paired data discussed in Section 6.3. By analogy with the formulae

$$\operatorname{cov}(x, y) = \frac{\sum(x - \bar{x})(y - \bar{y})}{n} = \frac{\sum xy}{n} - \left(\frac{\sum x}{n}\right)\left(\frac{\sum y}{n}\right)$$

we now define covariance for two random variables X and Y.
  If $E[X] = \mu$ and $E[Y] = \lambda$, we define the *covariance* of X and Y to be

$$\operatorname{cov}(X, Y) = E[(X - \mu)(Y - \lambda)]$$

We shall show later that

$$\operatorname{cov}(X, Y) = E[XY] - E[X]E[Y]$$

We also define the coefficient of correlation between X and Y to be

$$\rho = \frac{\operatorname{cov}(X, Y)}{\sigma_X \sigma_Y}$$

where σ_X and σ_Y are the standard deviations of X and Y, and we always have $-1 \leqslant \rho \leqslant 1$.

Example 2

For the two random variables X and Y defined in Example 1, find the covariance, cov (X, Y) and the coefficient of correlation between X and Y.

We have cov $(X, Y) = E[XY] - E[X]E[Y]$, and $E[X] = 1$, $E[Y] = \frac{3}{2}$.

For $E[XY]$, the mean value of XY, we consider all 12 positions in the probability table. For each one we multiply the value of XY at that position by the probability entered there, and add these up.

$$E[XY] = 0 \times \tfrac{1}{8} + 0 \times \tfrac{1}{8} + 0 \times 0 + 0 \times 0$$

$$+ 0 \times 0 + 1 \times \tfrac{2}{8} + 2 \times \tfrac{2}{8} + 3 \times 0$$

$$+ 0 \times 0 + 2 \times 0 + 4 \times \tfrac{1}{8} + 6 \times \tfrac{1}{8} = 2.$$

Hence cov $(X, Y) = E[XY] - E[X]E[Y] = 2 - 1 \times \tfrac{3}{2} = \tfrac{1}{2}$

Now $\sigma_X = \sqrt{\tfrac{1}{2}}$, $\sigma_Y = \sqrt{\tfrac{3}{4}}$, so the coefficient of correlation,

$$\rho = \frac{\frac{1}{2}}{\sqrt{\tfrac{1}{2}}\sqrt{\tfrac{3}{4}}} = \sqrt{\frac{2}{3}} \approx 0.816.$$

Independent random variables

Two random variables X and Y are said to be independent if the distribution of Y is unaffected by knowledge of the value of X.

We always have, for example, $P(X = 1 \text{ and } Y = 2) = P(X = 1) \times P(Y = 2 \mid X = 1)$, but if knowing that $X = 1$ does not alter the probability that $Y = 2$, then $P(Y = 2 \mid X = 1) = P(Y = 2)$ and so

$$P(X = 1 \text{ and } Y = 2) = P(X = 1) \times P(Y = 2)$$

We therefore say that

> X and Y are *independent* if
> $P(X = x_i \text{ and } Y = y_j) = P(X = x_i) \times P(Y = y_j)$
> for all possible values x_i of X and y_j of Y

This means that every entry in the probability table is obtained by multiplying the corresponding row and column totals. That is, if the probabilities are as shown in the table below we have $a = p_1q_1$, $b = p_1q_2$, $c = p_2q_1$, and so on.

X \ Y	y_1	y_2	$\ldots$	
x_1	a	b		p_1
x_2	c	d		p_2
$\vdots$				
	q_1	q_2	$\ldots$	

To show that X and Y are independent, we must check *all* the entries in the table.

To show that X and Y are not independent, it is sufficient to find *one* pair of values x_i, y_j for which

$$P(X = x_i \text{ and } Y = y_j) \neq P(X = x_i) \times P(Y = y_j).$$

The random variables X and Y defined in Example 1 are *not* independent, since, for example,
$$P(X = 1 \text{ and } Y = 2) = \tfrac{2}{8}$$

$$\text{but } P(X = 1) \times P(Y = 2) = \tfrac{2}{4} \times \tfrac{3}{8} = \tfrac{3}{16}$$

$$\text{so } P(X = 1 \text{ and } Y = 2) \neq P(X = 1) \times P(Y = 2)$$

Example 3
Two random variables X and Y have the following joint distribution:

X \ Y	0	1	2	3
0	$\frac{1}{32}$	$\frac{3}{32}$	$\frac{3}{32}$	$\frac{1}{32}$
1	$\frac{2}{32}$	$\frac{6}{32}$	$\frac{6}{32}$	$\frac{2}{32}$
2	$\frac{1}{32}$	$\frac{3}{32}$	$\frac{3}{32}$	$\frac{1}{32}$

(i) Show that X and Y are independent.
(ii) Find $E[X]$, $E[Y]$, $E[XY]$, and verify that $E[XY] = E[X]E[Y]$.

(i) We first add up the rows and columns to obtain the distributions of X and of Y.

X \ Y	0	1	2	3	
0	$\frac{1}{32}$	$\frac{3}{32}$	$\frac{3}{32}$	$\frac{1}{32}$	$\frac{1}{4}$
1	$\frac{2}{32}$	$\frac{6}{32}$	$\left(\frac{6}{32}\right)$	$\frac{2}{32}$	$\left(\frac{2}{4}\right)$
2	$\frac{1}{32}$	$\frac{3}{32}$	$\frac{3}{32}$	$\frac{1}{32}$	$\frac{1}{4}$
	$\frac{1}{8}$	$\frac{3}{8}$	$\left(\frac{3}{8}\right)$	$\frac{1}{8}$	

Note that the distributions of X and of Y are the same as in Example 1, although the joint distribution is different.

It is now a simple matter to check that all 12 entries in the table are obtained by multiplying the corresponding row and column totals; for example, the circled probability $\frac{6}{32}$ is equal to $\frac{2}{4} \times \frac{3}{8}$.

Hence X and Y are independent.

(ii) We have $E[X] = 1$, $E[Y] = \frac{3}{2}$

$$E[XY] = 0 \times \tfrac{1}{32} + 0 \times \tfrac{3}{32} + 0 \times \tfrac{3}{32} + 0 \times \tfrac{1}{32}$$

$$+ 0 \times \tfrac{2}{32} + 1 \times \tfrac{6}{32} + 2 \times \tfrac{6}{32} + 3 \times \tfrac{2}{32}$$

$$+ 0 \times \tfrac{1}{32} + 2 \times \tfrac{3}{32} + 4 \times \tfrac{3}{32} + 6 \times \tfrac{1}{32} = \tfrac{3}{2}$$

Hence $E[XY] = E[X]E[Y]$.

If X and Y are independent random variables, it always follows that

$$E[XY] = E[X]E[Y]$$

We shall prove this for a general (2×3) table, but it is clear that a similar proof will work for a table of any size.

Suppose that the joint distribution of X and Y is

X \ Y	y_1	y_2	y_3	
x_1	a	b	c	p_1
x_2	d	e	f	p_2
	q_1	q_2	q_3	

Since X and Y are independent, we have $a = p_1q_1$, $b = p_1q_2$ and so on.

$$
\begin{aligned}
E[XY] &= x_1y_1a + x_1y_2b + x_1y_3c \\
&\quad + x_2y_1d + x_2y_2e + x_2y_3f \\
&= x_1y_1p_1q_1 + x_1y_2p_1q_2 + x_1y_3p_1q_3 \\
&\quad + x_2y_1p_2q_1 + x_2y_2p_2q_2 + x_2y_3p_2q_3 \\
&= x_1p_1(y_1q_1 + y_2q_2 + y_3q_3) + x_2p_2(y_1q_1 + y_2q_2 + y_3q_3) \\
&= (x_1p_1 + x_2p_2)(y_1q_1 + y_2q_2 + y_3q_3) \\
&= E[X]E[Y]
\end{aligned}
$$

Hence

> If X, Y are independent,
> then $E[XY] = E[X]E[Y]$.

It then follows that cov $(X, Y) = E[XY] - E[X]E[Y] = 0$

$$\text{and hence } \rho = 0$$

However, covariance is only a measure of *linear* dependence; it is quite possible to find random variables X and Y for which cov $(X, Y) = 0$ but X and Y are *not* independent.

Example 4

Random variables X and Y have the following joint distribution

X \ Y	1	2	3	
1	$\frac{1}{3}$	0	$\frac{1}{3}$	$\frac{2}{3}$
2	0	$\frac{1}{3}$	0	$\frac{1}{3}$
	$\frac{1}{3}$	$\frac{1}{3}$	$\frac{1}{3}$	

Show that (i)   cov $(X, Y) = 0$

(ii) X and Y are *not* independent.

(i) We have $E[X] = \frac{4}{3}, E[Y] = 2$

$$E[XY] = 1 \times \frac{1}{3} + 2 \times 0 + 3 \times \frac{1}{3}$$
$$+ 2 \times 0 + 4 \times \frac{1}{3} + 6 \times 0 = \frac{8}{3}$$

and so cov $(X, Y) = \frac{8}{3} - \frac{4}{3} \times 2 = 0$

(ii) We have $P(X = 1 \text{ and } Y = 2) = 0$
but $P(X = 1) \times P(Y = 2) = \frac{2}{3} \times \frac{1}{3} = \frac{2}{9}$
Thus $P(X = 1 \text{ and } Y = 2) \neq P(X = 1) \times P(Y = 2)$,
and so X and Y are *not* independent.

Hence

> X, Y independent $\Rightarrow$ cov $(X, Y) = 0$
> cov $(X, Y) = 0 \not\Rightarrow X, Y$ independent.

The expectation of $X + Y$

Example 5
Two random variables X and Y (defined in Example 1) have the following joint distribution

X \ Y	0	1	2	3	
0	$\frac{1}{8}$	$\frac{1}{8}$	0	0	$\frac{1}{4}$
1	0	$\frac{2}{8}$	$\frac{2}{8}$	0	$\frac{2}{4}$
2	0	0	$\frac{1}{8}$	$\frac{1}{8}$	$\frac{1}{4}$
	$\frac{1}{8}$	$\frac{3}{8}$	$\frac{3}{8}$	$\frac{1}{8}$	

Calculate $E[X + Y]$ and verify that $E[X + Y] = E[X] + E[Y]$.

To find $E[X + Y]$ we multiply the value of $(X + Y)$ at each position by the corresponding probability and add these up.

$$E[X + Y] = 0 \times \frac{1}{8} + 1 \times \frac{1}{8} + 2 \times 0 + 3 \times 0$$
$$+ 1 \times 0 + 2 \times \frac{2}{8} + 3 \times \frac{2}{8} + 4 \times 0$$
$$+ 2 \times 0 + 3 \times 0 + 4 \times \frac{1}{8} + 5 \times \frac{1}{8} = \frac{5}{2}$$

Since $E[X] = 1$, $E[Y] = \frac{3}{2}$, we have $E[X + Y] = E[X] + E[Y]$.

We shall now prove the result $E[X + Y] = E[X] + E[Y]$ for a general (2×3) probability table.

X \ Y	y_1	y_2	y_3	
x_1	a	b	c	p_1
x_2	d	e	f	p_2
	q_1	q_2	q_3	

$$E[X+Y] = (x_1 + y_1)a + (x_1 + y_2)b + (x_1 + y_3)c$$
$$+ (x_2 + y_1)d + (x_2 + y_2)e + (x_2 + y_3)f$$
$$= x_1(a+b+c) + x_2(d+e+f) + y_1(a+d) + y_2(b+e) + y_3(c+f)$$
$$= x_1 p_1 + x_2 p_2 + y_1 q_1 + y_2 q_2 + y_3 q_3$$
$$= \quad E[X] \quad + \quad E[Y]$$

This proof can clearly be generalised to a probability table of any size, so we have:
For any two random variables X and Y (whether independent or not)

and similarly

$$\boxed{\begin{aligned} E[X+Y] &= E[X] + E[Y] \\ E[X-Y] &= E[X] - E[Y] \end{aligned}}$$

We can now prove the result $\operatorname{cov}(X, Y) = E[XY] - E[X]E[Y]$, which we used earlier. Covariance is defined as

$$\operatorname{cov}(X, Y) = E[(X - \mu)(Y - \lambda)], \text{ where } \mu = E[X] \text{ and } \lambda = E[Y]$$

Hence $\operatorname{cov}(X, Y) = E[XY - \lambda X - \mu Y + \mu\lambda]$

$$= E[XY] - E[\lambda X] - E[\mu Y] + E[\mu\lambda]$$
$$= E[XY] - \lambda E[X] - \mu E[Y] + \mu\lambda$$
$$= E[XY] - \lambda\mu - \mu\lambda + \mu\lambda$$
$$= E[XY] - \mu\lambda$$
$$= E[XY] - E[X]E[Y]$$

Exercise 9.1   Joint distributions

1   Two coins (one 50p and one 10p) are tossed, and random variables X and Y are defined by

X is the number of heads showing on the 50p coin (so $X = 0$ or 1)
Y is the number of heads showing on both coins (so $Y = 0$, 1 or 2)

(i)   Find the joint distribution of X and Y
(ii) Calculate $E[X]$, var (X), $E[Y]$ and var (Y)
(iii) Explain why X and Y are not independent
(iv) Calculate $E[XY]$
(v)   Find the covariance $\operatorname{cov}(X, Y)$
(vi) Find the coefficient of correlation between X and Y.

2   Two dice are thrown, and random variables X and Y are defined by:

X is the number of sixes showing (so $X = 0$, 1 or 2)
Y is the number of fives showing (so $Y = 0$, 1 or 2)

(i)   Show that P($X = 2$ and $Y = 0$) = $\frac{1}{36}$, P($X = 1$ and $Y = 1$) = $\frac{2}{36}$,
P($X = 2$ and $Y = 1$) = 0. Find the other probabilities and hence obtain the joint
distribution of X and Y.
(ii) Are X and Y independent ?
(iii) Calculate the covariance cov (X, Y)
(iv) Find the coefficient of correlation between X and Y

3   Two dice are thrown, and the random variables X and Y are defined by:

X is the higher of the two scores
Y is the lower of the two scores

(e.g. if the dice show 3 and 5, then $X = 5$ and $Y = 3$
if the dice show 4 and 4, then $X = 4$ and $Y = 4$)
Obtain the joint distribution of X and Y, and calculate the covariance cov (X, Y).

4   Random variables X and Y have the following joint distribution

X \ Y	1	2	3
1	0.12	0.15	0.03
2	0.28	0.35	0.07

(i)   Show that X and Y are independent.
(ii) Calculate E$[XY]$, and verify that E$[XY]$ = E$[X]$E$[Y]$
(iii) Write down the value of cov (X, Y)

5   The number of births per week in a village is a random variable X having the
distribution

X	0	1	2	3
probability	0.4	0.3	0.2	0.1

and the number of deaths per week is a random variable Y having the distribution

Y	0	1	2	3	4
probability	0.1	0.3	0.4	0.15	0.05

Assuming that X and Y are independent
(i)   obtain the joint distribution of X and Y
(ii) state the coefficient of correlation between X and Y
(iii) find the probability that there are more births than deaths in a given week.

6   A particle initially at the point $(1, 1)$ is moved at random to one of the eight adjacent
integer points so that if the co-ordinates of the particle after the move are (X, Y) the

joint distribution of X and Y is

X \ Y	0	1	2
0	$\frac{1}{8}$	$\frac{1}{8}$	$\frac{1}{8}$
1	$\frac{1}{8}$	0	$\frac{1}{8}$
2	$\frac{1}{8}$	$\frac{1}{8}$	$\frac{1}{8}$

(i)   Are X and Y independent?
(ii) Calculate the covariance cov (X, Y) and comment.

7   The coefficient of correlation between two random variables X and Y is zero. Does it follow that
   (i)   cov $(X, Y) = 0$
   (ii) $E[XY] = E[X]E[Y]$
   (iii) X and Y are independent?

8   Random variables X and Y (defined in Question 1) have the following joint distribution

X \ Y	0	1	2
0	$\frac{1}{4}$	$\frac{1}{4}$	0
1	0	$\frac{1}{4}$	$\frac{1}{4}$

(i)   Calculate $E[X + Y]$ and verify that $E[X + Y] = E[X] + E[Y]$
(ii) Calculate $E[(X + Y)^2]$, and hence find var $(X + Y)$
(iii) Verify that var $(X + Y) = $ var $(X) + 2$ cov $(X, Y) + $ var (Y).

9.2   Sum and difference of two random variables

Expectation

The following results, which we discovered in the previous section, are in fact true for continuous random variables as well as for discrete ones:
   For any random variables X, Y, $E[X \pm Y] = E[X] \pm E[Y]$
   If X, Y are independent, then $E[XY] = E[X]E[Y]$ and cov $(X, Y) = 0$.
The result $E[X + Y] = E[X] + E[Y]$ can be extended as follows:
   If $X_1, X_2, \ldots , X_n$ are any random variables

$$E[X_1 + X_2 + \cdots + X_n] = E[X_1] + E[X_2] + \cdots + E[X_n]$$

and if $a_1, a_2, \cdots , a_n$ are constants

$$E[a_1X_1 + a_2X_2 + \cdots + a_nX_n] = E[a_1X_1] + E[a_2X_2] + \cdots + E[a_nX_n]$$

$$= a_1E[X_1] + a_2E[X_2] + \cdots + a_nE[X_n]$$

Variance of $(X \pm Y)$ and $(aX + bY)$

Suppose that X and Y are two random variables: let $E[X] = \mu$, $E[Y] = \lambda$.
By definition we have $\text{var}(X) = E[(X - \mu)^2]$, $\text{var}(Y) = E[(Y - \lambda)^2]$

$$\text{cov}(X, Y) = E[(X - \mu)(Y - \lambda)]$$

We now consider the variance of $X + Y$. The mean of $X + Y$ is

$E[X + Y] = E[X] + E[Y] = \mu + \lambda$, and so the variance of $X + Y$ is

$$\begin{aligned}
\text{var}(X + Y) &= E[\{(X + Y) - (\mu + \lambda)\}^2] \\
&= E[\{(X - \mu) + (Y - \lambda)\}^2] \\
&= E[(X - \mu)^2 + 2(X - \mu)(Y - \lambda) + (Y - \lambda)^2] \\
&= E[(X - \mu)^2] + 2E[(X - \mu)(Y - \lambda)] + E[(Y - \lambda)^2] \\
&= \quad \text{var}(X) \quad + \quad 2\,\text{cov}(X, Y) \quad + \quad \text{var}(Y)
\end{aligned}$$

Similarly, the mean of $X - Y$ is $\mu - \lambda$, and so

$$\begin{aligned}
\text{var}(X - Y) &= E[\{(X - Y) - (\mu - \lambda)\}^2] \\
&= E[\{(X - \mu) - (Y - \lambda)\}^2] \\
&= E[(X - \mu)^2 - 2(X - \mu)(Y - \lambda) + (Y - \lambda)^2] \\
&= E[(X - \mu)^2] - 2E[(X - \mu)(Y - \lambda)] + E[(Y - \lambda)^2] \\
&= \quad \text{var}(X) \quad - \quad 2\,\text{cov}(X, Y) \quad + \quad \text{var}(Y)
\end{aligned}$$

Thus for any two random variables X and Y,

> $\text{var}(X + Y) = \text{var}(X) + 2\,\text{cov}(X, Y) + \text{var}(Y)$
> $\text{var}(X - Y) = \text{var}(X) - 2\,\text{cov}(X, Y) + \text{var}(Y)$

If the standard deviations of X and Y are σ_X and σ_Y, then $\text{var}(X) = \sigma_X^2$, $\text{var}(Y) = \sigma_Y^2$, and the coefficient of correlation

$$\rho = \frac{\text{cov}(X, Y)}{\sigma_X \sigma_Y} \text{ and so cov}(X, Y) = \rho \sigma_X \sigma_Y$$

Hence $\text{var}(X + Y) = \text{var}(X) + 2\,\text{cov}(X, Y) + \text{var}(Y)$

$$= \sigma_X^2 + 2\rho\sigma_X\sigma_Y + \sigma_Y^2$$

Also, if a and b are constants, then

$$\begin{aligned}
\text{var}(aX + bY) &= \text{var}(aX) + 2\,\text{cov}(aX, bY) + \text{var}(bY) \\
&= a^2\,\text{var}(X) + 2\,ab\,\text{cov}(X, Y) + b^2\,\text{var}(Y)
\end{aligned}$$

Example 1
The random variable X has mean 35 and standard deviation 6; the random variable Y has mean 20 and standard deviation 4; and the coefficient of correlation between X and Y is 0.8.

Find the mean and standard deviation of
(i) $X + Y$
(ii) $2X - 3Y$

We have cov $(X, Y) = \rho \sigma_X \sigma_Y = 0.8 \times 6 \times 4 = 19.2$

(i) $E[X + Y] = E[X] + E[Y] = 35 + 20 = 55$

var $(X + Y) = $ var $(X) + 2$ cov $(X, Y) + $ var $(Y) = 6^2 + 2 \times 19.2 + 4^2 = 90.4$ so the standard deviation of $X + Y$ is $\sqrt{90.4} = 9.508$

(ii) $E[2X - 3Y] = 2E[X] - 3E[Y] = 2 \times 35 - 3 \times 20 = 10$

var $(2X - 3Y) = $ var $(2X) - 2$ cov $(2X, 3Y) + $ var $(3Y)$

$$= 4 \text{ var } (X) - 12 \text{ cov } (X, Y) + 9 \text{ var } (Y)$$

$$= 4 \times 6^2 - 12 \times 19.2 + 9 \times 4^2 = 57.6$$

so the standard deviation of $2X - 3Y$ is $\sqrt{57.6} = 7.589$

Application to Normal variables

The above formulae enable us to find the mean and standard deviation of $X + Y$, but to calculate probabilities (such as the probability that $X + Y$ is greater than 60) we need to know the actual distribution of $X + Y$, and in general this can be a difficult problem.

However, if X and Y are Normally distributed, it can be shown that their sum $X + Y$, and indeed any linear combination $aX + bY$, also has a Normal distribution. Hence we can calculate probabilities in this case.

Example 2

A school cricket team finds that the time they spend fielding is Normally distributed with mean 2.2 hours and standard deviation 0.4 hours. The time they spend batting is Normally distributed with mean 2.5 hours and standard deviation 0.6 hours, and the coefficient of correlation between the times spent fielding and batting in the same match is -0.5.

Find the probability that, in a particular match,
(i)   the total time they spend fielding and batting exceeds 5 hours
(ii) the time they spend fielding is longer than the time spent batting.

Suppose that the team fields for X hours and bats for Y hours.
(i)   We require the probability that $X + Y$ is greater than 5.
We have $E[X + Y] = E[X] + E[Y] = 2.2 + 2.5 = 4.7$
and var $(X + Y) \quad = \sigma_X^2 + 2\rho \sigma_X \sigma_Y + \sigma_Y^2$

$$= 0.4^2 + 2 \times (-0.5) \times 0.4 \times 0.6 + 0.6^2$$

$$= 0.28$$

$(X + Y)$ is Normally distributed with mean 4.7 and standard deviation $\sqrt{0.28} = 0.5292$

so $P(X + Y > 5) = P\left(Z > \dfrac{5 - 4.7}{0.5292}\right) = P(Z > 0.567)$

$$= 1 - 0.7147$$

$$= 0.2853$$

(ii) We require the probability that $X > Y$, which is equivalent to $X - Y > 0$, so we consider the distribution of $X - Y$.

We have $E[X - Y] = E[X] - E[Y] = 2.2 - 2.5 = -0.3$

and var $(X - Y)$ $= \sigma_X^2 - 2\rho\sigma_X\sigma_Y + \sigma_Y^2$

$$= 0.4^2 - 2 \times (-0.5) \times 0.4 \times 0.6 + 0.6^2$$

$$= 0.76$$

$(X - Y)$ is Normally distributed with mean -0.3 and standard deviation $\sqrt{0.76} = 0.8718$,

so $P(X > Y) = P(X - Y > 0) = P\left(Z > \dfrac{0 - (-0.3)}{0.8718}\right)$

$$= P(Z > 0.344)$$

$$= 1 - 0.6346$$

$$= 0.3654$$

Exercise 9.2 Sum or difference of two random variables

1 For two random variables X and Y, express (i) var $(X + 3Y)$ (ii) var $(2X - 5Y)$ in terms of var (X), var (Y) and cov (X, Y).

2 The random variable X has mean 25 and standard deviation 7; the random variable Y has mean 35 and standard deviation 12, and the coefficient of correlation between X and Y is $\frac{1}{4}$.
Calculate the means and standard deviations of
(i) $X + Y$ (ii) $X - Y$ (iii) $3X + 5Y$

3 The random variable X has standard deviation 6, and the random variable Y has standard deviation 10.
(i)   Express var $(X + Y)$ in terms of the coefficient of correlation ρ between X and Y.
(ii) What are the maximum and minimum possible values of the standard deviation of $(X + Y)$, and when do they occur?
(iii) What is the standard deviation of $(X + Y)$ when X and Y are independent?

4 In a mathematics examination, the marks (X) in Paper 1 are Normally distributed with mean 48 and standard deviation 15. The marks (Y) in Paper 2 are also Normally distributed, but with mean 56 and standard deviation 12. The coefficient of correlation between X and Y is 0.7.
(i) Find the mean and standard deviation of $X + Y$.
     Find the probability that a candidate scores a total mark of
     (a) more than 130   (b) less than 90.
(ii) Find the mean and standard deviation of $X - Y$.
     Find the probability that a candidate scores a higher mark in Paper 1 than in Paper 2.

5 In a certain climate, the rainfall in any particular month may be assumed to be Normally distributed. For May the rainfall has mean 15 cm and standard deviation 4 cm and for June it has mean 18 cm and standard deviation 5 cm. The cocfficient of correlation between May and June rainfalls in the same year is -0.3.
Find the probability that, in a given year,
(i)   the May rainfall exceeds 20 cm
(ii) the June rainfall exceeds 20 cm
(iii) the total rainfall in May and June exceeds 40 cm
(iv) the May rainfall exceeds the June rainfall.

6 My time of arrival at work (measured in minutes past 8 o'clock) is Normally distributed with mean 52 and standard deviation 4. My boss' time of arrival has mean 61 and standard deviation 6. Our journeys are similarly affected by the weather and so on, and the coefficient of correlation between our arrival times is 0.4.
Find the probability that, on a particular morning, my boss arrives before I do.

7 It may be assumed that the weights of airline passengers are Normally distributed with mean 74 kg, and the weights of their baggage are Normally distributed with mean 20 kg.
(i) It is found that 10% of passengers weigh more than 85 kg and 20% of baggages weigh more than 24 kg. Find the standard deviation of the passenger weights and of the baggage weights.
(ii) What is the mean weight of a passenger and his baggage?
It is found that 10% of passenger and baggage combinations weigh more than 108 kg. Use

$$\text{var } (X + Y) = \sigma_X^2 + 2\rho\sigma_X\sigma_Y + \sigma_Y^2$$

to find the coefficient of correlation between the weights of a passenger and his baggage.

8 Random variables X and Y are such that $\text{var}\,(X) = 20$, $\text{var}\,(Y) = 12$ and $\text{var}\,(X + Y) = 26$.
Find $\text{cov}\,(X, Y)$ and $\text{var}\,(X - Y)$

9 Results such as $\text{var}\,(x + y) = \text{var}\,(x) + 2\,\text{cov}\,(x, y) + \text{var}\,(y)$ also apply to samples. Five pairs of values (x, y) are given as follows

x	3	5	4	10	8
y	2	7	6	8	9

(i)   List the 5 values of $(x + y)$
Find the standard deviations of x, y and $(x + y)$
(ii) Use $\text{var}\,(x + y) = s_x^2 + 2rs_xs_y + s_y^2$ to calculate the coefficient of correlation r between x and y
(iii) Check by calculating r in the usual way.

9.3 Sums of independent random variables

Mean and variance

If X and Y are *independent* random variables, then cov $(X, Y) = 0$ and the formulae var $(X \pm Y) = $ var $(X) \pm 2$ cov $(X, Y) + $ var (Y) simplify to give

$$\boxed{\begin{aligned} \text{var } (X + Y) &= \text{var } (X) + \text{var } (Y) \\ \text{var } (X - Y) &= \text{var } (X) + \text{var } (Y) \end{aligned}}$$

These results can be extended as follows:
If $X_1, X_2, \ldots, X_n$ are independent random variables

$$\text{var}(X_1 + X_2 + \cdots + X_n) = \text{var } (X_1) + \text{var } (X_2) + \cdots + \text{var } (X_n)$$

and if $a_1, a_2, \ldots, a_n$ are constants

$$\begin{aligned} \text{var } (a_1 X_1 + a_2 X_2 + \cdots + a_n X_n) &= \text{var } (a_1 X_1) + \text{var } (a_2 X_2) + \cdots + \text{var } (a_n X_n) \\ &= a_1^2 \text{ var } (X_1) + a_2^2 \text{ var } (X_2) + \cdots + a_n^2 \text{ var } (X_n) \end{aligned}$$

Thus the variance of a sum (or difference) of independent random variables is obtained by adding the variances of the individual variables. Note particularly that it is the *variances* which are added; if we are given the standard deviations, we must square these first, and if we require the standard deviation we must take the square root at the end.

The mean of a sum (or difference) of random variables is obtained, as before, by adding (or subtracting) the individual means.

The results $E[X \pm Y] = E[X] \pm E[Y]$

$$E[X_1 + X_2 + \cdots + X_n] = E[X_1] + E[X_2] + \cdots + E[X_n]$$

$$E[a_1 X_1 + a_2 X_2 + \cdots + a_n X_n] = a_1 E[X_1] + a_2 E[X_2] + \cdots + a_n E[X_n]$$

are true whether the variables are independent or not.

Example 1
The random variables X and Y are independent;
X has mean 25 and standard deviation 5
Y has mean 35 and standard deviation 3
Find the mean and standard deviation of
(i) $X + Y$
(ii) $X - Y$

(i) We have $E[X + Y] = E[X] + E[Y] = 25 + 35 = 60$
   and since X and Y are independent

   var $(X + Y) = $ var $(X) + $ var $(Y) = 5^2 + 3^2 = 34$

   Hence $(X + Y)$ has mean 60 and standard deviation $\sqrt{34} = 5.831$
(ii) We have $E[X - Y] = E[X] - E[Y] = 25 - 35 = -10$
   and since X and Y are independent

   var $(X - Y) = $ var $(X) + $ var $(Y) = 5^2 + 3^2 = 34$

Hence $(X - Y)$ has mean -10 and standard deviation $\sqrt{34} = 5.831$

Note that if X and Y are independent, then $X + Y$ and $X - Y$ have the same standard deviation.

Example 2

Find the mean and variance of the score on a single die.

(i)   A die is thrown and the random variable R is obtained by multiplying the number showing on the die by five.
Find the mean and variance of R.

(ii) Two dice, one red and one blue, are thrown, and the random variable S is calculated as three times the number showing on the red die plus twice the number showing on the blue die. Find the mean and variance of S.

(iii) Five dice are thrown, and T is the total score.
Find the mean and variance of T.

The distribution of the score X on a single die is as shown below

X	1	2	3	4	5	6
probability	$\frac{1}{6}$	$\frac{1}{6}$	$\frac{1}{6}$	$\frac{1}{6}$	$\frac{1}{6}$	$\frac{1}{6}$

$E[X] = \frac{21}{6} = 3.5$, $E[X^2] = \frac{91}{6}$, so var $(X) = \frac{91}{6} - (\frac{7}{2})^2 = \frac{35}{12} \approx 2.92$

(i)   If the score on the die is X, then $R = 5X$

$E[R] = E[5X] = 5E[X] = 5 \times 3.5 = 17.5$

var $(R) = $ var $(5X) = 25$ var $(X) = 25 \times \frac{35}{12} = \frac{875}{12} \approx 72.92$

(ii) If the scores on the red and blue dice are X and Y respectively, then X and Y are independent, and $S = 3X + 2Y$

$E[S] = E[3X + 2Y] = 3E[X] + 2E[Y] = 3 \times 3.5 + 2 \times 3.5 = 17.5$

var $(S) = $ var $(3X + 2Y) = 9$ var $(X) + 4$ var (Y) (since X and Y are independent)

$$= 9 \times \frac{35}{12} + 4 \times \frac{35}{12} = \frac{455}{12} \approx 37.92$$

(iii) If the scores on the dice are X_1, X_2, X_3, X_4, X_5, then $X_1, X_2, \ldots, X_5$ are independent and $T = X_1 + X_2 + \cdots + X_5$

$E[T] = E[X_1 + X_2 + \cdots + X_5] = E[X_1] + E[X_2] + \cdots + E[X_5]$

$$= 3.5 + 3.5 + \cdots + 3.5$$

$$= 3.5 \times 5$$

$$= 17.5$$

Since $X_1, X_2, \ldots, X_5$ are independent

var $(T) = $ var $(X_1 + X_2 + \cdots + X_5) = $ var $(X_1) + $ var $(X_2) + \cdots + $ var (X_5)

$$= \frac{35}{12} + \frac{35}{12} + \cdots + \frac{35}{12}$$

$$= \frac{35}{12} \times 5 = \frac{175}{12} \approx 14.58$$

Note that in (i), (ii) and (iii) the means are the same, but the variances are different. Throwing a die once and multiplying the score by five (as in (i)) is quite different from throwing five dice and adding the scores (as in (iii)).

Application to Normal variables

Any sum of independent Normal variables itself has a Normal distribution.

Example 3

A bag of garden compost is made by mixing peat with fertiliser. The peat and fertiliser are dispensed by different machines, and the weights of peat and fertiliser in a bag may be assumed to be independent. The weight of peat is Normally distributed with mean 7.0 kg and standard deviation 1.2 kg; and the weight of fertiliser is Normally distributed with mean 5.0 kg and standard deviation 0.9 kg.

Given that a bag of compost is selected at random, find the probability that
(i)   the total weight of the compost is less than 9.5 kg
(ii) the weight of the peat is less than the weight of the fertiliser
(iii) the weight of peat is more than double the weight of fertiliser.

Suppose the bag contains X kg of peat and Y kg of fertiliser; then X and Y are independent.
(i)   The total weight of the compost is $(X + Y)$ kg
We have $E[X + Y] = E[X] + E[Y] = 7.0 + 5.0 = 12.0$, and since X and Y are independent

$$\text{var } (X + Y) = \text{var } (X) + \text{var } (Y) = 1.2^2 + 0.9^2 = 2.25$$

$(X + Y)$ is Normally distributed with mean 12.0 and standard deviation $\sqrt{2.25} = 1.5$, and so

$$P(X + Y < 9.5) = P\left(Z < \frac{9.5 - 12.0}{1.5}\right) = P(Z < -1.667)$$

$$= 1 - 0.9522$$

$$= 0.0478$$

(ii) We require the probability that $X < Y$, which is equivalent to $X - Y < 0$.
We have $E[X - Y] = E[X] - E[Y] = 7.0 - 5.0 = 2.0$
and since X and Y are independent

$$\text{var } (X - Y) = \text{var } (X) + \text{var } (Y) = 1.2^2 + 0.9^2 = 2.25$$

$(X - Y)$ is Normally distributed with mean 2.0 and standard deviation 1.5, and so

$$P(X < Y) = P(X - Y < 0) = P\left(Z < \frac{0 - 2.0}{1.5}\right)$$

$$= P(Z < -1.333)$$

$$= 1 - 0.9087$$

$$= 0.0913$$

(iii) We require the probability that $X > 2Y$, which is equivalent to $X - 2Y > 0$, so we consider the distribution of $X - 2Y$
We have

$$E[X - 2Y] = E[X] - 2E[Y] = 7.0 - 2 \times 5.0 = -3.0$$

and since X and Y are independent

$$\text{var } (X - 2Y) = \text{var } (X) + 4 \text{ var } (Y) = 1.2^2 + 4 \times 0.9^2 = 4.68$$

$(X - 2Y)$ is Normally distributed with mean -3.0 and standard deviation $\sqrt{4.68} = 2.163$, so

$$P(X > 2Y) = P(X - 2Y > 0) = P\left(Z > \frac{0 - (-3.0)}{2.163}\right)$$

$$= P(Z > 1.387)$$

$$= 1 - 0.9173$$

$$= 0.0827$$

Example 4

A doctor leaves home at 9.00 a.m. He drives to a clinic, he sees three patients at the clinic, then he drives on to the hospital. The time taken to travel from home to the clinic is Normally distributed with mean 16 minutes and standard deviation 2 minutes; the time spent with each patient is Normally distributed with mean 10 minutes and standard deviation 2.5 minutes; and the time taken to drive to the hospital from the clinic is Normally distributed with mean 5 minutes and standard deviation 0.5 minutes.
  Find the probability that he arrives at the hospital before 10.00 a.m.

Suppose that the times taken to drive to the clinic, see the first patient, see the second patient, see the third patient, and drive to the hospital are W, X_1, X_2, X_3, and Y minutes respectively.
  The total time between leaving home and arriving at the hospital is then $T = W + X_1 + X_2 + X_3 + Y$ minutes; he arrives before 10.00 a.m. if $T < 60$.

$$\text{We have } E[T] = E[W] + E[X_1] + E[X_2] + E[X_3] + E[Y]$$

$$= 16 \quad + 10 \quad + 10 \quad + 10 \quad + 5 = 51$$

Assuming that the five times are independent, we have

$$\text{var } (T) = \text{var } (W) + \text{var } (X_1) + \text{var } (X_2) + \text{var } (X_3) + \text{var } (Y)$$

$$= 2^2 \quad + 2.5^2 \quad + 2.5^2 \quad + 2.5^2 \quad + 0.5^2 = 23$$

The total time T is Normally distributed with mean 51 minutes and standard deviation $\sqrt{23} = 4.796$ minutes, so

$$P(\text{he arrives before } 10.00 \text{ a.m.}) = P(T < 60)$$

$$= P\left(Z < \frac{60 - 51}{4.796}\right)$$

$$= P(Z < 1.877)$$

$$= 0.9698$$

The central limit theorem

It can be shown that if $X_1, X_2, ..., X_n$ are independent random variables, all having the same distribution, then their sum

$$X_1 + X_2 + \cdots + X_n$$

is approximately *Normally* distributed, whatever the distribution of the original variables $X_1, X_2, ..., X_n$, provided that n is large enough.

This is a remarkable result from advanced statistical theory, and it demonstrates the importance of the Normal distribution.

If $X_1, X_2, ..., X_n$ are themselves approximately Normal, then of course their sum $X_1 + X_2 + \cdots + X_n$ is approximately Normal for any value of n.

It is difficult to say in general how large n needs to be for the Normal approximation to be a good one. If $X_1, X_2, ..., X_n$ have a roughly symmetrical distribution, then n greater than about 20 is usually sufficient. The more skew the original distribution, the larger n needs to be.

Example 5

A die is thrown 30 times. Find the probability that the total score is more than 120.

Let $X_1, X_2, ..., X_{30}$ be the scores on the first throw, on the second throw and so on. Then $X_1, X_2, ..., X_{30}$ are independent random variables, all having the same distribution; in particular each one has mean 3.5 and variance $\frac{35}{12}$ (see Example 2).

By the Central Limit Theorem it follows that the total score $T = X_1 + X_2 + \cdots + X_{30}$ has an approximately Normal distribution.

$$\text{We have } E[T] = E[X_1] + E[X_2] + \cdots + E[X_{30}]$$

$$= 3.5 \quad + 3.5 \quad + \cdots + 3.5$$

$$= 30 \times 3.5 = 105$$

and since $X_1, X_2, ..., X_{30}$ are independent,

$$\text{var } (T) = \text{var } (X_1) + \text{var } (X_2) + \cdots + \text{var } (X_{30})$$

$$= \tfrac{35}{12} \quad + \tfrac{35}{12} \quad + \cdots + \tfrac{35}{12} = 30 \times \tfrac{35}{12} = 87.5$$

Hence T has mean 105 and standard deviation $\sqrt{87.5} = 9.354$.

The total score T can take only whole number values, so when we approximate it by a Normal distribution, we should use a continuity correction.

For a score of more than 120, 120 is not included, but 121 is, thus

$$P(T > 120) \approx P(T > 120.5 \text{ in Normal}) = P\left(Z > \frac{120.5 - 105}{9.354}\right)$$

$$= P(Z > 1.657)$$

$$= 1 - 0.9512$$

$$= 0.0488$$

Example 6

In a timber yard, 180 sheets of hardboard and 50 sheets of softboard are stacked in a pile. The thickness of a hardboard sheet has mean 0.6 cm and standard deviation 0.04 cm, and the thickness of a softboard sheet has mean 1.9 cm and standard deviation 0.14 cm. Find the probability that the complete pile is higher than 2.05 m.

The height of the pile is

$$H = X_1 + X_2 + \cdots + X_{180} + Y_1 + Y_2 + \cdots + Y_{50}$$

where $X_1, X_2, \ldots, X_{180}$ are the thicknesses of the hardboard sheets and $Y_1, Y_2, \ldots, Y_{50}$ are the thicknesses of the softboard sheets.

We have $E[H] = 0.6 + 0.6 + \cdots + 0.6 + 1.9 + 1.9 + \cdots + 1.9$

$$= 180 \times 0.6 + 50 \times 1.9$$

$$= 203$$

Assuming that the thicknesses of the 230 sheets in the pile are independent,

$$\text{var}(H) = 0.04^2 + 0.04^2 + \cdots + 0.04^2 + 0.14^2 + 0.14^2 + \cdots + 0.14^2$$

$$= 180 \times 0.04^2 + 50 \times 0.14^2$$

$$= 1.268$$

The height has mean 203 cm and standard deviation $\sqrt{1.268} = 1.126$ cm. By the Central Limit Theorem,

$$X_1 + X_2 + \cdots + X_{180} \qquad \text{and} \qquad Y_1 + Y_2 + \cdots + Y_{50}$$

are approximately Normally distributed, so H is approximately Normally distributed.

Hence $P(H > 205) = P\left(Z > \dfrac{205 - 203}{1.126}\right) = P(Z > 1.776)$

$$= 1 - 0.9621$$

$$= 0.0379$$

Exercise 9.3   Sums of independent random variables

1   X and Y are independent random variables, with $E[X] = 8$, $\text{var}(X) = 4$ and $E[Y] = 5$, $\text{var}(Y) = 9$
Find the mean and variance of
(i) $X + Y$ (ii) $X - Y$ (iii) $2X + 3Y$ (iv) $X - 4Y$ (v) $\frac{1}{2}(X + Y)$

2   Three dice, coloured red, white and blue, are thrown, and the numbers showing are R, W and B respectively. Find the mean and variance of the quantity $S = R + 3W - 2B$.

3   X and Y are independent random variables, each having a rectangular distribution

between 1 and 3, i.e. with pdf

$$f(x) = \begin{cases} \frac{1}{2} & \text{if } 1 \leqslant x \leqslant 3 \\ 0 & \text{otherwise.} \end{cases}$$

Find the mean and standard deviation of
(i) X (ii) $X + Y$ (iii) $X - Y$

4   The weights of men are Normally distributed with mean 78 kg and standard deviation 12 kg. The weights of women are Normally distributed with mean 70 kg and standard deviation 8 kg.
   (i)   A man and a woman are chosen at random. Find the mean and standard deviation of their total weight. Find the probability that
     (a) their total weight exceeds 160 kg
     (b) their average weight is less than 72 kg.
   (ii)   Given that a man and a woman are chosen at random, state the mean and standard deviation of the difference of their weights.
     Find the probability that
     (a) the man is more than 5 kg heavier than the woman
     (b) the woman is heavier than the man.
   (iii)   Five men and three women are chosen at random. Find the probability that their total weight is less than 560 kg.

5   In the manufacture of a certain article, a bar has to be fitted through a slot. The thicknesses of the bars are Normally distributed with mean 18 mm and standard deviation 1.5 mm; and the widths of the slots are Normally distributed with mean 20 mm and standard deviation 2 mm.
     Given that a bar and a slot are selected at random, find the probability that
   (i)   the bar is too thick to pass through the slot
   (ii)   the bar fits through the slot leaving a gap of more than 3 mm.

6   A train leaves the station at X minutes past 8 a.m., where X is Normally distributed with mean 46 and standard deviation 2. My journey from home to the station takes Y minutes where Y is Normally distributed with mean 30 and standard deviation 5.
   (i)   Find the mean and standard deviation of $X - Y$.
   (ii)   Find the probability that I catch the train if I leave home at 10 minutes past 8.
   (iii)   When should I leave home if the probability of catching the train is to be 0.95?

7   In a village there are 250 houses, 4 farms and 1 garage.
    At a certain time of the day the consumption of electricity
      by each house has mean 2.8 kW and standard deviation 0.8 kW
      by each farm has mean 15.5 kW and standard deviation 3.6 kW
      by the garage has mean 8.0 kW and standard deviation 1.5 kW.
    The consumptions may be assumed to be independently and Normally distributed.
   (i)   Find the mean and standard deviation of the total electrical consumption of the village.
   (ii)   Find the probability that the total consumption exceeds 800 kW.
   (iii)   Find the consumption which is exceeded with probability 0.99.

8   Find the mean and variance of a single digit (between 0 and 9 inclusive) chosen from a random number table.

40 digits are now taken from the table. Write down the mean and variance of the sum of these digits.

Assuming that this sum has an approximately Normal distribution and using a continuity correction, find the probability that the sum of the digits is greater than 180 but less than 200.

9   In Town A, 62% of the population vote Conservative, and in Town B 24% do so.

Samples of 30 people from Town A and 70 people from Town B are chosen. If the numbers of Conservative voters in these samples are X and Y respectively, state the distributions of X and Y, and deduce the mean and standard deviation of $X + Y$.

Using the Normal approximations to the distributions of X and Y, find the probability that the combined sample of 100 people contains more than 40 Conservative voters.

10   Random variables X and Y are such that var $(X + Y) =$ var $(X) +$ var (Y). Does it necessarily follow that

(i)   cov $(X, Y) = 0$

(ii) $E[XY] = E[X]E[Y]$

(iii) X and Y are independent

(iv)   var $(X - Y) =$ var $(X) +$ var (Y)?

10
Random Samples

10.1 Samples

Population and sample

In a given situation, the *population* consists of everything under consideration. This might be all the adult inhabitants of Britain, or all the plants in a meadow, or all the ropes produced by a factory in a given week, and so on, but equally it could be just the 24 children in Form 2A at a particular school.

It should always be made clear what the population is. For example, when we speak of the 'weights of people' we should specify whether we are considering

     all the people in a particular 'keep fit' group

or all people over 18 years old living in London

or all the people in the World (which would include babies) and so on.

   Suppose that we wish to find out something about a population, for example,

     the proportion of the European electorate who held a certain opinion

or the mean and standard deviation of the heights of trees in the New Forest

or the mean tension at which ropes produced in a certain factory will break,

then the only certain way of finding out is to study every single member of the population. However, if the population is a large one this is usually too expensive or too time-consuming. It might even defeat the object: for example if we tested every rope produced to measure the tension at which it breaks, we would have no ropes left to sell.

We therefore select a *sample,* consisting of some members of the population. We just study the sample, hoping that it is representative of the population as a whole.

Intuitively, if 1000 people are selected 'at random' from everyone in Britain, we would expect the proportion of women, the mean age, and so on, for this sample to be nearly the same as that in the whole population of Britain.

The people must be selected 'at random' because it is certainly not true that any sample of 1000 people will be representative of the population: for example, if we selected the first 1000 people in an anti-nuclear march, we would not expect their views on nuclear warfare to be similar to those of the whole population.

Random samples

A *random sample* is chosen in such a way that

(i)   every member of the population has an equal chance of being selected, and

(ii) the selections are made independently.

Notice that this is sampling with replacement (since the selections are made independently); it is possible for the same member of the population to be chosen more than once.

Example 1

Explain how to select a random sample of 250 people from a town of 45 346 inhabitants.

First number all the inhabitants of the town from 1 to 45 346. Using a random-number table, or a calculator (with a random number key) or a computer, obtain a sequence of random digits, for example,

| 90 | 32 | 18 | 85 | 42 | 77 | 37 | 32 | 50 | 51 | 40 | 19 | 59 | 99 | 02 |
| 91 | 25 | 17 | 41 | 08 | 04 | 76 | 62 | 60 | 33 | 92 | 98 | 62 | 70 | 83 |

Reading these in groups of 5 gives

| 90321 | 88542 | 77373 | 25051 | 40195 | 99902 |
| 91251 | 74108 | 04766 | 26033 | 92986 | 27083 |

We ignore any of these which are not in the range 00001 to 45346, so the first five people selected are those numbered 25051, 40195, 4766, 26033, 27083.

We continue in this way until 250 people have been selected.

Other methods of sampling

In practice it may be very difficult to select a random sample from a large population; the problem lies in numbering all the members of the population (imagine numbering all the plants in a large meadow). The following modifications to random sampling are often used.

Random sampling without replacement

This is the same as random sampling except that any repeats are ignored, so that the same member of the population cannot be chosen more than once. When a sample of (say) 250 members is chosen in this way, then every possible set of 250 members from the population has an equal chance of being selected as the sample.

When the population is very large compared with the size of the sample, the difference between sampling with, and without, replacement is negligible.

Stratified sampling

The population is first divided into groups called 'strata' (an example of a stratum might be 'men aged between 36 and 40 and earning between £10 000 and £11 000 per annum'). Each stratum is sampled separately and these samples are combined to give a stratified sample from the whole population.

For example, if a sample of 2000 people is required, and 3% of the population belong to a particular stratum, then 60 people will be selected from that stratum.

Systematic sampling

A starting point is chosen at random, but then the population is sampled at regular inter-

vals. For example, we might select every tenth name on a list, or we might test every fiftieth item on a production line.

Cluster sampling

One or more locations are chosen at random, then all the members of the population lying in these locations are taken to form the sample.

For example, to obtain a sample of plants from a meadow, a wire frame one metre square (called a quadrat) is thrown; wherever it lands, all the plants inside it are taken as part of the sample; the frame is then thrown again as often as necessary.

All these methods of sampling, when properly carried out, can give samples which are representative of the population (and in many respects a stratified sample is better than a random sample). It is usually possible (and desirable) to arrange that each member of the population has the same chance of selection, but in all four methods outlined above, the selections are not made independently. This makes the theoretical treatment of these modifications complicated.

We shall consider only random samples.

Estimating from a sample

Now suppose that we are interested in a numerical quantity associated with each member of the population, for example, the height of a person (or the weight of a pebble, the number of flowers on a plant, the current at which a fuse will blow, and so on).

The *population mean* μ and *population standard deviation* σ are calculated from the heights of all members of the population. If the population is large, it would be impractical to obtain all the heights, and so the values of μ and σ will not be known; but we can get a rough idea of their values by studying a random sample.

Suppose we have a random sample of n people, and suppose that their heights are $X_1, X_2, ..., X_n$. In this context we may say that the set of values $\{X_1, X_2, ..., X_n\}$ is a random sample of size n from a population having mean μ and standard deviation σ.

From the sample values $\{X_1, X_2, ..., X_n\}$, we may calculate the mean, median, standard deviation, inter-quartile range, and so on, as described in Chapters 1 and 2 (if n is large, the sample values may of course be grouped into classes first). In particular the sample mean is

$$\bar{X} = \frac{X_1 + X_2 + \cdots + X_n}{n}$$

and the sample variance is

$$s^2 = \frac{\sum_{i=1}^{n} (X_i - \bar{X})^2}{n}$$

$$= \frac{\sum X_i^2}{n} - \bar{X}^2$$

(where s is the standard deviation of the sample).

These values refer only to the particular sample for which they are calculated; if another

sample is chosen, it will generally have a different mean $\bar{X}$ and so on. However we hope that the sample is representative of the population as a whole, and that the sample mean $\bar{X}$ and standard deviation s will not be very different from the population values μ and σ.

If μ and σ are unknown, and we only have a sample of values, then we cannot calculate the exact values of μ and σ (we could only do this if we knew the values for the whole population), but

we can *estimate* μ by the sample mean $\bar{X}$

and   we can *estimate* σ by the sample standard deviation s.

(Note that the word 'estimate' has a technical meaning in statistics: it does not mean 'guess').

Example 2

From a large batch, a random sample of 10 nails was selected, and their lengths were measured (in mm) as

$$42.6, 38.4, 41.2, 41.4, 40.0, 39.5, 43.2, 40.8, 41.4, 39.0$$

Estimate the mean and standard deviation of the lengths of the nails in the batch.

The sample mean is $\bar{X} = \dfrac{\sum X_i}{n} = \dfrac{407.5}{10} = 40.75$

and the sample standard deviation is $s = \sqrt{\dfrac{\sum X_i^2}{n} - \bar{X}^2}$

$$= \sqrt{\dfrac{16626.81}{10} - \left(\dfrac{407.5}{10}\right)^2}$$

$$= 1.46$$

If the mean and standard deviation for the whole batch are μ and σ, we estimate $\mu \approx 40.75$ mm

and $\sigma \approx 1.46$ mm

Note that, if we now selected another random sample, we would obtain different estimates for μ and σ.

Obviously, the larger the sample, the more accurate we would expect our estimates to be. In this chapter we investigate how close $\bar{X}$ is to μ, and what can be deduced about the population mean μ from a random sample.

Exercise 10.1   Samples

1   A school has 1120 pupils. There are 70 'forms', of varying sizes. For sports competitions and so on, the pupils are divided into 35 'houses', each of which has 32 pupils. A sample of 70 pupils is required to fill in a questionnaire.
(i)   Explain why the following methods would *not* give a random sample

(a)   Select the first 70 pupils who enter the tuck-shop at break.
(b)   Put up a notice asking for volunteers, and select the first 70 pupils who volunteer.

(c)   Select every 16th name on the alphabetical list of pupils.
(d)   Select two pupils from each 'house'.
(e)   Select one pupil from each 'form'.

Which of the above methods would give a stratified sample; and which method would give a systematic sample?
(ii)   How would you select a random sample of 70 pupils?

2   {27.2, 13.8, 17.7, 14.9, 22.0} is a random sample taken from a population having mean 19.6 and standard deviation 8.2.
With the usual notation, state the values of
(i) μ (ii) σ (iii) $\bar{X}$ (iv) s

(v) $Z = \dfrac{\bar{X} - \mu}{\dfrac{\sigma}{\sqrt{n}}}$ (vi) $t = \dfrac{\bar{X} - \mu}{\dfrac{s}{\sqrt{n-1}}}$

3   A random sample of 120 students at a certain University had the following heights

Height (cm)	141–145	146–150	151–155	156–160	161–165	166–170
Number of students	2	1	3	7	15	24

Height (cm)	171–175	176–180	181–185	186–190	191–195	196–200
Number of students	23	20	15	8	1	1

Estimate the mean and standard deviation of the heights of all the students at that University.

10.2   The distribution of the sample means

Suppose we take a random sample of n values from a population having mean μ and standard deviation σ.
   We shall now investigate how the sample mean $\bar{X}$ varies from sample to sample.
   Consider a general random sample $\{X_1, X_2, ..., X_n\}$ of size n,
where X_1 is the first value selected
         X_2 is the second value selected, and so on.
Then X_1 is a random variable (varying from sample to sample); since the first value selected is equally likely to come from any member of the population, the distribution of X_1 is the same as the population;
in particular, $E[X_1] = \mu$ and var $(X_1) = \sigma^2$
   Similarly, X_2 is a random variable having the same distribution as the population, so $E[X_2] = \mu$ and var $(X_2) = \sigma^2$, and so on.

Since selections are made independently, the random variables $X_1, X_2, ..., X_n$ are independent.

We may therefore regard the random samples of size n as a set of independent random variables $X_1, X_2, ..., X_n$, where each one has the same distribution as the population (and so has mean μ and standard deviation σ).

The sample mean is

$$\bar{X} = \frac{1}{n} (X_1 + X_2 + \cdots + X_n)$$

this is also a random variable (varying from sample to sample). Applying the results of Section 9.3, we have

$$E[\bar{X}] = E\left[\frac{1}{n} (X_1 + X_2 + \cdots + X_n)\right]$$

$$= \frac{1}{n} E[X_1 + X_2 + \cdots + X_n]$$

$$= \frac{1}{n} \{E[X_1] + E[X_2] + \cdots + E[X_n]\}$$

$$= \frac{1}{n} \{\mu + \mu + \cdots + \mu\}$$

$$= \frac{1}{n} (n\mu)$$

$$= \mu$$

Thus although $\bar{X}$ varies (for some samples it will be greater than μ, and for other samples it will be less than μ), its average value over all random samples is equal to the population mean μ.

Also

$$\text{var} (\bar{X}) = \text{var} \left\{\frac{1}{n} (X_1 + X_2 + \cdots + X_n)\right\}$$

$$= \frac{1}{n^2} \text{var} (X_1 + X_2 + \cdots + X_n)$$

$$= \frac{1}{n^2} \{\text{var} (X_1) + \text{var} (X_2) + \cdots + \text{var} (X_n)\}$$

$$\text{(since } X_1, X_2, \cdots, X_n \text{ are independent)}$$

$$= \frac{1}{n^2} (\sigma^2 + \sigma^2 + \cdots + \sigma^2)$$

$$= \frac{1}{n^2} (n\sigma^2)$$

$$= \frac{\sigma^2}{n}$$

If the variance of $\bar{X}$ is small, then $\bar{X}$ is likely to be close to μ. We can see that, as we would expect, the larger the sample size n the smaller is the variance of $\bar{X}$.

Now $X_1 + X_2 + \cdots + X_n$ is a sum of independent random variables all having the same distribution, so by the Central Limit Theorem, provided that n is large enough, $X_1 + X_2 + \cdots + X_n$ will be approximately Normally distributed and hence the distribution of the sample means

$$\bar{X} = \frac{1}{n} (X_1 + X_2 + \cdots + X_n)$$

is also approximately Normal (whatever the distribution of the population).

We shall always assume that conditions are such that $\bar{X}$ has an approximately Normal distribution. This is true if

(i)   the population itself is approximately Normal

or (ii)   the sample size (n) is very large

or (iii)   the population has a fairly symmetrical distribution and the sample size is reasonably large (say $n > 20$)

We then have

> The sample mean $\bar{X}$ is approximately Normally distributed
>
> with mean μ and standard deviation $\dfrac{\sigma}{\sqrt{n}}$

The standard deviation of $\bar{X}$, $\dfrac{\sigma}{\sqrt{n}}$ is called the *standard error of the mean*. It gives an indication of how much $\bar{X}$ is likely to differ from μ (the difference between $\bar{X}$ and μ being the 'error' in the sample mean $\bar{X}$).

Example 1.
A random sample of 250 values is selected from a population having mean 36.0 and standard deviation 5.0.

Find the probability that the sample mean is greater than 36.4.

The sample mean $\bar{X}$ is approximately Normal, with mean $\mu = 36.0$ and standard deviation $\dfrac{\sigma}{\sqrt{n}} = \dfrac{5.0}{\sqrt{250}} = 0.3162$.

$$P(\bar{X} > 36.4) = P\left(Z > \frac{36.4 - 36.0}{0.3162}\right).$$

$$= P(Z > 1.265)$$

$$= 1 - 0.8971$$

$$= 0.1029$$

Example 2
Men have mean height 176 cm with standard deviation 11 cm.
Given that two (independent) random samples are taken, one of 80 men and the other of 120 men, find the probability that the mean heights of the two samples will differ by more than 1 cm.

Suppose that the sample of 80 men has mean height $\bar{X}$ and the sample of 120 men has mean height $\bar{Y}$:

then $\bar{X}$ is approximately Normal with mean 176 cm, standard deviation $\dfrac{11}{\sqrt{80}}$ cm

$\bar{Y}$ is approximately Normal with mean 176 cm, standard deviation $\dfrac{11}{\sqrt{120}}$ cm

We now consider $\bar{X} - \bar{Y}$;

$$\mathrm{E}[\bar{X} - \bar{Y}] = \mathrm{E}[\bar{X}] - \mathrm{E}[\bar{Y}] = 176 - 176 = 0,$$

and since $\bar{X}, \bar{Y}$ are independent,

$$\mathrm{var}\,(\bar{X} - \bar{Y}) = \mathrm{var}\,(\bar{X}) + \mathrm{var}\,(\bar{Y}) = \left(\frac{11}{\sqrt{80}}\right)^2 + \left(\frac{11}{\sqrt{120}}\right)^2$$

$$= 2.521$$

Thus $(\bar{X} - \bar{Y})$ is approximately Normal, with mean 0 and standard deviation $\sqrt{2.521} = 1.588$.

Hence P(sample means differ by more than 1 cm) $= \mathrm{P}(\bar{X} - \bar{Y} > 1) + \mathrm{P}(\bar{X} - \bar{Y} < -1)$

$$= \mathrm{P}(Z > 0.630) + \mathrm{P}(Z < -0.630)$$

$$= (1 - 0.7357) + (1 - 0.7357)$$

$$= 0.5286$$

Exercise 10.2   Distribution of the sample means

1 The weights of people have mean 78.5 kg and standard deviation 11.2 kg. Given that $\bar{X}$ is the mean weight of a random sample of 20 people, state the (approximate) distribution of $\bar{X}$.
(i)   Find the probability that $\bar{X}$ is greater than 82.0 kg
(ii)   Between which weights will $\bar{X}$ lie for 95% of samples?

2 The marks in an examination have mean 52 and standard deviation 18. If a random sample of 150 candidates is taken, find the probability that the sample mean will differ from the true mean (52) by more than 2 marks.

3 Two random samples, each of size 60, are taken from a population having mean 125 and standard deviation 12.
The sample means are $\bar{X}$ and $\bar{Y}$.
State the distributions of $\bar{X}$ and $\bar{Y}$, and deduce the distribution of $\bar{X} - \bar{Y}$.
Find the probability that $\bar{X}$ and $\bar{Y}$ differ by more than 2.5.

4 Scots have mean height 179 cm, with standard deviation 12 cm.
Welshmen have mean height 177 cm, with standard deviation 8 cm.
A random sample of 32 Scots has mean height $\bar{X}$, and a random sample of 75 Welshmen has mean height $\bar{Y}$.
Find the probability that $\bar{Y}$ is greater than $\bar{X}$.

10.3 Confidence intervals for the population mean

Suppose that a population has mean μ and standard deviation σ; and suppose that we wish to find out about the population mean μ. We cannot calculate the exact value of μ unless we study the whole population; so we take a random sample of size n and calculate the sample mean $\bar{X}$.

Then $\bar{X}$ is an estimate for μ, but this is of little use unless we know how accurate the estimate is likely to be.

If we consider all random samples of size n, the sample means $\bar{X}$ are approximately Normally distributed, with mean μ and standard deviation $\dfrac{\sigma}{\sqrt{n}}$.

For a standard Normal variable Z, there is probability 0.95 that Z lies between $-a$ and a where $\Phi(a) = 0.975$

$$a = 1.96$$

i.e. that Z lies between -1.96 and 1.96 (Fig. 10.1).

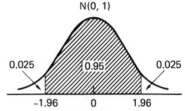

Fig. 10.1

Thus, for 95% of samples of size n

$$\mu - 1.96 \, \frac{\sigma}{\sqrt{n}} < \bar{X} < \mu + 1.96 \, \frac{\sigma}{\sqrt{n}}$$

or, on rearranging

$$\bar{X} - 1.96 \, \frac{\sigma}{\sqrt{n}} < \mu < \bar{X} + 1.96 \, \frac{\sigma}{\sqrt{n}}$$

This is a *95% confidence interval* for μ
The end-points

$$\boxed{\bar{X} \pm 1.96 \, \frac{\sigma}{\sqrt{n}} \text{ are 95\% confidence limits for } \mu}$$

The confidence limits vary from sample to sample (since $\bar{X}$ does).
The population mean μ lies between these confidence limits for 95% of samples.

Example 1

The weights of apples from an orchard have unknown mean μ and standard deviation 15 g; and a random sample of 400 apples has mean weight 82.36 g.

Find 95% confidence limits for the mean weight of all the apples in the orchard.

95% confidence limits for μ are $\bar{X} \pm 1.96 \, \dfrac{\sigma}{\sqrt{n}} = \bar{X} \pm 1.96 \times \dfrac{15}{\sqrt{400}}$

$$= \bar{X} \pm 1.47 = 82.36 \pm 1.47$$

$$= 80.89, \ 83.83 \text{ g}$$

Considering all random samples of 400 apples, the sample mean $\bar{X}$ will vary, and so the confidence limits $\bar{X} \pm 1.47$ differ for each sample; the population mean μ lies between these confidence limits for 95% of samples.

We have only one sample (for which $\bar{X} = 82.36$ and the confidence limits are 80.89 and 83.83). We do not know whether this particular sample is one of the 95% for which μ lies between the confidence limits or whether it is one of the 5% for which μ does not lie between the limits. However, we can be '95% confident' that μ does lie in the interval $80.89 < \mu < 83.83$.

While it is true to say that $P(\bar{X} - 1.47 < \mu < \bar{X} + 1.47) = 0.95$, the variable quantity here is $\bar{X}$; if we substitute a particular value $\bar{X} = 82.36$, it is *not* right to say that $P(80.89 < \mu < 83.83) = 0.95$; the statement '$80.89 < \mu < 83.83$' is either true or false, depending on the true (although unknown) value of μ.

It is of course possible to work at confidence levels other than 95%.

Example 2

A random sample of 120 bullets was selected. These bullets were fired from a gun; their muzzle velocities were measured and found to have a mean of 634 m s^{-1} and standard deviation 8 m s^{-1}. Find

(i)   a 90% confidence interval   (ii)   a 95% confidence interval
(iii)   a 99% confidence interval   (iv)   a 99.8% confidence interval

for the mean velocity of all the bullets.

We first consider the standard Normal distribution N(0, 1).

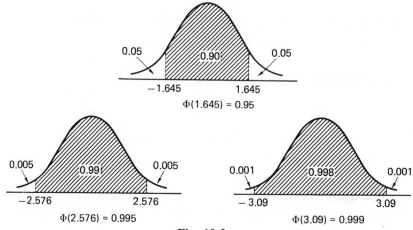

Fig. 10.2

Suppose that the muzzle velocity for all bullets has mean μ and standard deviation σ. We estimate $\sigma \approx 8$.

(i)   90% confidence limits for μ are $\bar{X} \pm 1.645 \dfrac{\sigma}{\sqrt{n}} = 634 \pm 1.645 \times \dfrac{8}{\sqrt{120}}$ (Fig. 10.2(a))

$$= 634 \pm 1.20$$

$$= 632.80, 635.20$$

90% confidence interval is $632.80 < \mu < 635.20$

(ii)   95% confidence limits are $\bar{X} \pm 1.96 \dfrac{\sigma}{\sqrt{n}} = 634 \pm 1.96 \times \dfrac{8}{\sqrt{120}}$

$$= 634 \pm 1.43$$

95% confidence interval is $632.57 < \mu < 635.43$

(iii)   99% confidence limits are $\bar{X} \pm 2.576 \dfrac{\sigma}{\sqrt{n}} = 634 \pm 1.88$ (Fig. 10.2 (b)).

99% confidence interval is $632.12 < \mu < 635.88$

(iv)   99.8% confidence limits are $\bar{X} \pm 3.09 \dfrac{\sigma}{\sqrt{n}} = 634 \pm 2.26$ (Fig. 10.2(c)).

99.8% confidence interval is $631.74 < \mu < 636.26$

Clearly the more confident we wish to be, the wider is the confidence interval, so the less we can say about μ.

We need to compromise between, on the one hand

making a precise statement about μ, in which we have little confidence

and on the other hand

making a statement in which we have great confidence, but which gives imprecise information about μ.

The generally accepted compromise is to use a confidence level of 95% (for which there is of course a 5% chance of being wrong), unless the consequences of making a mistake would be serious, in which case a higher confidence level (e.g. 99%) should be used.

Note   In Example 2, the population standard deviation σ was unknown, so this was estimated from the sample. It is reasonable to do this if the sample is large (say $n > 50$); but for small samples, the sample standard deviation s might differ considerably from σ, and so using s to estimate σ would introduce an unacceptably large error.

Example 3

The heights of people have a standard deviation of 12 cm. What size sample should be taken if it is required to estimate the mean height, with 95% confidence, to within $\pm\, 0.5$ cm?

Suppose that the population has mean height μ.

If a sample of n people has mean height $\bar{X}$, then the 95% confidence limits for μ are

$\bar{X} \pm 1.96 \dfrac{\sigma}{\sqrt{n}}$.

To estimate μ to within $\pm\, 0.5$ cm we require $1.96 \dfrac{\sigma}{\sqrt{n}} = 0.5$

$$\sqrt{n} = \frac{1.96\,\sigma}{0.5} = \frac{1.96 \times 12}{0.5}$$

$$= 47.04$$

$$n = 2212.8$$

A sample of at least 2213 people will be needed.

One-sided confidence limits

Sometimes we are only interested in showing that the population mean μ is, for example, greater than some value.
For 95% of samples (Fig. 10.3),

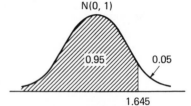

$$\bar{X} < \mu + 1.645 \,\frac{\sigma}{\sqrt{n}}$$

i.e. $\mu > \bar{X} - 1.645 \,\dfrac{\sigma}{\sqrt{n}}$;

this is a 95% lower confidence limit for μ.

Fig. 10.3

Example 4

For a random sample of 850 tomato plants, the mean yield was 27.2 fruit per plant, with a standard deviation of 3.8.
  Give a 95% lower confidence limit for the true mean number of fruit per plant.

Suppose that all tomato plants have mean yield μ with standard deviation σ. We estimate $\sigma \approx 3.8$.

95% lower confidence limit for μ is $\bar{X} - 1.645 \,\dfrac{\sigma}{\sqrt{n}}$

$$= 27.2 - 1.645 \times \frac{3.8}{\sqrt{850}}$$

$$= 26.99$$

This means that we are 95% confident that $\mu > 26.99$, in the sense that μ is greater than the lower limit obtained in this way for 95% of samples.

Example 5

An insurance company found that, for a random sample of 90 accidents of a certain category, the mean claim was £1265 with a standard deviation of £205.
  Give a 99% upper confidence limit for the mean claim for all accidents in this category.

Suppose that, for all such accidents, the mean claim is μ and the standard deviation is σ.
For 99% of samples,

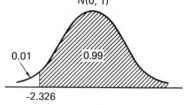

$$\bar{X} > \mu - 2.326 \,\frac{\sigma}{\sqrt{n}} \qquad \text{(Fig. 10.4)}$$

and thus $\mu < \bar{X} + 2.326 \,\dfrac{\sigma}{\sqrt{n}}$

We have $n = 90$, $\bar{X} = 1265$, and we estimate $\sigma \approx 205$, so the 99% upper confidence limit for μ is

Fig. 10.4

$$\bar{X} + 2.326 \,\frac{\sigma}{\sqrt{n}} = 1265 + 2.326 \times \frac{205}{\sqrt{90}}$$

$$= 1315$$

The company can be 99% confident that the mean claim for accidents in this category is less than £1315.

Exercise 10.3   Confidence intervals (for μ)

1   A population has mean μ and standard deviation 10.
A random sample of size 784 from this population has mean $\bar{X}$.
Find (in terms of $\bar{X}$)
(i)   95% confidence limits for μ
(ii) 98% confidence limits for μ
(iii) 99% confidence limits for μ
(iv) 99.8% confidence limits for μ.

Given $\bar{X} = 58.2$, find the corresponding 95% confidence interval for μ.

2   The weights of cattle are known to have standard deviation 38.2 kg.
A random sample of 250 cattle has mean weight 215.4 kg.
Give 95% confidence limits for the mean weight of all cattle.

3   A bank selected a random sample of 600 of its customers, and found that these accounts had a mean credit of £125, with standard deviation £360.
Give 99% confidence limits for the mean credit in all the bank's accounts.

4   A population has mean μ and standard deviation σ, and a random sample of size n has mean $\bar{X}$. Find general expressions (in terms of $\bar{X}$, σ and n) for
(i)   a 95% upper confidence limit for μ
(ii) a 99% lower confidence limit for μ
(iii) a 99.8% upper confidence limit for μ.

5   The tension at which a certain type of lift cable will break is known to have a standard deviation of 18 kN. Ten such cables broke at the following tensions:

$$224, 230, 182, 185, 206, 229, 191, 177, 200, 196 \quad \text{(kN)}.$$

Give a 95% lower confidence limit for the mean breaking tension, and explain the meaning of your value.

6   The times taken for a certain train journey on 250 occasions were as follows:

Journey time (minutes)	118–120	120–122	122–124	124–126
Number of occasions	3	15	36	58

Journey time (minutes)	126–128	128–130	130–132	132–134
Number of occasions	63	42	20	13

Assuming that these journeys are a random sample of all journeys on this route, find a time T for which it may be said, with 90% confidence, that the mean journey time for this route is less than T.

7   A random sample of 180 cars had mean age 4.2 years with standard deviation 2.7 years.
   (a)   Give a 95% confidence interval for the mean age of all cars.
   (b)   What size sample would be needed in order to
         (i)   estimate the mean age, with 95% confidence, to within ± 0.1 year
         (ii)   find 95% confidence limits which differ by 0.1 year?
   (c)   What size sample would be needed in order to estimate the mean age, with 99% confidence, to within ± 0.1 year?

8   The heights of people are known to have a standard deviation of 11.2 cm. What size of sample should be taken in order to estimate the mean height (in cm) correct to one decimal place, with 95% confidence?

9   A random sample of 400 chickens had mean weight 2.08 kg, with standard deviation 0.22 kg. With what confidence may it be said that
   (i)   the true mean weight is between 2.06 and 2.10 kg
   (ii)   the true mean weight is greater than 2.07 kg?

10   Given that $Z = \dfrac{\bar{X} - \mu}{\dfrac{\sigma}{\sqrt{n}}}$ is standard Normal, find numbers α and β such that

   $P(Z < \alpha) = 0.01$ and $P(Z > \beta) = 0.04$. There is probability 0.95 that $\alpha < Z < \beta$; obtain the corresponding 95% confidence interval for μ, in terms of $\bar{X}$, σ and n.
      Find this confidence interval when $n = 50$, $\sigma = 1.4$ and $\bar{X} = 57.2$. Find also, for these values, the usual (symmetric) 95% confidence interval. Which of these two confidence intervals do you prefer, and why?

10.4   Hypothesis testing for the population mean

We now consider situations in which we have some preconceived idea about the value of the population mean μ. This might be based on past experience or on the original intention or design. For example, suppose a machine was designed to produce metal rods having a mean length of 120 mm, with standard deviation 4 mm.
   We define two rival 'hypotheses',
the *null hypothesis,* usually written H_0, which asserts that μ is equal to the particular value suggested (in this case 120 mm)
and the *alternative hypothesis,* usually written H_1, which asserts that μ does not take this value.
For the example above we have H_0: $\mu = 120$

$$H_1: \mu \neq 120$$

Suppose that a random sample of n rods has mean length $\bar{X}$.
   Using the value of $\bar{X}$, we must decide
      either to accept H_0
      or to reject H_0 (and hence accept H_1)

Clearly if $\bar{X}$ is close to 120 we would favour H_0 and if $\bar{X}$ is far from 120 we would favour H_1; the problem is to find exactly where to take the dividing line.

We might make the correct decision (to accept H_0 when it is actually true, or to reject H_0 when it is actually false), but there are two possible kinds of mistake:

Type I error   To reject H_0 when it is actually true.
We conclude that μ is not equal to the design value of 120 when, in fact, it is. This is an unfortunate mistake, as we might then make adjustments to the machine, quite unnecessarily.
The probability of making a Type I error, often denoted by α, is called the *significance level* of the test.
Type II error   To accept H_0 when it is actually false.
We conclude that μ is equal to 120 (i.e. the machine is performing satisfactorily) when in fact it is not.
The probability of making a Type II error is often denoted by β.
The probability of correctly rejecting H_0 when it is false is then $(1 - \beta)$; this is called the *power* of the test.

It is desirable that both α and β should be as small as possible. However for a fixed sample size (n), if we decrease α we shall increase β, so some compromise is necessary.

We usually fix the significance level α. The probability β, of accepting H_0 when it is false, is generally difficult to calculate, and it may be quite large.

We shall devise a test having a significance level of 5%, i.e. the probability of rejecting H_0, when it is actually true, is to be 5%.

Suppose that all the rods produced by the machine have a true mean length μ, with standard deviation σ.

$$\text{We test} \quad H_0 : \mu = 120$$

$$\text{against} \quad H_1 : \mu \neq 120$$

If a random sample of n rods has mean length $\bar{X}$, then

$\bar{X}$ is approximately Normal, with mean μ and standard deviation $\dfrac{\sigma}{\sqrt{n}}$

and so $\dfrac{\bar{X} - \mu}{\dfrac{\sigma}{\sqrt{n}}}$ has the standard Normal distribution $N(0, 1)$

Consider $Z = \dfrac{\bar{X} - 120}{\dfrac{\sigma}{\sqrt{n}}}$

H_1 is favoured when $\bar{X}$ is far from 120, i.e. when Z is far from zero, so we shall reject H_0 when Z is sufficiently far from zero.

If H_0 is true, then $Z = \dfrac{\bar{X} - 120}{\dfrac{\sigma}{\sqrt{n}}} = \dfrac{\bar{X} - \mu}{\dfrac{\sigma}{\sqrt{n}}}$, so Z is standard Normal.

Thus if we decide to

Accept H_0 if $-1.96 \leqslant Z \leqslant 1.96$ (Fig. 10.5)

Reject H_0 otherwise (i.e. $Z > 1.96$ or $Z < -1.96$

i.e. $|Z| > 1.96$)

then the probability of rejecting H_0 when it is true will be 5%; so this test has a significance level of 5% as required.

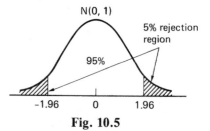

Fig. 10.5

The set of values of Z for which H_0 will be rejected is sometimes called the *critical region* for the test.

This consists of two parts ($Z > 1.96$ and $Z < -1.96$), so the above test is called a *two-tailed test*.

Example 1

Using the above test, what conclusion would be made about the machine if

(i)   a random sample of 50 rods had mean length 120.97 mm

(ii) a random sample of 300 rods had mean length 119.50 mm?

(i)   We have $n = 50$, $\bar{X} = 120.97$, and we assume $\sigma = 4$

so $Z = \dfrac{\bar{X} - 120}{\sigma/\sqrt{n}} = \dfrac{120.97 - 120}{4/\sqrt{50}} = 1.715$

This is between -1.96 and $+1.96$, so we

Accept H_0

The machine appears to be functioning satisfactorily.

(ii) We have $n = 300$, $\bar{X} = 119.50$, $\sigma = 4$

so $Z = \dfrac{119.50 - 120}{4/\sqrt{300}} = -2.165$

This lies in the rejection region, so we

Reject H_0

This sample suggests that the true mean length of the rods is not 120 mm.

Suppose that, as the result of the test, H_0 is rejected.

Since the probability of rejecting H_0 when it is actually true is fairly small (5%), we can be reasonably confident that H_0 is indeed false. This is a positive result; the sample has provided evidence against the null hypothesis.

Now suppose that H_0 is accepted.

This does *not* show that H_0 is probably true. It means that the sample mean $\bar{X}$ is a reasonable value under the assumption of the null hypothesis, so there is insufficient evidence to reject the null hypothesis.

Example 2

A farmer has found that, over many years, the yields from his potato plants have had mean 1.82 kg per plant, with a standard deviation 0.34 kg per plant.

One year he planted 750 plants of a new variety, and these gave a total yield of 1395 kg.
Test, at the 5% level of significance, whether the mean yield per plant for the new
variety is different from that of his usual variety.

Suppose that the yield per plant for the new variety has mean μ and standard deviation σ.
The null hypothesis must give a precise value for μ, so this will state that the new variety
has the same mean yield as the original.

$$\text{We test} \quad H_0 : \mu = 1.82$$

$$\text{against} \quad H_1 : \mu \neq 1.82$$

If a random sample of n plants gives mean yield $\bar{X}$, then $\bar{X}$ is approximately Normal
with mean μ and standard deviation $\sigma/\sqrt{n}$, so

$$\frac{\bar{X} - \mu}{\sigma/\sqrt{n}}$$

is standard Normal.

$$\text{Let } Z = \frac{\bar{X} - 1.82}{\dfrac{\sigma}{\sqrt{n}}}$$

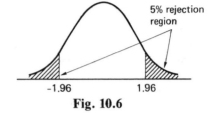

Fig. 10.6

If H_0 is true, then Z is standard Normal.

We have $n = 750, \bar{X} = \dfrac{1395}{750} = 1.86, \sigma = 0.34$

$$\text{thus } Z = \frac{1.86 - 1.82}{\dfrac{0.34}{\sqrt{750}}} = 3.222$$

Reject H_0.

The value of Z is considerably greater than 1.96, so there is very strong evidence that
the mean yield for the new variety is different.

It is of course possible to devise tests with significance levels other than 5%. For
example a 1% level of significance might be used; in this case we shall require stronger
evidence before we reject the null hypothesis.

Example 3
A certain type of electric motor is claimed to have a power output of 3.25 kW. A random
sample of 200 of these motors was selected, and when tested individually the power out-
puts were found to have mean 3.21 kW and standard deviation 0.22 kW.
  Is there evidence, at the 1% level of significance, that the true mean power output is
different from the specified value?

Suppose that the power output from these motors has mean μ and standard deviation σ.

$$\text{We test} \quad H_0 : \mu = 3.25$$

$$\text{against} \quad H_1 : \mu \neq 3.25$$

If a sample of n motors has mean power output $\bar{X}$, then $\bar{X}$ is approximately Normal

with mean μ and standard deviation $\sigma/\sqrt{n}$

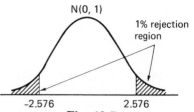

so $\dfrac{\bar{X} - \mu}{\sigma/\sqrt{n}}$ is standard Normal.

Let $Z = \dfrac{\bar{X} - 3.25}{\sigma/\sqrt{n}}$

If H_0 is true, then Z is standard Normal.
We have $n = 200$, $\bar{X} = 3.21$ and we estimate $\sigma \approx 0.22$,

thus $Z = \dfrac{3.21 - 3.25}{0.22/\sqrt{200}} = -2.571$

Fig. 10.7

Accept H_0.

There is *not* quite enough evidence (at the 1% level of significance) to say that the mean power output is different from 3.25 kW.

Example 4
The lifetimes of mice kept under laboratory conditions have mean 258 days, and standard deviation 45 days. 36 of these mice, selected at random, were each given a measured dose of a certain drug every day, and the mean lifetime for this group was 274 days.

At what level of significance does this indicate that the drug has altered the mean lifetime of the mice?

Suppose that, when given the drug, the lifetimes of the mice have mean μ and standard deviation σ.

We test $H_0 : \mu = 258$

against $H_1 : \mu \neq 258$

If a random sample of n mice has mean lifetime $\bar{X}$, then $\bar{X}$ is approximately Normal

with mean μ and standard deviation $\dfrac{\sigma}{\sqrt{n}}$,

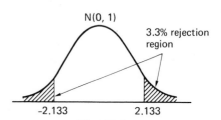

so $\dfrac{\bar{X} - \mu}{\sigma/\sqrt{n}}$ is standard Normal.

Let $Z = \dfrac{\bar{X} - 258}{\sigma/\sqrt{n}}$

If H_0 is true, then Z is standard Normal.
We have $n = 36$, $\bar{X} = 274$, $\sigma = 45$,

Fig. 10.8

thus $Z = \dfrac{274 - 258}{45/\sqrt{36}} = 2.133$.

Now $P(Z > 2.133) = 1 - \Phi(2.133)$

$= 1 - 0.9835$

$= 0.0165$

and $P(Z < -2.133) = 0.0165$,

so H_0 would just be rejected at a significance level of 3.3%.

As this is less than 5%, we would usually regard this as evidence that the drug has altered the mean lifetime of the mice.

If the level of significance is not specified, we normally use 5%. A value of Z which is significant at about 5% is regarded as giving some evidence against the null hypothesis; a value of Z which is significant at about 1% is regarded as giving strong evidence against the null hypothesis.

One-tailed tests

Sometimes we are interested in showing only that the population mean μ lies on one particular side of the value specified in the null hypothesis.
  For example, suppose that over a long period, the alcohol content in barrels of a certain wine has had mean 15% and standard deviation 3%. The process by which the wine is made is now modified with the intention of increasing the alcohol content. A random sample of barrels made by the new process is taken, and used to test the effectiveness of the new process.
  Suppose that for barrels made by the new process, the alcohol content has mean μ% and standard deviation σ%.
  The null hypothesis will assert that the new process has made no difference (i.e. μ is still 15); but the new process will have been effective only if $\mu > 15$.

$$\text{So we test} \quad H_0 : \mu = 15$$

$$\text{against} \quad H_1 : \mu > 15$$

Suppose that a random sample of n barrels has mean alcohol content $\bar{X}$%, then $\bar{X}$ is approximately Normal with mean μ and standard deviation $\sigma/\sqrt{n}$, so

$$\frac{\bar{X} - \mu}{\sigma/\sqrt{n}} \text{ is standard Normal.}$$

$$\text{Let } Z = \frac{\bar{X} - 15}{\sigma/\sqrt{n}}$$

Now H_1 is favoured only when $\bar{X} > 15$, i.e. when Z is positive, so we shall reject H_0 when Z is sufficiently large and positive.
  When H_0 is true, Z is standard Normal.
  If we decide to

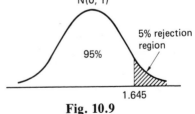

$$\text{Accept} \quad H_0 \text{ if } Z \leqslant 1.645$$

$$\text{Reject} \quad H_0 \text{ if } Z > 1.645$$

then this test has a significance level of 5%

Fig. 10.9

The rejection region now consists of just one part ($Z > 1.645$); this is a *one-tailed test*.

Example 5

Using the above test, what would you deduce if
(i) a random sample of 30 barrels had mean alcohol content 16%?
(ii) a random sample of 275 barrels had mean alcohol content 14.4%?

(i) We have $n = 30, \bar{X} = 16, \sigma = 3,$

$$\text{thus } Z = \frac{16 - 15}{3/\sqrt{30}} = 1.826$$

Reject H_0.

There is evidence that $\mu > 15$, i.e. that the new process is effective.

(ii) We have $n = 275, \bar{X} = 14.4, \sigma = 3$

$$\text{thus } Z = \frac{14.4 - 15}{3/\sqrt{275}} = -3.317$$

Accept H_0.

Although H_0 seems unlikely to be true, the evidence does not favour H_1. There is certainly no evidence that the new process is effective.

Example 6

Up until January 1983, the daily number of deaths and serious injuries due to road accidents in a certain county had mean 3.16 and standard deviation 0.75. In January 1983, the wearing of seat-belts became compulsory, with the intention of reducing casualties.

During a period of 100 days after that there were a total of 297 deaths and serious injuries. Does this support the view that the seat-belt law has reduced the casualty rate? (Use the 1% level of significance).

Suppose that, after January 1983, the daily number of deaths and serious injuries has mean μ and standard deviation σ.

$$\text{We test} \quad H_0 : \mu = 3.16$$

$$\text{against} \quad H_1 : \mu < 3.16$$

If, for a random sample of n days, the mean number is $\bar{X}$, then $\bar{X}$ is approximately Normal with mean μ and standard deviation $\sigma/\sqrt{n}$, so

$$\frac{\bar{X} - \mu}{\sigma/\sqrt{n}} \text{ is standard Normal. Let } Z = \frac{\bar{X} - 3.16}{\sigma/\sqrt{n}}$$

H_1 is favoured when $\bar{X} < 3.16$, i.e. when Z is negative. If H_0 is true, then Z is standard Normal.

We have $n = 100, \bar{X} = \dfrac{297}{100} = 2.97, \sigma = 0.75$

$$\text{thus } Z = \frac{2.97 - 3.16}{0.75/\sqrt{100}} = -2.533$$

Reject H_0.

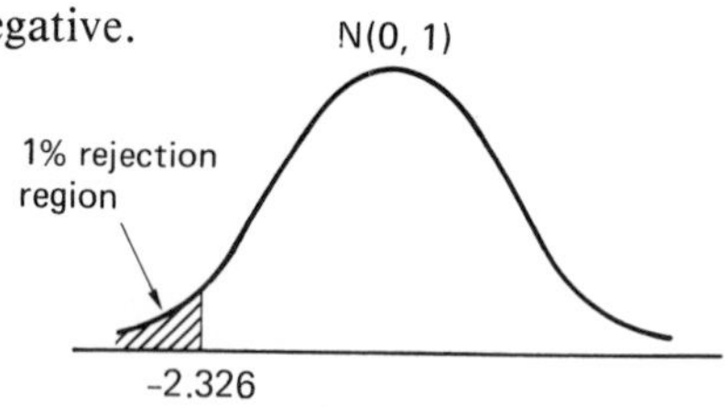

Fig. 10.10

There *is* evidence (at the 1% level of significance) that the casualty rate has been reduced.

It is not always easy to decide whether to carry out a one-tailed or a two-tailed test. Remember that (in theory at least) the hypotheses should be formulated and the test devised before the random sample is taken; so that the decision to use a one-tailed or a two-tailed test must not be influenced by the results of the sample.

Consider the following two situations:

(A)   The mean weight of people in the UK is known to be 76.4 kg. A social scientist believes that people living in urban areas tend to be heavier than those living in rural areas. To test his theory he takes a random sample of 500 people living in London, and finds that this sample has mean weight 77.1 kg with standard deviation 9.2 kg. What conclusion should he reach?

(B)   The mean weight of people in the UK is known to be 76.4 kg. A random sample of 500 people living in London was found to have mean weight 77.1 kg and standard deviation 9.2 kg.

Does this provide evidence that people living in London are heavier than the national average?

Suppose that μ is the mean weight of people living in London.

In (A), it is quite clear, before the sample was taken, that the social scientist is only interested in showing that $\mu > 76.4$. He would use a one-tailed test, of

$$H_0 : \mu = 76.4$$

against $H_1 : \mu > 76.4$

In fact $Z = 1.701$, so at the 5% level of significance he would reject H_0 and conclude that his theory is correct.

In (B), the situation is less clear, and we need to consider why the word 'heavier' is used in the last sentence. If it is 'obviously expected' that people living in London are heavier than average, the situation is essentially the same as in (A) and we would use a one-tailed test. On the other hand, if the sample was taken first and seeing that the sample mean was greater than 76.4 it was thought that people living in London might be heavier, this is based on the results of the sample and so we cannot use a one-tailed test.

Whenever there is doubt, as in this case, it is best to use a two-tailed test. We test first whether the mean weight of people living in London is different from the national average (then if we do find there is evidence of a difference, it is clear from the sample that the London people are heavier).

We test $H_0 : \mu = 76.4$

against $H_1 : \mu \neq 76.4$

Since $Z = 1.701$, at the 5% level of significance, we would accept H_0; there is no evidence that the mean weight of people living in London is different from the national average.

Exercise 10.4   Hypothesis testing (for μ)

Always state the null hypothesis (H_0) and the alternative hypothesis (H_1); use the 5% level of significance (unless stated otherwise); show the rejection region on a sketch of the standard Normal curve; make a decision and comment on the strength of the evidence.

1   A population has (unknown) mean μ and (known) standard deviation σ. A random sample of size n from this population has mean $\bar{X}$.

(i)   State the approximate distribution of (a) $\bar{X}$ (b) $\dfrac{\bar{X} - \mu}{\sigma/\sqrt{n}}$

The null hypothesis $H_0 : \mu = 60$ is to be tested against $H_1 : \mu \neq 60$

   (ii) Devise a test which has a significance level of 5%.
        What decision would be reached in the following cases:
        (a) $\sigma = 15, n = 20, \bar{X} = 62.3$
        (b) $\sigma = 15, n = 850, \bar{X} = 58.7$
        (c) $\sigma = 15, n = 120, \bar{X} = 64.0$?
  (iii) Devise a test which has a significance level of 2%
  (iv) Devise a test which has a significance level of 1%
   (v) Devise a test which has a significance level of 0.2%.

2 Over a long period, the lengths of bolts produced by a machine had a mean of 78.36 mm and standard deviation 0.80 mm. After a major overhaul, a random sample of 150 bolts had mean length 78.57 mm. Does this provide evidence that the mean length has changed?

3 A population has standard deviation 12.5 and unknown mean μ.
Test the null hypothesis that $\mu = 50$ against the alternative $\mu \neq 50$, given that a sample of 2000 values has mean 49.6.

4 Cans of lager are supposed to have mean contents 0.7 litres, and experience has shown that the standard deviation of the contents is $0.012\ \ell$.
The contents of 10 cans were measured as follows:

$0.704, 0.672, 0.680, 0.710, 0.666, 0.678, 0.685, 0.671, 0.681, 0.690$ (ℓ)

Test whether the true mean contents is different from the stated value of $0.7\ \ell$.

5 A manufacturer claims that his light bulbs have a mean lifetime of 650 hours with standard deviation 90 hours. A random sample of 500 bulbs had a mean lifetime of 660.3 hours. Test, at the 1% level of significance, whether the mean lifetime is different from the value quoted by the manufacturer.

6 British students are known to have mean height 176 cm with standard deviation 11 cm. A random sample of 60 American students had a mean height of 179 cm. Does this suggest that the mean height of American students is different from that of British students?

7 A random sample of 400 bags of sugar had mean weight 0.998 kg, with standard deviation 0.021 kg. Test the hypothesis that the true mean weight is 1 kg.

8 The mean mark for an examination taken by several thousand candidates was 48.7. One particular examiner marked 265 scripts, which may be assumed to be a random sample of the candidates. His marks had mean 44.5 and standard deviation 22.5. Is there any evidence that this examiner is marking differently from the others?

9 In 1960, the average number of children per family was 2.4, with standard deviation 0.7. In 1983, a random sample of 30 families had a total of 66 children. At what level of significance does this indicate a change in the average number of children?

10 A population has mean μ and standard deviation σ.

A random sample of size n from this population has mean $\bar{X}$.

  (i)   Devise a test for $H_0 : \mu = 80$ against $H_1 : \mu > 80$ at the 5% level of significance. What decision would be reached in the following cases

   (a) $\sigma = 25, n = 16, \bar{X} = 88.7$

   (b) $\sigma = 25, n = 3500, \bar{X} = 80.7$

   (c) $\sigma = 25, n = 450, \bar{X} = 76$?

  (ii) Devise a test for $H_0 : \mu = 45$ against $H_1 : \mu < 45$ at the 5% level of significance.

  (iii) Devise a test for $H_0 : \mu = 120$ against $H_1 : \mu > 120$ at the 1% level of significance.

  (iv) Devise a test for $H_0 : \mu = 2.8$ against $H_1 : \mu < 2.8$ at the 0.2% level of significance.

11 The weekly weight gain for chickens fed on a normal diet has mean 75 g with standard deviation 15 g. A new diet is introduced with the intention of increasing the weight gain, and a random sample of 200 chickens fed on this new diet had a mean weekly weight gain of 76.9 g. Is the new diet effective?

12 A farmer sells potatoes in 25 kg sacks. An inspector suspects that the farmer is selling underweight sacks; he weighs a random sample of 150 sacks and finds that the mean weight is 24.6 kg, with standard deviation 0.9 kg. What does he conclude?

13 A swimmer's times for 300 m have had mean 192 s with standard deviation 2 s. After a fortnight's holiday, his first 25 swims had a mean time of 191.3 seconds. Has the holiday altered his performance?

14 The tourist office of a seaside resort claims that the resort is sunnier than the national average of 5.2 hours per day. A sample of 85 days at the resort gave mean daily sunshine of 5.6 hours with standard deviation 2.6 hours. Is the claim justified?

15 People in Britain have mean height 176 cm with standard deviation 11 cm. A sample of 60 teachers had mean height 173.4 cm, and therefore it is claimed that teachers are shorter than average. Is this claim justified?

16 'Standard' light bulbs have mean lifetime 850 hours with standard deviation 120 hours. A sample of 300 'long life' bulbs had mean lifetime 867 hours. Does this provide evidence (at the 1% level of significance) that 'long life' bulbs are better than the 'standard' ones?

11

The Binomial Distribution

11.1 Revision of Binomial distributions

Consider a sequence of n 'trials', each of which can result in

> either 'success', with probability p
> or   'failure', with probability $q = 1 - p$

Then the total number of successes X, has the Binomial distribution $B(n, p)$ provided that
    (i) the probability p of success is the same for each trial
and (ii) the trials are independent.
We now derive the probability that there are exactly r successes.
   If there are r successes, there must be $(n - r)$ failures.
Each arrangement of r successes and $(n - r)$ failures is an arrangement of n items, of which r are successes and $(n - r)$ are failures, so the number of such arrangements is

$$\frac{n!}{r!(n-r)!} = \binom{n}{r}$$

   The probability of obtaining r successes and $(n - r)$ failures in a specified arrangement is $p^r q^{n-r}$ (since the n trials are independent).
Hence the probability that there are exactly r successes is

$$P(X = r) = \binom{n}{r} p^r q^{n-r}$$

so that the distribution of X is

X	0	1	2	$\dots r \dots$	n
Probability	q^n	$\binom{n}{1}pq^{n-1}$	$\binom{n}{2}p^2q^{n-2}$ $\binom{n}{r}p^rq^{n-r}$		p^n

Note that

$$q^n + \binom{n}{1}pq^{n-1} + \binom{n}{2}p^2q^{n-2} + \cdots + \binom{n}{r}p^rq^{n-r} + \cdots + p^n$$

is the binomial expansion of $(q + p)^n$; and since $q + p = 1$, this confirms that the probabilities add up to one.

Mean and variance

The Binomial distribution $B(n, p)$ has mean np

and variance npq

We shall now prove these results (which we have assumed and used before).
 Define the random variables $Y_1, Y_2, \ldots, Y_n$ as follows

$$Y_1 = \begin{cases} 1 \text{ if the first trial is a success} \\ 0 \text{ if the first trial is a failure} \end{cases}$$

$$Y_2 = \begin{cases} 1 \text{ if the second trial is a success} \\ 0 \text{ if the second trial is a failure} \end{cases}$$

and so on.

The distribution of Y_1 is

Y_1	0	1
Probability	q	p

Thus $E[Y_1] = p$ and $E[Y_1^2] = p$

so $\text{var}(Y_1) = p - p^2 = p(1 - p) = pq$

Now $Y_2, Y_3, \ldots, Y_n$ all have the same distribution as Y_1 and $Y_1, Y_2, \ldots, Y_n$ are independent (since the trials are independent).
 $X = Y_1 + Y_2 + \cdots + Y_n$ is the total number of successes, and so X is $B(n, p)$

$$E[X] = E[Y_1 + Y_2 + \cdots + Y_n]$$
$$= E[Y_1] + E[Y_2] + \cdots + E[Y_n]$$
$$= p + p + \cdots + p$$
$$= np$$

$$\text{var}(X) = \text{var}(Y_1 + Y_2 + \cdots + Y_n)$$
$$= \text{var}(Y_1) + \text{var}(Y_2) + \cdots + \text{var}(Y_n)$$

(since $Y_1, Y_2, \ldots, Y_n$ are independent)

$$= pq + pq + \cdots + pq$$
$$= npq$$

Fitting a Binomial distribution to data

Suppose we have a sample of values from a population which is thought to have a Binomial distribution $B(n, p)$, where p is unknown. We can use the sample mean to estimate the population mean (np), and hence to estimate the value of p. We then calculate the theoretical probabilities, and compare the expected frequencies with the actual ones.

Example 1

Eggs are packed in boxes of six. 1000 boxes were inspected and the number of cracked eggs in each box was noted, with the following results.

Number of cracked eggs	0	1	2	3	4	5	6
Number of boxes	527	358	101	12	2	0	0

Fit a Binomial distribution to this data.

If the probability that an individual egg is cracked is p, we would expect the number of cracked eggs in a box to have the $B(6, p)$ distribution, which has mean $6p$.

The sample mean is $\dfrac{(0 \times 527) + (1 \times 358) + \cdots}{1000} = 0.604$

so we estimate $6p \approx 0.604$

i.e. $\qquad\qquad p \approx 0.1007$

We calculate the probabilities from $B(6, 0.1007)$, for example

$$P(X = 2) = \binom{6}{2}(0.1007)^2(0.8993)^4 = 0.0994$$

and so the expected number of boxes having 2 cracked eggs is

$$0.0994 \times 1000 = 99.4$$

We obtain

Number of cracked eggs	0	1	2	3	4	5	6
Probability from B(6, 0.1007)	0.5291	0.3553	0.0994	0.0148	0.0012	0.0001	0
Expected frequency	529.1	355.3	99.4	14.8	1.2	0.1	0.0

The expected frequencies are very close to the actual frequencies; the Binomial distribution fits the data well.

The Normal approximation to the Binomial distribution

In the proof of the mean and variance formulae, we have seen that $Y_1 + Y_2 + \cdots + Y_n$ has the Binomial distribution $B(n, p)$.

Now $Y_1, Y_2, \ldots, Y_n$ are independent random variables, all having the same distribution, so by the Central Limit Theorem, $Y_1 + Y_2 + \cdots + Y_n$ will have an approximately Normal distribution provided that n is large enough. Hence the Binomial distribution $B(n, p)$ can be approximated by a Normal distribution having mean np and standard deviation $\sqrt{npq}$.

We have seen that this approximation is a good one if we can go three standard deviations each side of the mean without leaving the range of possible values (0 to n). Since we are approximating a discrete variable (the Binomial) by a continuous one (the Normal) a continuity correction should be used.

Example 2

650 people attend a performance at a theatre, and the probability that a particular person buys a programme is 0.7.

(i) Find the probability that fewer than 440 programmes will be sold.

(ii) How many programmes should be available for sale if the probability of having too few available is to be less than 0.01?

If X people wish to buy a programme then,

X is B(650, 0.7) which is approximately Normal,
with mean $\mu = 650 \times 0.7 = 455$
and standard deviation $\sigma = \sqrt{650 \times 0.7 \times 0.3} = 11.68$

(Note that $\mu \pm 3\sigma = 455 \pm 35 = 420, 490$, which are both well within the range 0–650; so the Normal approximation will be a good one.)

(i) $P(X < 440 \text{ in Binomial}) \approx P(X < 439.5 \text{ in Normal})$

$$= P\left(Z < \frac{439.5 - 455}{11.68}\right)$$

$$= P(Z < -1.327)$$

$$= 1 - 0.9077$$

$$= 0.0923$$

(ii) If N programmes are available, then the probability that there will be too few is
$P(X > N) \approx P(X > N + 0.5 \text{ in Normal})$
We require $P(X > N + 0.5) = 0.01$
so $N + 0.5$ corresponds to z, where $\Phi(z) = 0.99$
$$z = 2.326$$

Thus $$\frac{(N + 0.5) - 455}{11.68} = 2.326$$

$$N = 481.7$$

Hence 482 programmes should be available for sale.

The mode

Suppose X is B(n, p), and let $p_r = P(X = r) = \binom{n}{r} p^r q^{n-r}$

$$\text{Consider } \frac{p_r}{p_{r-1}} = \frac{\binom{n}{r} p^r q^{n-r}}{\binom{n}{r-1} p^{r-1} q^{n-r+1}}$$

$$= \frac{n!}{(n-r)!\, r!}\, p \left/ \frac{n!}{(n-r+1)!\, (r-1)!}\, q \right.$$

$$= \frac{n!\, (n-r+1)!\, (r-1)!\, p}{n!\, (n-r)!\, r!q}$$

$$= \frac{(n-r+1)\, p}{rq}$$

Thus
$$\frac{p_r}{p_{r-1}} > 1 \Leftrightarrow \frac{(n-r+1)p}{rq} > 1$$

$$\Leftrightarrow (n-r+1)p > rq$$

$$\Leftrightarrow (n-r+1)p > r(1-p)$$

$$\Leftrightarrow np - rp + p > r - rp$$

$$\Leftrightarrow (n+1)p > r$$

Consider the sequence of probabilities $p_0, p_1, p_2, \ldots, p_{r-1}, p_r, \ldots, p_n$.

So long as $r < (n+1)p$, we have $\dfrac{p_r}{p_{r-1}} > 1$

i.e. p_r is greater than the preceding probability p_{r-1}

but when $r > (n+1)p$, we have $\dfrac{p_r}{p_{r-1}} < 1$,

so p_r is less than the preceding probability.

Hence the maximum probability p_r occurs when r is just less than $(n+1)p$ i.e. when r is the integer part of $(n+1)p$.

If X is B(n, p), then the most likely value of X i.e. the mode, is the integer part of $(n+1)p$

Note that, if $(n+1)p$ is an integer, then when $r = (n+1)p$ we have

$$\frac{p_r}{p_{r-1}} = 1, \text{ so } p_{r-1} = p_r, \text{ and there are two modes}$$

$$(n+1)p - 1 \quad \text{and} \quad (n+1)p$$

Example 3
The probability that a person chosen at random will be left-handed is 0.24. If 40 people are selected at random, what is the most likely number of left-handed people in the sample?

The number of left-handed people in the sample is B(40, 0.24).
We have $(n+1)p = 41 \times 0.24 = 9.84$, so the mode is 9.
The most likely number is 9.

Example 4
Find the mode of the Binomial distribution B$(800, \frac{2}{3})$.

We have $(n+1)p = 801 \times \frac{2}{3} = 534$.
Since this is an integer, there are two modes, 533 and 534.

Exercise 11.1 Binomial distributions (revision)

1   For the following situations, does X have a Binomial distribution? If it does, state the values of n and p if possible.

(i)   Throw a die 20 times; X is the number of sixes
(ii) X is the number of boys in a family of five children
(iii) X is the number of Queens in a bridge hand (13 cards selected from a normal pack of 52 cards)
(iv)   A multiple choice test contains 20 questions, which start with easy questions and become progressively more difficult. For each question the candidate has to choose one out of four possible answers. An entirely clueless candidate guesses randomly; X is his number of correct answers.
(v)   Consider the same test as in (iv); but for a candidate who does know something about the subject, and makes a sensible attempt at each question.
(vi)   Select 12 people at random; X is the number of left-handed people selected.
(vii) Toss a coin until the third head is obtained; X is the number of tosses.
(viii) 1% of items produced by a machine are defective; X is the number of defective items in a random sample of 150 items.
(ix)   A car-hire firm has 4 cars for hire; X is the number of cars out on hire.

2   A corridor is lit by 9 light-bulbs. After 5 weeks the probability that any particular bulb has failed is 0.3. Find the probability that after 5 weeks
(i)   no bulbs have failed
(ii)   between 3 and 5 bulbs have failed (inclusive)
(iii) at least 7 bulbs have failed.

3   The probability that a certain type of seed will grow is 0.6, and 12 seeds are planted. Find the largest value of N for which the probability that at least N seeds will grow is greater than 95%.

4   For mass-produced silicon chips the probability that a chip will work correctly is $\frac{1}{4}$. Given that n chips are taken, find the probability that at least one chip will work. How many chips need to be taken so as to be 99% confident that at least one of them will work?

5   When asked to donate £1 towards a worthy cause, 70% of people will do so (the other 30% giving nothing). A collector asks 8 people to donate £1. State the mean and variance of the amount collected. Find the probability that this amount is within one standard deviation of its mean.

6   The random variable X has mean 7.2 and standard deviation 1.2. Given that X has a Binomial distribution, find the probability that $X = 6$.

7   A commuter train runs 5 times a week. A record was kept of the number of times that the train was late arriving at its destination each week. Over a period of 3 years (156 weeks), the following figures were obtained:

Number of times late during a week X	0	1	2	3	4	5
Number of weeks	8	16	30	41	36	25

Calculate the mean value of X. Assuming that X has a Binomial distribution B(5, p), estimate the value of p; calculate the probabilities, and hence the expected frequencies, for this Binomial distribution. Does it appear to fit the data well?

8   Glasses are packed in boxes of eight. After a journey, 100 boxes were inspected and the number of broken glasses were counted, with the following results

Number of broken glasses in a box	0	1	2	3	4	5	6	7	8
Number of boxes	27	39	25	7	1	0	1	0	0

Fit a Binomial distribution to this data, and comment on how well it fits.

9   Given that 30% of people have blue eyes, and a random sample of 200 people is chosen, use the Normal approximation to find the probability that between 50 and 65 of these people (inclusive) have blue eyes.

10   An unbiased die is thrown 150 times and the number of sixes is recorded. Given that X is the number of sixes, find the probability that $20 \leqslant X \leqslant 30$.

11   10 000 supporters want to travel from Chelsea to Madrid for a football match. Thin-air is one of the airlines operating on this route. Each supporter acts independently and the probability that he chooses to travel by Thin-air is $\frac{1}{2}$. How many seats should the airline provide if it is prepared to run a 1% risk that it will have too few seats?
   Suppose now that Thin-air reduces its price so that there is a probability of $\frac{3}{5}$ that a supporter will choose to travel with them. Given that Thin-air puts on 6100 seats, find the probability that they will have too few seats.

12   Find the mode(s) of the following Binomial distributions:
   (i) B(12, 0.8)   (ii) B(7, 0.37)   (iii) B(19, 0.45)   (iv) B(365, 0.09)

13   Given that 3.2% of the items produced by a factory are defective, what is the most likely number of defectives in a crate containing 300 items?

11.2   The sum of Binomial distributions

Suppose a coin is tossed 30 times. If X is the number of heads on the first 20 tosses, and Y is the number of heads on the last 10 tosses, then X is B(20, $\frac{1}{2}$), Y is B(10, $\frac{1}{2}$) and X, Y are independent.
   Their sum $(X + Y)$ is the total number of heads on the 30 tosses, so $(X + Y)$ is B(30, $\frac{1}{2}$).
   Generally, if we consider a sequence of $(m + n)$ independent trials with probability of success p,

   and let X be the number of successes on the first m trials
   and Y be the number of successes on the last n trials,

so that $(X + Y)$ is the total number of successes, we have

> If X is B(m, p), Y is B(n, p) and X, Y are independent, then $(X + Y)$ is B$(m + n, p)$

Similarly, if X_1 is B(n_1, p), X_2 is B$(n_2, p), \ldots, X_k$ is B(n_k, p) and $X_1, X_2, \ldots, X_k$ are independent,

$$\text{then } (X_1 + X_2 + \cdots + X_k) \text{ is B}(n_1 + n_2 + \cdots + n_k, p)$$

When adding Binomial distributions in this way, it is essential that the probability of success (p) is the same for each distribution.

Example 1
One Saturday, a man enters 14 homing pigeons in a race, and the probability that any individual pigeon is lost during the race is 0.1. Next week he enters 9 pigeons in a race, and the week after that he enters 7 pigeons. Assuming that the probability of losing a pigeon remains at 0.1, find the probability that he loses a total of exactly 4 pigeons during these three weeks.

If he loses W pigeons in the first race, X in the second and Y in the third,

$$\text{then } W \text{ is B}(14, 0.1), \ X \text{ is B}(9, 0.1) \text{ and } Y \text{ is B}(7, 0.1)$$

so the total $T = W + X + Y$ is B$(30, 0.1)$

$$P(T = 4) = \binom{30}{4} (0.1)^4 (0.9)^{26} = 0.1771$$

If X and Y have Binomial distributions with different probabilities of success, then $(X + Y)$ does not have a Binomial distribution. However, if conditions are such that X and Y have approximately Normal distributions, then $(X + Y)$ will also be approximately Normal.

Example 2
X and Y are independent random variables; X is B$(2, 0.4)$ and Y is B$(2, 0.7)$.
Obtain the distribution of $T = X + Y$ and show that this is not a Binomial distribution.

The distributions of X and Y are given below.

X	0	1	2
Probability	$(0.6)^2$ $= 0.36$	$2 \times 0.4 \times 0.6$ $= 0.48$	$(0.4)^2$ $= 0.16$

Y	0	1	2
Probability	0.09	0.42	0.49

$(X + Y)$ has possible values 0, 1, 2, 3, 4

$P(X + Y = 0) = P(X = 0 \text{ and } Y = 0) = P(X = 0) \times P(Y = 0)$ since X, Y are independent
$= 0.36 \times 0.09$
$= 0.0324$

$P(X + Y = 1) = P(X = 1 \text{ and } Y = 0) + P(X = 0 \text{ and } Y = 1)$
$= 0.48 \times 0.09 + 0.36 \times 0.42$
$= 0.1944$

$P(X + Y = 2) = P(X = 2 \text{ and } Y = 0) + P(X = 1 \text{ and } Y = 1) + P(X = 0 \text{ and } Y = 2)$
$= 0.16 \times 0.09 + 0.48 \times 0.42 + 0.36 \times 0.49$
$= 0.3924$

$P(X + Y = 3) = P(X = 2 \text{ and } Y = 1) + P(X = 1 \text{ and } Y = 2)$
$= 0.16 \times 0.42 + 0.48 \times 0.49$
$= 0.3024$

$P(X + Y = 4) = P(X = 2 \text{ and } Y = 2)$
$= 0.16 \times 0.49$
$= 0.0784$

Thus the distribution of $T = X + Y$ is

T	0	1	2	3	4
Probability	0.0324	0.1944	0.3924	0.3024	0.0784

We now show that this cannot be a Binomial distribution.
If it *is*, then it must be B(4, p) for some p,
then from $P(T = 4) = p^4$ we have $p^4 = 0.0784$

$$p \approx 0.529$$

but from $P(T = 0) = (1 - p)^4$ we have $(1 - p)^4 = 0.0324$
$$1 - p \approx 0.424$$
$$p \approx 0.576$$

so we cannot find a value of p which gives the correct probabilities for both $T = 0$ and $T = 4$.

Hence T does *not* have a Binomial distribution.

Example 3
In town A, 42% of the population are Conservative voters and in town B, 18% of the population are Conservative voters. Given that 350 people are selected from town A, and 150 people are selected from town B, find the probability that more than 180 Conservative voters are selected altogether.

If X and Y are the numbers of Conservative voters selected from towns A and B

respectively, then

X is B(350, 0.42), which is approximately Normal,
with mean $350 \times 0.42 = 147$
and variance $350 \times 0.42 \times 0.58 = 85.26$

Y is B(150, 0.18), which is approximately Normal,
with mean $150 \times 0.18 = 27$
and variance $150 \times 0.18 \times 0.82 = 22.14$

Hence $(X + Y)$ is approximately Normal, with mean $147 + 27 = 174$
and variance $85.26 + 22.14 = 107.4$

(assuming that X and Y are independent).

$$P(X + Y > 180) \approx P(X + Y > 180.5 \text{ in Normal})$$

$$= P\left(Z > \frac{180.5 - 174}{\sqrt{107.4}}\right)$$

$$= P(Z > 0.627)$$

$$= 0.2653$$

Exercise 11.2   The sum of Binomial distributions

1   W, X, Y are independent random variables;
W is B(3, 0.2), X is B(5, 0.2) and Y is B(8, 0.2)
State the distribution of $W + X + Y$ and find $P(W + X + Y = 4)$

2   Tickets for an important football match and an important cricket match are issued randomly and independently to one quarter of those who apply for them. A group of 10 people all apply for a ticket for each match. What is the distribution of
(i)   the number of football match tickets obtained by this group?
(ii)   the number of cricket match tickets obtained by the group?
(iii)   the total number of tickets obtained by the group?
(a)   Calculate the probability that exactly four tickets are obtained altogether.
(b)   Calculate the probability that at least two people obtain tickets for both matches.
   (*Hint: This part of the question has nothing to do with parts (i), (ii), (iii) or (a).*)

3   The University of Swindletown Examination Board awards O level passes to 70% of its candidates, quite randomly and without regard to performance in the examination (in fact it does not even bother to get the scripts marked!). For a class of 12 pupils 8 enter O level mathematics, 5 enter O level English, and 2 enter O level history.
State the distribution of
(i)   the number of passes in mathematics
(ii)   the number of passes in English
(iii)   the number of passes in history
(iv)   the total number of passes.
Calculate the probability that the class will obtain at least 13 O level passes.

4 X and Y are independent random variables.
  (i) Suppose that X is B$(1,\frac{1}{5})$ and Y is B$(2,\frac{1}{5})$

  Write out the distributions (i.e. list the possible values, and their probabilities) of X and of Y. From these distributions, obtain the distribution of $(X + Y)$.
  Write out the distribution B$(3,\frac{1}{5})$ and hence verify that $(X + Y)$ is B$(3,\frac{1}{5})$
  (ii) Suppose now that X is B$(1,\frac{1}{2})$ and Y is B$(2,\frac{1}{5})$.
  Obtain the distribution of $(X + Y)$, and show that this is *not* a Binomial distribution.

5 The probability that a man is left-handed is 0.2, and the corresponding probability for a woman to be left-handed is 0.15. If 300 men and 200 women are selected at random, find the mean and standard deviation of
  (i)   the number of left-handed men
  (ii) the number of left-handed women
  (iii) the total number of left-handed people.
  Using the Normal approximation, estimate the probability that there are more than 100 left-handed people altogether.

11.3   Distribution of the sample proportion

Consider a population in which a (generally unknown) proportion θ possesses some characteristic. If a random sample of size n is selected from the population, and X members of the sample possess the characteristic, then the sample proportion, X/n, may be used to estimate the value of θ.
  Now X has the Binomial distribution B(n,θ) so

$$E[X] = n\theta \text{ and var } (X) = n\theta(1 - \theta)$$

hence $\dfrac{X}{n}$ has mean $E\left[\dfrac{X}{n}\right] = \dfrac{1}{n} E[X] = \dfrac{1}{n} n\theta = \theta$

and variance $\text{var}\left(\dfrac{X}{n}\right) = \dfrac{1}{n^2} \text{var}(X) = \dfrac{1}{n^2} n\theta(1 - \theta) = \dfrac{\theta(1 - \theta)}{n}$

We shall assume that conditions are such that X has an approximately Normal distribution; then so does the sample proportion $\dfrac{X}{n}$, i.e.

> The sample proportion $\dfrac{X}{n}$ is approximately Normal,
> with mean θ and standard deviation $\sqrt{\dfrac{\theta(1 - \theta)}{n}}$

$\sqrt{\dfrac{\theta(1 - \theta)}{n}}$ is called the 'standard error of proportion'.

Example 1
In a certain country, 54% of the population are women.

(i) If a random sample of 450 people is selected, find the probability that the proportion of women in the sample is less than $\frac{1}{2}$.

(ii) For what size of sample is there a probability 0.01 that the proportion of women in the sample is less than $\frac{1}{2}$?

If a random sample of n people contains X women, then the proportion of women in the sample, X/n, is approximately Normal, with mean $\theta = 0.54$ and standard deviation

$$\sqrt{\frac{\theta(1-\theta)}{n}} = \sqrt{\frac{0.54 \times 0.46}{n}}$$

(i) If $n = 450$, X/n has mean 0.54

and standard deviation $\sqrt{\dfrac{0.54 \times 0.46}{450}} = 0.02349$

$$P\left(\frac{X}{n} < 0.5\right) = P\left(Z < \frac{0.5 - 0.54}{0.02349}\right)$$

$$= P(Z < -1.7025)$$

$$= 1 - 0.9557$$

$$= 0.0443$$

(ii) If $P(X/n < 0.5) = 0.01$, then 0.5 corresponds to $z = -2.326$. Thus

$$\frac{0.5 - 0.54}{\sqrt{\dfrac{0.54 \times 0.46}{n}}} = -2.326$$

$$\frac{n}{0.54 \times 0.46} = \left(\frac{2.326}{0.04}\right)^2$$

$$n = 839.9$$

i.e. a sample of 840 people is needed.

Now suppose that the proportion θ in the population is not known. For 95% of samples, we have

$$\theta - 1.96\sqrt{\frac{\theta(1-\theta)}{n}} < \frac{X}{n} < \theta + 1.96\sqrt{\frac{\theta(1-\theta)}{n}}$$

i.e. $\dfrac{X}{n}$ differs from θ by less than $1.96\sqrt{\dfrac{\theta(1-\theta)}{n}}$

i.e. $\left|\dfrac{X}{n} - \theta\right| < 1.96\sqrt{\dfrac{\theta(1-\theta)}{n}}$

Rearranging the inequalities, we obtain a 95% confidence interval for θ as:

$$\frac{X}{n} - 1.96\sqrt{\frac{\theta(1-\theta)}{n}} < \theta < \frac{X}{n} + 1.96\sqrt{\frac{\theta(1-\theta)}{n}}$$

Unfortunately the confidence limits

$$\frac{X}{n} \pm 1.96 \sqrt{\frac{\theta(1-\theta)}{n}}$$

contain the unknown quantity θ. However, it is reasonable to estimate $\theta \approx X/n$ to obtain approximate 95% confidence limits.

Example 2
In a random sample of 250 cars, 185 were British. Give
(i)   approximate 95% confidence limits, (ii) an approximate 99% lower confidence limit, for the true proportion of British cars.
(iii) What size sample would be needed in order to estimate the proportion of British cars, with 95% confidence, to within ± 0.02?

(i)   95% confidence limits for the true proportion θ are $\dfrac{X}{n} \pm 1.96 \sqrt{\dfrac{\theta(1-\theta)}{n}}$

We have $n = 250$, $X = 185$, and estimating $\theta \approx \dfrac{X}{n} = \dfrac{185}{250} = 0.74$, approximate 95% confidence limits are

$$0.74 \pm 1.96 \sqrt{\frac{0.74 \times 0.26}{250}} = 0.74 \pm 0.054$$

$$= 0.686, \ 0.794.$$

We can be 95% confident that $0.686 < \theta < 0.794$
(ii) The 99% lower confidence limit for θ is

$$\frac{X}{n} - 2.326 \sqrt{\frac{\theta(1-\theta)}{n}}$$

$$\approx 0.74 - 2.326 \sqrt{\frac{0.74 \times 0.26}{250}}$$

$$= 0.675$$

We can be 99% confident that $\theta > 0.675$
(iii) The 95% confidence limits for θ are $\dfrac{X}{n} \pm 1.96 \sqrt{\dfrac{\theta(1-\theta)}{n}}$

so we require

$$1.96 \sqrt{\frac{\theta(1-\theta)}{n}} = 0.02$$

Estimating $\theta \approx 0.74$,

$$1.96 \sqrt{\frac{0.74 \times 0.26}{n}} = 0.02$$

$$\frac{n}{0.74 \times 0.26} = \left(\frac{1.96}{0.02}\right)^2$$

$$n = 1847.8$$

A sample of about 1850 cars would be needed.

Example 3

It is required to estimate the proportion θ of homes having a video recorder, to within ± 0.01 with 95% confidence.

In a random sample of n homes, there were X homes with a video recorder.

Show that X/n differs from θ by less than 0.01, with probability 95%, provided that $n > 38416\,\theta(1-\theta)$.

Find the smallest sample size for which this condition will be satisfied whatever the value of θ.

With probability 95%, X/n differs from θ by less than $1.96\sqrt{\dfrac{\theta(1-\theta)}{n}}$

so we need $1.96\sqrt{\dfrac{\theta(1-\theta)}{n}} < 0.01$

$$\frac{n}{\theta(1-\theta)} > \left(\frac{1.96}{0.01}\right)^2$$

$$n > 38416\,\theta(1-\theta)$$

Now consider $f(\theta) = \theta(1-\theta) = \theta - \theta^2$

$$f'(\theta) = 1 - 2\theta$$

$$f''(\theta) = -2$$

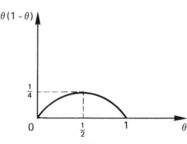

Fig. 11.1

so $f(\theta) = \theta(1-\theta)$ has a maximum value when $1 - 2\theta = 0$

$$\theta = \tfrac{1}{2}$$

and the maximum value is $\theta(1-\theta) = \tfrac{1}{2} \times \tfrac{1}{2} = \tfrac{1}{4}$

Thus, if $n > 38416\,\theta(1-\theta)$ for all possible values of θ, then

$$n > 38416 \times \tfrac{1}{4} = 9604$$

A sample of at least 9604 homes is needed.

Exercise 11.3   Confidence intervals for a proportion

1   A certain species of moth occurs in light and dark forms, and it is known that 30% of moths are light. Given that a sample of n moths contains X light ones state the mean and standard deviation of the proportion X/n of light moths in the sample.
  (a) For a sample of 400 moths
    (i)   find the probability that the proportion of light moths in the sample is more than 0.33;
    (ii) find the two limits between which the sample proportion will lie with probability 0.95.
  (b) For a sample of n moths, state the two limits between which the sample proportion X/n will lie with probability 0.95. Deduce the size of sample which will ensure, with probability 0.95, that X/n differs from 0.3 by less than 0.02.

2 A random sample of 200 people contains 42 smokers.
Obtain approximate 95% confidence limits for the true proportion of smokers in the population.

3 A manufacturer found that in a random sample of 600 watches, 54 required to be replaced during the guarantee period. Give approximate 99% confidence limits for the proportion of watches which need to be replaced under guarantee. What size sample is needed if the manufacturer wants to find this proportion, with 99% confidence, to within ± 0.01.

4 In a random sample of 1400 items produced by a machine, there were 140 defective items. Obtain an approximate 95% upper confidence limit for the proportion of defective items.

5 An opinion poll is carried out by interviewing n people, and it is desired to give, with 95% confidence, error bounds for the proportion of people holding certain views. Using the fact that, with 95% probability, the sample proportion X/n differs from the true proportion θ by less than $1.96\sqrt{\dfrac{\theta(1-\theta)}{n}}$ and by considering the maximum value of $\theta(1-\theta)$, show that these error bounds can be given as $\pm 1.96\sqrt{\dfrac{1}{4n}}$, whatever the value of θ.
(i) Find the error bounds when $n = 1000$
(ii) What size sample is required if the error bounds are to be ± 0.005?

12
The Poisson Distribution

12.1 A limit of Binomial distributions

The following table shows the probabilities that $X = 0, 1, 2, 3, 4, 5$ in various Binomial distributions $B(n, p)$, each of which has mean $np = 2$. For example, in $B(250, 0.008)$,

$$P(X = 3) = \binom{250}{3} (0.008)^3 (0.992)^{247}$$

$$= \frac{250 \times 249 \times 248}{1 \times 2 \times 3} (0.008)^3 (0.992)^{247}$$

$$= 0.1812$$

X	0	1	2	3	4	5
probability in						
B(50, 0.04)	0.1299	0.2706	0.2762	0.1842	0.0902	0.0346
B(100, 0.02)	0.1326	0.2707	0.2734	0.1823	0.0902	0.0353
B(250, 0.008)	0.1343	0.2707	0.2718	0.1812	0.0902	0.0358
B(1000, 0.002)	0.1351	0.2707	0.2709	0.1806	0.0902	0.0360
B(5000, 0.0004)	0.1353	0.2707	0.2707	0.1805	0.0902	0.0361
B(20000, 0.0001)	0.1353	0.2707	0.2707	0.1805	0.0902	0.0361
B(200000, 0.00001)	0.1353	0.2707	0.2707	0.1804	0.0902	0.0361

The probabilities appear to approach limiting values as n becomes larger and larger.

In general, consider the limit of the Binomial distribution $B(n, p)$ as $n \to \infty$ in such a way that the mean $\mu = np$ remains constant.

$$\text{Then } p = \frac{\mu}{n} \to 0. \text{ Let } p_r = P(X = r).$$

We have

$$p_0 = P(X = 0) = (1 - p)^n = \left(1 - \frac{\mu}{n}\right)^n \to e^{-\mu} \text{ as } n \to \infty$$

$$\text{(using the result that } \left(1 + \frac{x}{n}\right)^n \to e^x \text{ as } n \to \infty \text{ with } x = -\mu)$$

We have seen that, for the Binomial distribution B(n, p),

$$\frac{p_r}{p_{r-1}} = \frac{(n-r+1)p}{r(1-p)}$$

$$= \frac{np-(r-1)p}{r(1-p)}$$

$$= \frac{\mu-(r-1)p}{r(1-p)}$$

$$\to \frac{\mu}{r} \qquad \text{as } n \to \infty \text{ and } p \to 0$$

Thus the limiting distribution is given by

$$p_0 = e^{-\mu} \qquad \text{and } p_r = p_{r-1} \times \frac{\mu}{r},$$

i.e.

$$p_1 = p_0 \times \frac{\mu}{1} = e^{-\mu}\mu$$

$$p_2 = p_1 \times \frac{\mu}{2} = e^{-\mu}\mu \times \frac{\mu}{2} = e^{-\mu}\frac{\mu^2}{2!}$$

$$p_3 = p_2 \times \frac{\mu}{3} = e^{-\mu}\frac{\mu^2}{2!} \times \frac{\mu}{3} = e^{-\mu}\frac{\mu^3}{3!} \text{ and so on.}$$

Since B(n, p) has mean $np = \mu$
and variance $np(1-p) = \mu(1-p) \to \mu$,
we would expect this limiting distribution to have mean μ and variance μ.
We shall prove this later.

We say that a random variable X has the *Poisson distribution* with mean μ, (and written Poisson (μ)), if X has possible values $0, 1, 2, 3, \ldots$ and

$$\boxed{P(X = r) = e^{-\mu}\frac{\mu^r}{r!}}$$

i.e. the distribution is

X	0	1	2	3	...	r	...
Probability	$e^{-\mu}$	$e^{-\mu}\mu$	$e^{-\mu}\frac{\mu^2}{2!}$	$e^{-\mu}\frac{\mu^3}{3!}$		$e^{-\mu}\frac{\mu^r}{r!}$	

Example 1
The number of letters received at a house on a weekday has the Poisson distribution with mean 1.5. Find the probability that on a particular day,
(i) no letters are received
(ii) exactly two letters are received

(iii) less than three letters are received
(iv) at least four letters are received.

The number of letters, X, is Poisson (1.5)
(i) $P(X = 0) = e^{-1.5} = 0.2231$

(ii) $P(X = 2) = e^{-1.5} \times \dfrac{1.5^2}{2!} = 0.2510$

(iii) $P(X < 3) = P(X = 0) + P(X = 1) + P(X = 2)$

$$= 0.2231 + e^{-1.5} \times 1.5 + 0.2510$$

$$= 0.2231 + 0.3347 + 0.2510$$

$$= 0.8088$$

(iv) $P(X \geqslant 4) = 1 - \{P(X = 0) + P(X = 1) + P(X = 2) + P(X = 3)\}$

$$= 1 - \left(0.2231 + 0.3347 + 0.2510 + e^{-1.5} \times \frac{1.5^3}{3!} \right)$$

$$= 0.0656$$

Example 2
A garage finds that the number, X, of customers wishing to hire a car on any particular day has the Poisson distribution with mean 2.
  Find the probabilities that $X = 0, 1, 2, 3, 4, 5, 6$ and calculate the cumulative probabilities

$$P(X \leqslant 0), P(X \leqslant 1), P(X \leqslant 2), ..., P(X \leqslant 6).$$

(a) If the garage has 4 cars available for hire, find the probability that, on one particular day,
   (i)   not all the cars are hired
   (ii) all the cars are hired
   (iii) the garage is unable to satisfy all the demands for hired cars.
(b) If the garage has 4 cars available, calculate the mean number of cars out on hire.
(c) How many cars should be available for hire if the probability of being unable to satisfy the demand is to be less than 2% ?

X is Poisson (2)

$P(X = 0)$	$= e^{-2}$		$= 0.1353$	$P(X \leqslant 0)$	$= 0.1353$
$P(X = 1)$	$= e^{-2} \times 2$	$= P(X = 0) \times 2$	$= 0.2707$	$P(X \leqslant 1)$	$= 0.4060$
$P(X = 2)$	$= e^{-2} \times \dfrac{2^2}{2!}$	$= P(X = 1) \times \dfrac{2}{2}$	$= 0.2707$	$P(X \leqslant 2)$	$= 0.6767$
$P(X = 3)$	$= e^{-2} \times \dfrac{2^3}{3!}$	$= P(X = 2) \times \dfrac{2}{3}$	$= 0.1804$	$P(X \leqslant 3)$	$= 0.8571$
$P(X = 4)$	$= e^{-2} \times \dfrac{2^4}{4!}$	$= P(X = 3) \times \dfrac{2}{4}$	$= 0.0902$	$P(X \leqslant 4)$	$= 0.9473$

$$P(X = 5) \qquad = e^{-2} \times \frac{2^5}{5!} \qquad = P(X = 4) \times \frac{2}{5} \qquad = 0.0361 \qquad P(X \leqslant 5) \quad = 0.9834$$

$$P(X = 6) \qquad = e^{-2} \times \frac{2^6}{6!} \qquad = P(X = 5) \times \frac{2}{6} \qquad = 0.0120 \qquad P(X \leqslant 6) \quad = 0.9955$$

Note   Although we have recorded these values correct to 4 decimal places, each probability has been calculated from the previous one without clearing the calculator.

The cumulative probabilities are obtained by summing the probabilities in the memory of the calculator, i.e. by pressing the $\boxed{M+}$ key after calculating each probability. The memory may then be read, and the latest probability retrieved, by pressing the memory exchange key $\boxed{M \leftrightarrow X}$ twice, or, if this key is not available, by a sequence such as $\boxed{X \leftrightarrow Y}$, $\boxed{RM}$, $\boxed{X \leftrightarrow Y}$

Note that the probabilities that $X = 0, 1, 2, 3, 4, 5$ agree with the limiting values obtained for the Binomial distributions with mean 2.

(a) (i)   P(not all cars are hired) = $P(X \leqslant 3) = 0.8571$
   (ii)   P(all cars are hired) = $P(X \geqslant 4) = 1 - 0.8571 = 0.1429$
   (iii) P(unable to satisfy demand) = $P(X > 4) = 1 - P(X \leqslant 4)$

$$= 1 - 0.9473$$

$$= 0.0527$$

(b) If Y cars are out on hire, then the distribution of Y is

Y	0	1	2	3	4
Probability	0.1353	0.2707	0.2707	0.1804	0.1429

$$E[Y] = 0 \times 0.1353 + 1 \times 0.2707 + \ldots + 4 \times 0.1429 = 1.925$$

(c) If the garage has n cars available, then it is unable to satisfy the demand when $X > n$; so we require $P(X > n) < 0.02$
   i.e. $P(X \leqslant n) > 0.98$

Since $P(X \leqslant 4) = 0.9473$ and $P(X \leqslant 5) = 0.9834$, the garage should have at least 5 cars available for hire.

The Poisson approximation to the Binomial distribution

We have seen that the Poisson distribution, Poisson (μ) arises as the limiting distribution of $B(n, p)$, where $np = \mu$, as $n \to \infty$.

Hence, if n is sufficiently large,

> $B(n, p)$ can be approximated by Poisson (np).
> The approximation will be fairly good if $n > 50$ and $p < 0.1$

Example 3

A manufacturer finds that 3% of items produced by a machine are defective. If a random

sample of 150 items is taken, find the probability that it contains more than 5 defective items.

The number of defective items, X, is B(150, 0.03)
This has mean $150 \times 0.03 = 4.5$, so X is approximately Poisson (4.5)

$$P(X > 5) \approx 1 - \{P(X = 0) + P(X = 1) + \cdots + P(X = 5)\}$$

$$= 1 - \left\{ e^{-4.5} + e^{-4.5} \times 4.5 + \cdots + e^{-4.5} \times \frac{4.5^5}{5!} \right\}$$

$$= 1 - \{0.0111 + 0.0500 + 0.1125 + 0.1687 + 0.1898 + 0.1708\}$$

$$= 1 - 0.7029$$

$$= 0.2971$$

Note   The true value is

$$P(X > 5) = 1 - \left\{ (0.97)^{150} + 150 \, (0.03)(0.97)^{149} + \cdots + \binom{150}{5}(0.03)^5 (0.97)^{145} \right\}$$

$$= 1 - \{0.0104 + 0.0481 + 0.1108 + 0.1691 + 0.1922 + 0.1736\}$$

$$= 1 - 0.7043$$

$$= 0.2957$$

but the Poisson approximation is easier to calculate.

Exercise 12.1   The Poisson distribution

1   The number of pips in an orange has a Poisson distribution with mean 3. Find the probability that an orange contains:
  (i)   no pips   (ii)   one pip   (iii)   two pips   (iv)   three pips
  (v)   four pips   (vi)   less than two pips   (vii)   at least three pips.

2   In a large book, the number of misprints per page is a Poisson variable with mean 1. If a page is selected at random, find the probability that it contains:
  (i)   exactly two misprints   (ii)   at least two misprints.

3   A shop finds that, in one day, the number of customers wishing to buy a certain colour television set is a Poisson variable with mean 1.2. Find the probabilities that, on some particular day
  (i)   no customers wish to buy that model television
  (ii) exactly three customers wish to buy one.
    There are always 3 models available in the shop at the beginning of the day, and no more can be obtained until the beginning of the next day. Find the probability that the shop will be able to satisfy all the demands for that set on any particular day.
    A new manager takes over, and he decides that he must have a large enough stock at the beginning of each day so that there is at least 99.9% probability that he can satisfy all the demands for that set.

By calculating the cumulative probabilities $P(X \leq 0)$, $P(X \leq 1)$, $P(X \leq 2),\ldots$ (where X is the number of customers wishing to buy the set), determine how many sets should be stocked each day.

4 Find the probability that $X = 3$ in the following Binomial distributions

$$B(50, 0.08), \quad B(100, 0.04), \quad B(400, 0.01), \quad B(2000, 0.002), \quad B(10000, 0.0004)$$

What is the mean of these Binomial distributions? Find the probability that $X = 3$ in the Poisson distribution with this mean.

5 An insurance company insures 750 oil-tankers for one year, and estimates that for any particular ship, the probability that a claim will arise is 0.0032. Use the Poisson approximation to find the probability that the company receives (i) no claims (ii) exactly two claims.

6 A manufacturer of calculators finds that 2% of the calculators produced are defective. Given that a random sample of 250 calculators is selected, find the probability that this sample contains
(i) exactly two defective calculators
(ii) less than three defective calculators.

7 A school has 1151 pupils (none of whom has a birthday on 29th February). If a date (e.g. 17th September) is chosen at random, find the probability that at least 5 pupils will have their birthday on that date.

8 The same school employs 158 teachers. The probability that any particular teacher is away on a Monday is 0.02. Find the probability that the number of teachers away on a Monday is:
(i) exactly four   (ii) between two and five (inclusive).

9 Eggs are supplied to a supermarket in boxes of 60, and on average 5% of the eggs are cracked. Find the probability that a box contains exactly 4 cracked eggs
(i) directly from the Binomial distribution $B(60, 0.05)$
(ii) using the Poisson approximation to the Binomial distribution.

12.2 Poisson distributions for random events

Consider events which occur randomly and independently, with a constant average rate.

This means that the probability that an event occurs during a given time interval is unaffected by what has occurred previously; and the mean number of occurrences in a time interval is proportional to the length of the interval.

For example, suppose that a small mass of radioactive material emits α-particles at random with average rate 120 α-particles per minute. Let X be the number of α-particles emitted during a given 2 second interval.

Since the mean number of α-particles emitted is proportional to the length of the interval, and 2 seconds is $\frac{1}{30}$ minute, the mean number emitted during a 2 second interval is $\frac{1}{30} \times 120 = 4$. Thus $E[X] = 4$.

Divide the 2 second interval into 10 separate parts, each of length $\frac{1}{5}$ second. The diagram illustrates 3 emissions, occurring during the second, fifth and sixth parts.

If we ignore the possibility that more than one α-particle may be emitted during a $\frac{1}{5}$ second part, then the total number of emissions, X, is approximately B(10, p) where p is the probability that an α-particle is emitted during any one of these $\frac{1}{5}$ second parts.

(Note that the conditions for random events ensure that p is constant for all ten parts, and that emissions during the 10 parts occur independently).

Since E[X] = 4, we have $p = 0.4$, and X is approximately B(10, 0.4). This is only an approximation for the distribution of X, because we have neglected the possibility that two (or more) emissions might occur during a $\frac{1}{5}$ second interval.

If we now divide the 2 second interval into 100 separate parts, each of length $\frac{1}{50}$ second, we shall find that X is approximately B(100, 0.04). This will be a better approximation since it is clearly less likely that there will be two (or more) emissions during a $\frac{1}{50}$ second interval.

Similarly, by dividing the 2 second interval into 1000 parts, we shall find that X is approximately B(1000, 0.004), and this is an even better approximation. Hence the true distribution of X is the limit of the Binomial distributions B(10, 0.4), B(100, 0.04), B(1000, 0.004), ..., which is the distribution Poisson (4).

In general, for a random process (as defined above)

> The number of events occurring in a fixed time interval has a Poisson distribution with appropriate mean

(i.e. the mean number of occurrences corresponding to the length of the given interval).

Example 1

Telephone calls are received by a switchboard at a constant average rate of 2 calls per minute. Calls may be considered to occur at random times and independently of each other. Find the probability that

(i)   exactly 5 calls are received in a 2 minute interval
(ii) no calls are received in a 30 second interval
(iii) at least one call is received in a 10 second interval.

(i)   For a 2 minute interval, the mean number of calls is 4, so the actual number of calls received, X, is Poisson (4)

$$P(X = 5) = e^{-4} \times \frac{4^5}{5!} = 0.1563$$

(ii) For a 30 second interval, the mean number of calls is 1, so the actual number of calls, X, is Poisson (1)

$$P(X = 0) = e^{-1} = 0.3679$$

(iii) For a 10 second interval, the mean number of calls is $\frac{1}{3}$, so the actual number of calls, X, is Poisson $(\frac{1}{3})$

$$P(X \geqslant 1) = 1 - P(X = 0) = 1 - e^{-1/3} = 0.2835$$

Exercise 12.2   Poisson distributions for random events

In the following questions it may be assumed that the events described occur randomly and independently with a constant average rate, so that the Poisson distribution may be used.

1   At a certain cross-roads, accidents occur at an average rate of one every two days. Find the probability that
(i)   there is more than one accident on a particular day
(ii) there are less than three accidents in a given week.

2   Meteorites strike the Earth's surface at an average rate of 73 per year (365 days). Find the probability that at least 4 meteorites will strike the surface during a given 14 day period.

3   In a small town, deaths occur at an average rate of 2 per week. Find the probability that
(i)   no-one dies on a given day
(ii) more than two deaths occur on a given week-end (Saturday and Sunday).

4   Customers enter a bank at an average rate of 40 per hour. Find the probability that
(i)   exactly three customers enter during a given minute
(ii) the interval between two successive customers is more than 5 minutes.
   Hint: This is the same as saying that no customers enter during a five minute period.

5   A cathode ray gun emits electrons at a rate of 4.2×10^9 electrons per second. Find the probability that, in 1 nanosecond (10^{-9} second), less than two electrons are emitted.

6   During the manufacture of sheet metal, imperfections occur randomly at an average rate of one per 5 square metres. Find the probability that
(i)   a sheet with area 12 m^2 is free of imperfections
(ii) a sheet with area 8 m^2 has more than three imperfections.

7   Cars pass a certain point on a road at an average rate of 2 per minute.
(i)   Find the probability that exactly 6 cars pass in a 3 minute interval.
(ii) For an interval of t minutes, write down in terms of t
   (a) the mean number of cars which pass the point;
   (b) the probability that at least one car passes by. Deduce the length of the interval for which the probability that at least one car passes by is 0.99.

12.3 Mean, variance and mode of the Poisson distribution

The Poisson (μ) distribution is defined as follows:

X	0	1	2	...	r	...
Probability	$e^{-\mu}$	$e^{-\mu}\mu$	$e^{-\mu}\dfrac{\mu^2}{2!}$		$e^{-\mu}\dfrac{\mu^r}{r!}$	
	p_0	p_1	p_2	...	p_r	...

We shall now prove formally that this does define a probability distribution, with mean $E[X] = \mu$ and variance $\mathrm{var}(X) = \mu$

Remember that the series expansion for e^x is

$$e^x = 1 + x + \frac{x^2}{2!} + \frac{x^3}{3!} + \cdots \qquad \text{(which is valid for all values of } x\text{)}.$$

The conditions necessary for a probability distribution are $p_r \geqslant 0$ (which is clearly true) and $\sum\limits_{r=0}^{\infty} p_r = 1$

$$\text{Now } \sum_{r=0}^{\infty} p_r = e^{-\mu} + e^{-\mu}\mu + e^{-\mu}\frac{\mu^2}{2!} + e^{-\mu}\frac{\mu^3}{3!} + \cdots$$

$$= e^{-\mu}\left(1 + \mu + \frac{\mu^2}{2!} + \frac{\mu^3}{3!} + \cdots\right)$$

$$= e^{-\mu}e^{\mu}$$

$$= 1$$

hence we do have a proper probability distribution.

The mean, $E[X] = \sum\limits_{r=0}^{\infty} rp_r$

$$= 0 \times p_0 + 1 \times p_1 + 2 \times p_2 + 3 \times p_3 + \cdots$$

$$= 0 + e^{-\mu}\mu + 2e^{-\mu}\frac{\mu^2}{2!} + 3 e^{-\mu}\frac{\mu^3}{3!} + \cdots$$

$$= \mu\, e^{-\mu}\left(1 + \mu + \frac{\mu^2}{2!} + \cdots\right)$$

$$= \mu e^{-\mu}e^{\mu}$$

$$= \mu$$

For the variance, we first consider

$$\sum_{r=0}^{\infty} r(r-1)p_r = 0 + 0 + 2 \times 1 \times p_2 + 3 \times 2 \times p_3 + 4 \times 3 \times p_4 + \cdots$$

$$= 2 \times 1 \times e^{-\mu} \frac{\mu^2}{2!} + 3 \times 2 \times e^{-\mu} \frac{\mu^3}{3!} + 4 \times 3 \times e^{-\mu} \frac{\mu^4}{4!} + \cdots$$

$$= \mu^2 e^{-\mu} \left(1 + \mu + \frac{\mu^2}{2!} + \cdots \right)$$

$$= \mu^2 e^{-\mu} e^{\mu}$$

$$= \mu^2$$

Then $E[X^2] = \sum_{r=0}^{\infty} r^2 p_r = \sum_{r=0}^{\infty} \{r(r-1) + r\} p_r$

$$= \sum_{r=0}^{\infty} r(r-1)p_r + \sum_{r=0}^{\infty} rp_r$$

$$= \mu^2 + \mu$$

and $\text{var}(X) = E[X^2] - \mu^2$

$$= (\mu^2 + \mu) - \mu^2$$

$$= \mu$$

The mode

Consider

$$\frac{p_r}{p_{r-1}} = \frac{e^{-\mu} \dfrac{\mu^r}{r!}}{e^{-\mu} \dfrac{\mu^{r-1}}{(r-1)!}} = \frac{\mu^r}{r!} \times \frac{(r-1)!}{\mu^{r-1}} = \frac{\mu}{r}$$

Thus

$$p_r = p_{r-1} \times \frac{\mu}{r}$$

While $r < \mu$, p_r is greater than the previous probability p_{r-1}, and the sequence $p_0, p_1, p_2, \ldots$, is increasing; when $r > \mu$, p_r is less than p_{r-1}.

Hence the maximum probability p_r occurs when r is just less than μ i.e.

> The mode of Poisson (μ) is the greatest integer less than μ.

If μ is an integer, then when $r = \mu$ we have $p_\mu = p_{\mu-1}$, so there are two modes, $\mu - 1$ and μ.

Example 1
The mode of Poisson (7.3) is 7.
The mode of Poisson (2.98) is 2.
The modes of Poisson (6) are 5 and 6.

Fitting a Poisson distribution to data

If we have a sample which is thought to come from a Poisson distribution, we can check this by calculating the probabilities, and hence the expected frequencies, for a Poisson distribution having the same mean as that of the sample.

For a Poisson distribution, the mean and variance are equal. So, as an additional check, we usually calculate the variance of the sample. If the mean and variance of the sample are approximately equal, this supports the view that the sample comes from a Poisson distribution (although it is important to realize that this does not, in itself, prove that it does). However, if the mean and variance of the sample are very different then it is unlikely that a Poisson distribution will fit the data very well.

Example 2

The numbers of life insurance policies sold by a salesman over a period of 50 weeks were as follows

Number of policies sold per week, X	0	1	2	3	4	5	6	7
Number of weeks	3	7	13	11	8	5	2	1

Fit a Poisson distribution to this data.

The sample has mean $\bar{x} = 2.84$ and variance $s^2 = 2.5744$.
Since the mean and variance are approximately equal, this sample could be taken from a Poisson distribution.

We now calculate the expected frequencies for Poisson (2.84); for example,

$$P(X = 2) = e^{-2.84} \frac{(2.84)^2}{2!} = 0.2356$$

so the expected frequency for $X = 2$ is $0.2356 \times 50 = 11.8$.

Although the sample contained no values higher than $X = 7$, the theoretical distribution Poisson (2.84) can take values higher than 7. We therefore change the final 'cell' to $X \geq 7$, so that the probabilities add up to one and the expected frequencies add up to 50. $(\text{P} X \geq 7)$ is calculated as 'one minus the sum of the other probabilities').

X	0	1	2	3	4	5	6	≥ 7
Probability from Poisson (2.84)	0.0584	0.1659	0.2356	0.2231	0.1584	0.0900	0.0426	0.0261
Expected frequency	2.9	8.3	11.8	11.2	7.9	4.5	2.1	1.3

These expected frequencies are close to the actual frequencies; the Poisson distribution fits the data well.

Exercise 12.3   The mean, variance and mode of a
·   Poisson distribution

1   Find the probability that X lies within one standard deviation of its mean when
    (i) X is Poisson (6)   (ii) X is Poisson (1.5)

2   Write down the modes of the distributions
    (i)   Poisson (7.2)   (ii)   Poisson (3.95)   (iii)   Poisson (5)

3   At a large illumination display, an average of 32 lamps have to be replaced each week.
    What is the most likely number of lamps needing replacement
    (i)   in one day   (ii)   in July (31 days)?

4   The following table gives the number of goals scored in 80 football matches.

Number of goals	0	1	2	3	4	5	6	7
Number of matches	8	18	22	14	8	5	4	1

Find the mean number of goals. For a Poisson distribution with this mean calculate the probabilities of obtaining 0, 1, 2, 3, 4, 5, 6 and $\geqslant 7$ goals.
    Calculate the 'expected frequencies' for this Poisson distribution (i.e. the probabilities multiplied by 80), and compare these with the actual frequencies.

5   The number of minor faults detected at the final factory inspection of 250 new cars were as follows

Number of faults	0	1	2	3	4	5	6	7	8	9	10
Number of cars	1	15	25	43	50	41	35	18	12	6	4

Calculate the mean and variance of the number of faults. Why does this suggest that a Poisson distribution might fit the distribution?
    Find the expected frequencies for a Poisson distribution having the same mean.

12.4   The sum of two independent Poisson distributions

Suppose that, on a Sunday afternoon, an AA office receives requests for assistance from motorists at an average rate of 4 per hour. Let X be the number of requests received between 1 p.m. and 2 p.m., and let Y be the number of requests received between 2 p.m. and 4 p.m.

If we can regard requests for assistance as random events, the distribution of X is Poisson (4), and the distribution of Y is Poisson (8).

X and Y are independent random variables. Their sum $X + Y$ is the number of requests received between 1 p.m. and 4 p.m., so the distribution of $(X + Y)$ is Poisson (12).

This suggests the following general result:

> If X is Poisson (μ), Y is Poisson (λ), and X, Y are independent, then $(X + Y)$ is Poisson $(\mu + \lambda)$

We shall now prove this directly, working from the distributions of X and Y. X and Y have possible values $0, 1, 2, 3, \ldots$

Let $W = X + Y$. Then W has possible values $0, 1, 2, 3, \ldots$

$P(W = 0) = P(X = 0 \text{ and } Y = 0) = P(X = 0) \times P(Y = 0)$ since X, Y are independent.

$$= e^{-\mu} e^{-\lambda}$$

$$= e^{-(\mu+\lambda)}$$

$P(W = 1) = P(X = 1 \text{ and } Y = 0) + P(X = 0 \text{ and } Y = 1)$

$$= e^{-\mu}\mu\, e^{-\lambda} + e^{-\mu}\, e^{-\lambda}\lambda$$

$$= e^{-(\mu+\lambda)}(\mu + \lambda)$$

$P(W = 2) = P(X = 2 \text{ and } Y = 0) + P(X = 1 \text{ and } Y = 1) + P(X = 0 \text{ and } Y = 2)$

$$= e^{-\mu}\frac{\mu^2}{2!}\, e^{-\lambda} + e^{-\mu}\mu\, e^{-\lambda}\lambda + e^{-\mu}e^{-\lambda}\frac{\lambda^2}{2!}$$

$$= e^{-(\mu+\lambda)}\frac{1}{2!}\,(\mu^2 + 2\mu\lambda + \lambda^2)$$

$$= e^{-(\mu+\lambda)}\frac{(\mu + \lambda)^2}{2!}$$

$P(W = r) = P(X = r \text{ and } Y = 0) + P(X = r - 1 \text{ and } Y = 1) + P(X = r - 2 \text{ and } Y = 2) + \cdots$
$$+ P(X = 0 \text{ and } Y = r)$$

$$= e^{-\mu}\frac{\mu^r}{r!}\, e^{-\lambda} + e^{-\mu}\frac{\mu^{r-1}}{(r-1)!}\, e^{-\lambda}\lambda + e^{-\mu}\frac{\mu^{r-2}}{(r-2)!}\, e^{-\lambda}\frac{\lambda^2}{2!} + \cdots + e^{-\mu}e^{-\lambda}\frac{\lambda^r}{r!}$$

$$= e^{-(\mu+\lambda)}\frac{1}{r!}\left\{ \mu^r + \frac{r!}{(r-1)!1!}\,\mu^{r-1}\,\lambda + \frac{r!}{(r-2)!2!}\,\mu^{r-2}\,\lambda^2 + \cdots + \lambda^r \right\}$$

$$= e^{-(\mu+\lambda)}\frac{1}{r!}\left\{ \mu^r + \binom{r}{1}\mu^{r-1}\lambda + \binom{r}{2}\mu^{r-2}\lambda^2 + \cdots + \lambda^r \right\}$$

$$= e^{-(\mu+\lambda)}\frac{(\mu + \lambda)^r}{r!}$$

The probabilities that $W = 0, 1, 2, \ldots, r, \ldots$ are those given by a Poisson distribution with mean $(\mu + \lambda)$. Hence W is Poisson $(\mu + \lambda)$.

It follows that if $X_1, X_2, \ldots, X_n$ are independent Poisson variables with means $\mu_1, \mu_2, \ldots, \mu_n$ then their sum $(X_1 + X_2 + \cdots + X_n)$ has a Poisson distribution with mean $(\mu_1 + \mu_2 + \cdots + \mu_n)$

Example 1

An electrical shop sells television sets and video cassette recorders. The number of television sets sold per day has a Poisson distribution with mean 1, and the number of video cassette recorders sold per day has a Poisson distribution with mean 2.

Find the probability that, on a particular day, more than 3 items are sold altogether.

If X television sets and Y video cassette recorders are sold, then X is Poisson (1) and Y is Poisson (2); assuming that X, Y are independent, the total number of items sold, $T = X + Y$, is Poisson (3).

$$P(T > 3) = 1 - \{P(T = 0) + P(T = 1) + P(T = 2) + P(T = 3)\}$$

$$= 1 - \left\{ e^{-3} + e^{-3} \times 3 + e^{-3} \times \frac{3^2}{2!} + e^{-3} \times \frac{3^3}{3!} \right\}$$

$$= 0.3528$$

Exercise 12.4   The sum of two independent Poisson distributions

1   Chris and Leslie share a flat. The daily number of letters received by Chris is a Poisson variable with mean 2, and the number received by Leslie is an independent Poisson variable with mean 3. What is the distribution of the total number of letters delivered to the flat? On a given day find the probability that less than three letters are delivered.

2   At a certain point on a road, cars pass at the rate of 4 per minute, motorcycles at the rate of one every 5 minutes, and other vehicles (lorries, tractors etc) at the rate of one every 2 minutes. For a given 30 second interval, state the distributions of the numbers of cars, motorcycles and other vehicles, and deduce the distribution of the total number of vehicles.

Find the probability that at least 4 vehicles will pass during this interval.

3   The number of pearls in an oyster is a Poisson variable with mean 0.02. Given that n oysters are taken, write down in terms of n
(i)   the distribution of the total number of pearls
(ii)   the probability that at least one pearl is found.
How many oysters need be taken to ensure that the probability of finding at least one pearl is 0.95? With this number of oysters, calculate the probability of finding more than 4 pearls altogether.

12.5   The Normal approximation to the Poisson distribution

The Poisson distribution can be regarded as the limit of Binomial distributions. Since a Binomial distribution can sometimes be approximated by a Normal distribution, it is reasonable to expect that, under suitable conditions,

> Poisson (μ) can be approximated by a Normal distribution having the same mean μ and standard deviation $\sigma = \sqrt{\mu}$

We adopt the same criterion as for the Normal approximation to the Binomial distribution—that we should be able to go 3 standard deviations on each side of the mean without leaving the range of possible values, which for Poisson (μ) is 0 to infinity. Hence the approximation should be a good one if $\mu - 3\sqrt{\mu} > 0$

$$\text{i.e.} \qquad \mu > 3\sqrt{\mu}$$
$$\text{i.e.} \qquad \mu^2 > 9\mu$$
$$\text{i.e.} \qquad \mu > 9$$

The higher the value of the mean μ, the better will be the Normal approximation. The approximation can be used for lower values of μ, down to about $\mu = 5$, if the accuracy is not too critical.

Since the Poisson distribution is discrete, *the continuity correction must be used.*

Example 1

If X is Poisson (10), use the Normal approximation to find
(i) $P(X = 8)$
(ii) $P(8 \leqslant X \leqslant 11)$

Poisson (10) is approximately Normal, with mean 10 and standard deviation $\sqrt{10}$.
(i) $P(X = 8) \approx P(7.5 < X < 8.5 \text{ in Normal})$

$$= P\left(\frac{7.5 - 10}{\sqrt{10}} < Z < \frac{8.5 - 10}{\sqrt{10}}\right)$$
$$= P(-0.791 < Z < -0.474)$$
$$= (1 - 0.6822) - (1 - 0.7855)$$
$$= 0.1033$$

(ii) $P(8 \leqslant X \leqslant 11) \approx P(7.5 < X < 11.5 \text{ in Normal})$
$$= P(-0.791 < Z < 0.474)$$
$$= 0.6822 - (1 - 0.7855)$$
$$= 0.4677$$

Note The true values are

(i) $P(X = 8) = e^{-10}\dfrac{10^8}{8!} = 0.1126$

(ii) $P(8 \leqslant X \leqslant 11) = e^{-10}\dfrac{10^8}{8!} + \cdots + e^{-10}\dfrac{10^{11}}{11!}$

$$= 0.4766$$

Example 2

A tax office receives enquiries at an average rate of 270 per week of 5 working days. Assuming that enquiries may be regarded as random events, find the probability that
(i) more than 300 enquiries are received in a given week
(ii) more than 60 enquiries are received in a given day.

(i) The number, X, of enquiries received during a week is Poisson (270), which is approximately Normal with mean 270 and standard deviation $\sqrt{270}$

$P(X > 300) \approx P(X > 300.5$ in Normal)

$$= P\left(Z > \frac{300.5 - 270}{\sqrt{270}}\right)$$

$$= P(Z > 1.856)$$
$$= 0.0318$$

(ii) The number, Y, of enquiries received during a day is Poisson (54) which is approximately Normal with mean 54 and standard deviation $\sqrt{54}$

$P(Y > 60) \approx P(Y > 60.5$ in Normal)

$$= P\left(Z > \frac{60.5 - 54}{\sqrt{54}}\right)$$

$$= P(Z > 0.885)$$
$$= 0.1880$$

Note that these two probabilities are *not* the same, although 300 enquiries per week might be thought to be 'equivalent' to 60 enquiries per day.

Figure 12.1 summarises the conditions under which certain distributions may be approximated by others.

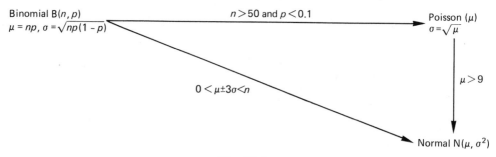

Fig. 12.1

Exercise 12.5   The Normal approximation to the Poisson

1 The number of people killed in road accidents each day has a Poisson distribution with mean 26. State the mean and the standard deviation of the approximating Normal distribution and use it to find the probability that, on a given day
   (i) more than 30 people are killed
   (ii) between 20 and 30 people (inclusive) are killed.

2 Given that X is Poisson (6), find the probability that $4 \leqslant X \leqslant 7$
   (i) using the Poisson probabilities
   (ii) using the Normal approximation.

3   Customers enter a supermarket at an average rate of 144 per hour. Find the probability that
    (i) more than 160 customers enter during a given hour
    (ii) less than 10 customers enter during a given 5 minute period.

4   A fire station receives an average of 10 emergency calls each week. Find the probability that in a four week period it receives at least 55 emergency calls.

12.6   Miscellaneous questions

The following Exercise contains slightly harder questions involving Binomial and Poisson distributions.

Exercise 12.6   Miscellaneous questions

1   The emission of α-particles from a radioactive source may be assumed to be a random process. In 250 intervals of 1 ms it was found that no α-particles were emitted in 41 of these intervals. By considering $P(X = 0)$ in the distribution Poisson (μ), estimate the mean number of α-particles emitted per millisecond. Hence calculate the probability that exactly 2 α-particles will be emitted in a given 3 ms interval.

2   A car-hire firm finds that the daily demand for its cars follows a Poisson distribution with mean 2.8. Each car owned by the firm costs the firm £8 per day regardless of whether it is hired or not, and the firm charges £20 for the hire of a car for one day.
    (i) Suppose the firm owns 3 cars. Find the probabilities that the daily demand is 0, 1, 2, $\geqslant 3$, and state the total profit in each case (e.g. if there are no demands, the profit is $-£24$; if there are $\geqslant 3$ demands the profit is £60 − £24 = £36). Hence find the expected daily profit.
    (ii) Find the expected daily profit if the firm owns 4 cars.

3   In a factory, 'incidents' occur at an average rate of one 'incident' every three working days.
    (i)   For a given day, find the probability that at least one 'incident' occurs.
    (ii) For a week of 5 working days, what is the distribution of the number of days on which 'incidents' occur? Find the probability that 'incidents' occur on at least 3 days.
    (iii) If two or more 'incidents' occur during one day, a report must be sent to Head Office. Find the mean and the standard deviation of the number of reports sent during a month of 22 working days.

4   A die is thrown until the third 6 appears. Find the probability that this takes exactly 8 throws. (*Hint:* how many sixes are there in the first 7 throws? What must the 8th throw be?)
    In a sequence of independent trials, the probability of success is p. Show that the

probability that the kth success occurs on the nth trial is $\binom{n-1}{k-1} p^k (1-p)^{n-k}$

(*Hint:* there must be $k-1$ successes in the first $n-1$ trials, and the nth trial must be a success.)

5   In the proofs of a new textbook, the number of misprints per page has a Poisson distribution with mean 1.5. Only pages containing at least one misprint are returned to the printer for correction. Considering only the pages which are returned, what do the Poisson probabilities of 1, 2, 3, 4, ... misprints add up to? How should these probabilities be adjusted so that they add up to 1? Find the probability that a page returned contains

(i) 1   (ii) 2   (iii) 3   (iv) 4 misprints.

6   The number of flowers on a certain pot-plant follows a Poisson distribution with mean 3, but only plants containing between 2 and 5 flowers (inclusive) are sold. For plants which are sold, list the distribution of the number of flowers, and calculate its mean and standard deviation.

13
Continuous Random Variables

13.1 The probability density function

If a continuous random variable X has probability density function (pdf) $f(x)$, then we have

$$f(x) \geqslant 0 \text{ and } \int_{\text{all } x} f(x) \, dx = 1$$

The probability that X is between a and b is given by the area under the pdf curve, i.e.

$$P(a < X < b) = \int_a^b f(x) \, dx$$

The mean and variance are found by integration

$$\text{mean, } \mu = E[X] = \int_{\text{all } x} x f(x) \, dx$$

$$E[X^2] = \int_{\text{all } x} x^2 f(x) \, dx$$

$$\text{variance, } \sigma^2 = \text{var}(X) = E[X^2] - \mu^2$$

$$\text{standard deviation, } \sigma = \sqrt{\text{var}(X)}$$

Example 1

A random variable X has an *exponential distribution* with pdf (Fig. 13.1)

$$f(x) = \begin{cases} \lambda e^{-\lambda x} & \text{if } x \geqslant 0 \\ 0 & \text{if } x < 0 \end{cases}$$

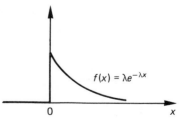

$$f(x) = \lambda e^{-\lambda x}$$

Fig. 13.1

where λ is a positive constant.

(i) Verify that $f(x)$ satisfies the conditions for a pdf.

(ii) Show that $E[X] = \dfrac{1}{\lambda}$ and $\text{var}(X) = \dfrac{1}{\lambda^2}$

(iii) Find the median.

(iv) Find the probability that X is greater than its mean.

(i) $f(x) \geq 0$ and $\displaystyle\int_0^\infty \lambda e^{-\lambda x}\, dx = \left[-e^{-\lambda x} \right]_0^\infty = (0) - (-1) = 1$

(ii) To find the mean and variance, we use integration by parts. We also need to use the results that $xe^{-\lambda x}$ and $x^2 e^{-\lambda x}$ tend to zero as $x \to +\infty$

$$\mathrm{E}[X] = \int_0^\infty x\lambda e^{-\lambda x}\, dx = \left[x(-e^{-\lambda x}) \right]_0^\infty - \int_0^\infty (1)(-e^{-\lambda x})\, dx$$

$$= \left[-xe^{-\lambda x} - \frac{1}{\lambda} e^{-\lambda x} \right]_0^\infty = (0-0) - \left(0 - \frac{1}{\lambda} \right)$$

$$= \frac{1}{\lambda}$$

$$\mathrm{E}[X^2] = \int_0^\infty x^2 \lambda e^{-\lambda x}\, dx = \left[x^2(-e^{-\lambda x}) \right]_0^\infty - \int_0^\infty (2x)(-e^{-\lambda x})\, dx$$

$$= \left[-x^2 e^{-\lambda x} - 2x\left(\frac{1}{\lambda} e^{-\lambda x} \right) \right]_0^\infty + \int_0^\infty (2)\left(\frac{1}{\lambda} e^{-\lambda x} \right) dx$$

$$= \left[-x^2 e^{-\lambda x} - \frac{2x}{\lambda} e^{-\lambda x} - \frac{2}{\lambda^2} e^{-\lambda x} \right]_0^\infty$$

$$= (0 - 0 - 0) - \left(0 - 0 - \frac{2}{\lambda^2} \right)$$

$$= \frac{2}{\lambda^2}$$

$$\mathrm{var}(X) = \frac{2}{\lambda^2} - \left(\frac{1}{\lambda} \right)^2 = \frac{1}{\lambda^2}$$

(iii) If m is the median, then $P(X > m) = \frac{1}{2}$,

so $\dfrac{1}{2} = \displaystyle\int_m^\infty \lambda e^{-\lambda x}\, dx = \left[-e^{-\lambda x} \right]_m^\infty = (0) - (-e^{-\lambda m}) = e^{-\lambda m}$

i.e. $e^{\lambda m} = 2$

$\lambda m = \ln 2$

The median, $m = \dfrac{\ln 2}{\lambda}$

(iv) $P\left(X > \dfrac{1}{\lambda} \right) = \displaystyle\int_{1/\lambda}^\infty \lambda e^{-\lambda x}\, dx = \left[-e^{-\lambda x} \right]_{1/\lambda}^\infty = (0) - (-e^{-1})$

$= e^{-1} \approx 0.3679$

Example 2

A random variable X has pdf (Fig. 13.2)

$$f(x) = \begin{cases} 0 & \text{if } x < 0 \\ kx & \text{if } 0 \leqslant x \leqslant 1 \\ \dfrac{k}{x^4} & \text{if } x > 1 \end{cases}$$

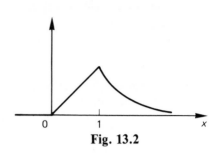

Fig. 13.2

Find (i) the value of the constant k

  (ii) the mean and the standard deviation of X

(i) We have $1 = \displaystyle\int_0^\infty f(x)\,dx = \int_0^1 kx\,dx + \int_1^\infty \frac{k}{x^4}\,dx$

$$= \left[\frac{1}{2}kx^2\right]_0^1 + \left[-\frac{k}{3x^3}\right]_1^\infty$$

$$= \tfrac{1}{2}k + \tfrac{1}{3}k = \tfrac{5}{6}k$$

Hence $k = \dfrac{6}{5}$

(ii) $E[X] = \displaystyle\int_0^\infty xf(x)\,dx = \int_0^1 xf(x)\,dx + \int_1^\infty xf(x)\,dx$

$$= \int_0^1 x\frac{6}{5}x\,dx + \int_1^\infty x\frac{6}{5x^4}\,dx$$

$$= \left[\frac{2}{5}x^3\right]_0^1 + \left[-\frac{3}{5x^2}\right]_1^\infty$$

$$= \frac{2}{5} + \frac{3}{5}$$

$$= 1$$

$E[X^2] = \displaystyle\int_0^\infty x^2 f(x)\,dx = \int_0^1 x^2\frac{6}{5}x\,dx + \int_1^\infty x^2\frac{6}{5x^4}\,dx$

$$= \left[\frac{3}{10}x^4\right]_0^1 + \left[-\frac{6}{5x}\right]_1^\infty$$

$$= \frac{3}{10} + \frac{6}{5}$$

$$= 1.5$$

$\text{var}\,(X) = 1.5 - 1^2 = 0.5$

standard deviation $\sigma = \sqrt{0.5} \approx 0.7071$

Exercise 13.1   Probability density functions

1   A random variable X has probability density function (pdf)

$$f(x) = \begin{cases} \dfrac{k}{x^2} & \text{for } 1 \leqslant x \leqslant 4 \\ 0 & \text{otherwise.} \end{cases}$$

(i)   Sketch the distribution, and state the mode of X.
(ii)   Find the value of k.
(iii)   Find the mean $E[X]$ and the variance, var (X).
(iv)   Find $P(2 < X < 3)$
(v)   Find $P(X > 2.5)$
(vi)   Find the median.

2   Show that $f(x) = \dfrac{2}{x^3}$ for $x \geqslant 1$ is a pdf for some random variable X.

Find (i) $P(X < 2)$ (ii) $P(X \geqslant 5)$ (iii) the mean.

3   During a thunderstorm, the time X minutes between two successive lightning flashes may be assumed to be a continuous random variable with pdf

$$f(x) = e^{-x} \text{ for } x \geqslant 0$$

Sketch this distribution.

(i)   Verify that $\displaystyle\int_0^\infty f(x)\, \mathrm{d}x = 1$
(ii)   Find the probability that the time between successive flashes is more than 2 minutes.
(iii)   Find the time T between flashes which is exceeded in 95% of cases (i.e. $P(X > T) = 0.95$).
(iv)   Find the median time between flashes.
(v)   Show that $\int xe^{-x}\, \mathrm{d}x = -xe^{-x} - e^{-x} + C$
        and $\int x^2 e^{-x}\, \mathrm{d}x = -x^2 e^{-x} - 2xe^{-x} - 2e^{-x} + C$
        and hence find the mean and standard deviation of the time between flashes.
        (You may assume that $xe^{-x} \to 0$ and $x^2 e^{-x} \to 0$ as $x \to \infty$)

4   A random variable X has an exponential distribution with pdf

$$f(x) = \lambda e^{-\lambda x} \qquad \text{for } x \geqslant 0$$

where λ is a positive constant. You may assume that $E[X] = \dfrac{1}{\lambda}$ and var $(X) = \dfrac{1}{\lambda^2}$.

(i)   Find the probability that X lies within one standard deviation of its mean.
(ii)   Find a in terms of λ if $P(X > a) = 0.1$

5   The random variable X has the rectangular distribution between 0 and a, (Fig. 13.3), i.e. the pdf is

$$f(x) = \begin{cases} \dfrac{1}{a} & \text{if } 0 \leqslant x \leqslant a \\ 0 & \text{otherwise} \end{cases}$$

Show that $E[X] = \tfrac{1}{2} a$ and var $(X) = \tfrac{1}{12} a^2$

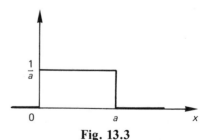

Fig. 13.3

6 The random variable X has a triangular distribution with pdf as shown in Fig. 13.4.

(a) State the value of h.

(b) State the mean of X.

(c) Find the equations for the pdf $f(x)$
   when (i) $-1 \leqslant x \leqslant 0$
   (ii) $0 \leqslant x \leqslant 1$

(d) Calculate the variance of X.

(e) Find the probability that X differs from its mean by more than two standard deviations.

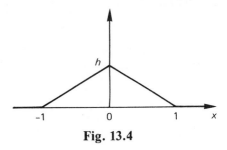

Fig. 13.4

7 A random variable X has pdf

$$f(x) = \begin{cases} A\sqrt{x} & \text{if} & 0 \leqslant x \leqslant 1 \\ A & \text{if} & 1 \leqslant x \leqslant 3 \\ 0 & \text{otherwise} \end{cases}$$

Sketch the pdf.

Find (i) the value of A (ii) the median   (iii) the mean   (iv) the variance.

13.2 The cumulative distribution function

The probability density function corresponds to a histogram; we now consider a function which corresponds to a cumulative frequency curve.

For a value x, a cumulative frequency curve gives the number of observations which are less than or equal to x; and a cumulative probability curve gives the probability that an observation is less than or equal to x.

For a continuous random variable X, we define the *cumulative distribution function* (cdf), $F(x)$, to be the probability that X is less than or equal to x, i.e.

$$\boxed{F(x) = \text{P}(X \leqslant x)}$$

The graph of the cdf (Fig. 13.5) corresponds to a cumulative probability curve.

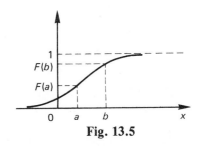

Fig. 13.5

Probabilities may be found as

$$\text{P}(X < a) = F(a)$$
$$\text{P}(X > a) = 1 - F(a)$$
$$\text{P}(a < X < b) = F(b) - F(a)$$

(Remember that, for a continuous variable, $\text{P}(X < a)$ is equal to $\text{P}(X \leqslant a)$, since $\text{P}(X = a) = 0$).

The median M, and the lower and upper quartiles Q_1 and Q_3 (Fig. 13.6) are such that

$$F(M) = \tfrac{1}{2}, \; F(Q_1) = \tfrac{1}{4}, \; F(Q_3) = \tfrac{3}{4}$$

and, for example, the 57th percentile is the value x for which $F(x) = 0.57$.

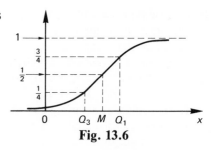
Fig. 13.6

The cdf $F(x) = P(X \leqslant x)$ is the area under the pdf curve to the left of x, and this may be found by integration,

$$F(x) = \int f(x) \, dx$$

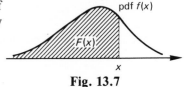
Fig. 13.7

(The constant of integration is found by considering the end-points, as shown in Examples 2, 3 and 4 below).

Example 1

A random variable X has the rectangular distribution between 2 and 6. Find, and sketch, the cdf $F(x)$.

The pdf is $f(x) = \tfrac{1}{4}$ (for $2 \leqslant x \leqslant 6$)
Clearly if $x < 2$, then $P(X \leqslant x) = 0$
   and if $x > 6$, then $P(X \leqslant x) = 1$
If $2 \leqslant x \leqslant 6$, then $P(X \leqslant x)$ is the shaded area, which is $\tfrac{1}{4}(x - 2)$.

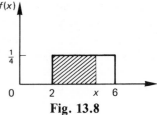
Fig. 13.8

Hence the cdf is

$$F(x) = \begin{cases} 0 & \text{if } x < 2 \\ \tfrac{1}{4}(x - 2) & \text{if } 2 \leqslant x \leqslant 6 \\ 1 & \text{if } x > 6 \end{cases}$$

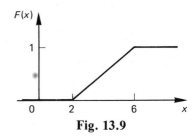
Fig. 13.9

Example 2

A random variable X has pdf

$$f(x) = \begin{cases} \tfrac{2}{9}x(3 - x) & \text{if } 0 \leqslant x \leqslant 3 \\ 0 & \text{otherwise} \end{cases}$$

Find the cdf $F(x)$, and use it to find $P(X > 2)$.

If $x < 0$, then $F(x) = 0$; and if $x > 3$, $F(x) = 1$.

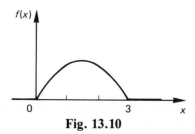
Fig. 13.10

If $0 \leqslant x \leqslant 3$,

$$F(x) = \int f(x)\, dx = \int \frac{2}{9}(3x - x^2)\, dx$$

$$= \frac{1}{3}x^2 - \frac{2}{27}x^3 + C$$

Since $F(0) = P(X \leqslant 0) = 0$, we have

$$0 = 0 - 0 + C, \text{ so } C = 0$$

Thus $F(x) = \frac{1}{3}x^2 - \frac{2}{27}x^3$
Note that $F(3) = 3 - 2 = 1$
The cdf is

$$F(x) = \begin{cases} 0 & \text{if } x < 0 \\ \dfrac{1}{3}x^2 - \dfrac{2}{27}x^3 & \text{if } 0 \leqslant x \leqslant 3 \\ 1 & \text{if } x > 3 \end{cases}$$

Then $P(X > 2) = 1 - F(2) = 1 - \left(\dfrac{1}{3} \times 4 - \dfrac{2}{27} \times 8\right) = \dfrac{7}{27}$

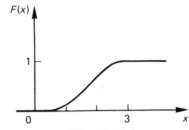
Fig. 13.11

Example 3
The random variable X has an exponential distribution with pdf

$$f(x) = \begin{cases} \lambda e^{-\lambda x} & \text{if } x \geqslant 0 \\ 0 & \text{if } x < 0 \end{cases}$$

Show that the cdf is $F(x) = 1 - e^{-\lambda x}$ (for $x \geqslant 0$), and hence find the median and the 90th percentile.

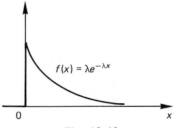
Fig. 13.12

If $x \geqslant 0$, the cdf is

$$F(x) = \int \lambda e^{-\lambda x}\, dx = -e^{-\lambda x} + C$$

Now $F(0) = P(X \leqslant 0) = 0$, so $0 = -1 + C$, i.e. $C = 1$
Hence the cdf is

$$F(x) = \begin{cases} 0 & \text{if } x < 0 \\ 1 - e^{-\lambda x} & \text{if } x \geqslant 0 \end{cases}$$

Note that as $x \to \infty$, $F(x) \to 1 - 0 = 1$

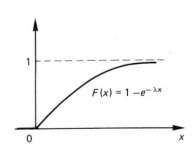
Fig. 13.13

For the median, we need $F(x) = \frac{1}{2}$, i.e. $1 - e^{-\lambda x} = \frac{1}{2}$

$$\text{i.e. } x = \frac{\ln 2}{\lambda}$$

For the 90th percentile, we need $F(x) = 0.9$ i.e. $1 - e^{-\lambda x} = 0.9$

$$\text{i.e.} \qquad e^{\lambda x} = 10$$

$$x = \frac{\ln 10}{\lambda}$$

Hence the median is $(\ln 2)/\lambda$, and the 90th percentile is $(\ln 10)/\lambda$.

Example 4

A random variable X has pdf

$$f(x) = \begin{cases} 0 & \text{if } x < 0 \\ \dfrac{6}{5}x & \text{if } 0 \leqslant x \leqslant 1 \\ \dfrac{6}{5x^4} & \text{if } x > 1 \end{cases}$$

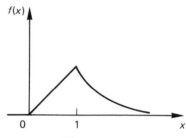

Fig. 13.14

Find (i) the cdf $F(x)$
    (ii) $P(\frac{1}{2} < X < 2)$
    (iii) the median and the semi-interquartile range.

(i) If $0 \leqslant x \leqslant 1$, $F(x) = \displaystyle\int \frac{6}{5} x \, dx = \frac{3}{5} x^2 + C$

Now $F(0) = 0$, and so $C = 0$

Thus $F(x) = \dfrac{3}{5} x^2$; we have $F(1) = \dfrac{3}{5}$

If $x > 1$, $F(x) = \displaystyle\int \frac{6}{5x^4} \, dx = -\frac{2}{5x^3} + D$

Now $F(1) = \dfrac{3}{5}$, and so $\dfrac{3}{5} = -\dfrac{2}{5} + D$, i.e. $D = 1$

Hence the cdf is

$$F(x) = \begin{cases} 0 & \text{if } x < 0 \\ \dfrac{3}{5}x^2 & \text{if } 0 \leqslant x \leqslant 1 \\ 1 - \dfrac{2}{5x^3} & \text{if } x > 1 \end{cases}$$

We check that as $x \to \infty$, $F(x) \to 1 - 0 = 1$.

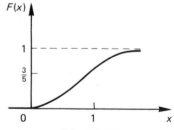

Fig. 13.15

(ii) $P(\frac{1}{2} < X < 2) = F(2) - F(\frac{1}{2}) = \left(1 - \dfrac{2}{5 \times 8}\right) - \dfrac{3}{5} \times \left(\dfrac{1}{2}\right)^2 = \dfrac{4}{5}$

(iii) For the median, we need $F(x) = \frac{1}{2}$, i.e. $\frac{3}{5}x^2 = \frac{1}{2}$, i.e. $x^2 = \frac{5}{6}$

Hence the median is $\sqrt{\frac{5}{6}} \approx 0.9129$

For the upper quartile, we need $F(x) = \frac{3}{4}$, i.e. $1 - \frac{2}{5x^3} = \frac{3}{4}$,

so $x^3 = \frac{8}{5}$ i.e. $x \approx 1.1696$

For the lower quartile, we need $F(x) = \frac{1}{4}$, i.e. $\frac{3}{5}x^2 = \frac{1}{4}$,

so $x^2 = \frac{5}{12}$ i.e. $x \approx 0.6455$

Hence the semi-interquartile range is $\frac{1}{2}(1.1696 - 0.6455) = 0.2621$
(Compare these values with the mean 1 and standard deviation 0.7071; See Example 2 in Section 13.1).

Since $F(x)$ is the integral of $f(x)$, the pdf $f(x)$ is the derivative of the cdf $F(x)$ i.e.

$$\boxed{f(x) = F'(x)}$$

Example 5
A continuous random variable X takes values between 2 and 5, and for $2 \leqslant x \leqslant 5$, the
probability that $X \leqslant x$ is $ax - \frac{b}{x}$, where a and b are constants.
Find (i)   the constants a and b.
  (ii) P$(3 < X < 4)$
  (iii) the pdf $f(x)$
  (iv) the mean of X.

(i)   We are given the cdf $F(x) = P(X \leqslant x) = ax - \frac{b}{x}$ (for $2 \leqslant x \leqslant 5$)

Since $F(2) = P(X \leqslant 2) = 0$, $2a - \frac{b}{2} = 0$, i.e. $b = 4a$

Since $F(5) = P(X \leqslant 5) = 1$, $5a - \frac{b}{5} = 1$,

thus $5a - \frac{4a}{5} = 1$, so $a = \frac{5}{21}$

and $b = \frac{20}{21}$

(ii) The cdf is

$$F(x) = \begin{cases} 0 & \text{if } x < 2 \\ \frac{5}{21}\left(x - \frac{4}{x}\right) & \text{if } 2 \leqslant x \leqslant 5 \\ 1 & \text{if } x > 5 \end{cases}$$

Hence P$(3 < X < 4) = F(4) - F(3) = \frac{5}{21}\left(4 - \frac{4}{4}\right) - \frac{5}{21}\left(3 - \frac{4}{3}\right) = \frac{20}{63}$

(iii) The pdf is $f(x) = F'(x)$, so

$$f(x) = \begin{cases} 0 & \text{if } x < 2 \\ \dfrac{5}{21}\left(1 + \dfrac{4}{x^2}\right) & \text{if } 2 \leqslant x \leqslant 5 \\ 0 & \text{if } x > 5 \end{cases}$$

(iv) $E[X] = \displaystyle\int_2^5 xf(x)\,dx = \int_2^5 \dfrac{5}{21}\left(x + \dfrac{4}{x}\right)dx = \dfrac{5}{21}\left[\dfrac{1}{2}x^2 + 4\ln x\right]_2^5$

$$= \dfrac{5}{2} + \dfrac{20}{21}\ln\dfrac{5}{2}$$

$$\approx 3.373$$

Exercise 13.2   The cumulative distribution function

1   The random variable X has the rectangular distribution on $[0, 2]$ i.e. the pdf is
$f(x) = \frac{1}{2}$ for $0 \leqslant x \leqslant 2$
State the value of $P(X \leqslant x)$ when
(i) $x < 0$ (ii) $0 \leqslant x \leqslant 2$ (iii) $x > 2$
and hence sketch the graph of the cdf $F(x)$

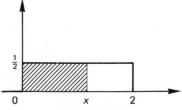

Fig. 13.16

2   The random variable X has the rectangular distribution on $[3, 4]$ i.e. the pdf is
$f(x) = 1$ for $3 \leqslant x \leqslant 4$
Find and sketch the cdf $F(x)$

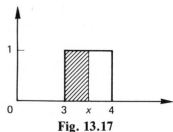

Fig. 13.17

3   The random variable X has pdf $f(x) = 2x$
for $0 \leqslant x \leqslant 1$.
Find and sketch the cdf $F(x)$

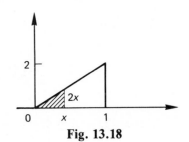

Fig. 13.18

4 The random variable X has pdf

$$f(x) = \begin{cases} \dfrac{3}{7}x^2 & \text{for } 1 \leqslant x \leqslant 2 \\ 0 & \text{otherwise} \end{cases}$$

Integrate to obtain the cdf $F(x)$ for $1 \leqslant x \leqslant 2$. (Use the fact that $F(1) = 0$ to obtain the constant of integration.)
Check that $F(2) = 1$.
State the value of $F(x)$ when (i) $x < 1$ (ii) $x > 2$
Sketch the pdf $f(x)$ and the cdf $F(x)$

5 The random variable X has pdf

$$f(x) = \begin{cases} \dfrac{2}{x^2} & \text{for } 1 \leqslant x \leqslant 2 \\ 0 & \text{otherwise} \end{cases}$$

Find and sketch the cdf $F(x)$.
Use the cdf to find
(i) $P\left(X < \dfrac{3}{2}\right)$ (ii) $P(X > 1.8)$ (iii) $P(1.4 < X < 1.6)$

6 A random variable X has pdf

$$f(x) = \begin{cases} 2e^{-2x} & \text{for } x \geqslant 0 \\ 0 & \text{otherwise} \end{cases}$$

Find and sketch the cdf $F(x)$.
   Hence find the median and the semi-interquartile range. Compare these values with the mean and the standard deviation (you may quote the results for a general exponential distribution).

7 A random variable has pdf

$$f(x) = \begin{cases} 6(x-2)(3-x) & \text{for } 2 \leqslant x \leqslant 3 \\ 0 & \text{otherwise} \end{cases}$$

Find and sketch the cdf $F(x)$.

8 A random variable X has a triangular
distribution as shown.
(i)   Find the value of h
(ii) Find the equations of the pdf $f(x)$ for
      $-2 \leqslant x \leqslant 0$ and $0 \leqslant x \leqslant 2$
(iii) Find the cdf $F(x)$ for $-2 \leqslant x \leqslant 0$, and
      find the value of $F(0)$.
(iv) Find the cdf $F(x)$ for $0 \leqslant x \leqslant 2$.
(v)   Sketch the cdf $F(x)$.

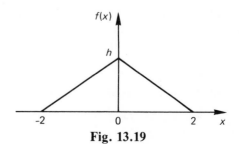

Fig. 13.19

9   A random variable X has pdf

$$f(x) = \begin{cases} \frac{1}{3} & \text{for } 0 \leqslant x < 1 \\ \frac{1}{2} & \text{for } 1 \leqslant x < 2 \\ \frac{1}{6} & \text{for } 2 \leqslant x < 3 \\ 0 & \text{otherwise} \end{cases}$$

Find and sketch the cdf $F(x)$.

10 A sample of 750 light-bulbs had the following lifetimes:

Lifetime (hours)	0–25	25–50	50–75	75–100	100–125
Number of bulbs	289	185	112	67	37

Lifetime (hours)	125–150	150–175	175–200	200–225	225–250
Number of bulbs	25	16	10	5	4

Estimate the mean lifetime.
It is thought that the lifetimes, X hours, have an exponential distribution with pdf
$f(x) = \lambda e^{-\lambda x}$ for $x \geqslant 0$ (and 0 otherwise), for which $E[X] = \dfrac{1}{\lambda}$.
What value of λ will give the same mean as the sample?
For this value of λ, tabulate the cdf $F(x) = 1 - e^{-\lambda x}$ for $x = 0, 25, 50, 75, \ldots, 250$.
Hence calculate the probabilities
$P(0 < X < 25)$, $P(25 < X < 50)$, $\ldots$, $P(200 < X < 225)$, $P(X > 225)$, and the expected frequencies.

11 A random variable X has cdf

$$F(x) = \begin{cases} 0 & \text{for } x < 0 \\ x - \frac{1}{4}x^2 & \text{for } 0 \leqslant x \leqslant 2 \\ 1 & \text{for } x > 2 \end{cases}$$

Find the pdf $f(x)$ and calculate the mean and variance of X.

12 A random variable X takes values between 0 and π, and $P(X \leqslant x) = \sin \frac{1}{2}x$ for $0 \leqslant x \leqslant \pi$. Find
(i) $P(1 < X < 2)$ (ii) the pdf $f(x)$ (iii) the mean of X

13 A random variable X takes values between 5 and 10, and $P(X > x) = \dfrac{10}{x} - 1$ for $5 \leqslant x \leqslant 10$. Find
(i) the cdf $F(x)$ (ii) the pdf $f(x)$ (iii) the mean of X.

14 A random variable X has cdf

$$F(x) = \begin{cases} 0 & \text{for } x < 0 \\ \dfrac{1}{5}x^2 & \text{for } 0 \leqslant x \leqslant 1 \\ \dfrac{2}{5}x - \dfrac{1}{5} & \text{for } 1 \leqslant x \leqslant 2 \\ \dfrac{4}{5}x - \dfrac{1}{10}x^2 - \dfrac{3}{5} & \text{for } 2 \leqslant x \leqslant 4 \\ 1 & \text{for } x > 4 \end{cases}$$

Find (i) $P\left(\dfrac{3}{2} < X < \dfrac{5}{2}\right)$ (ii) the pdf, $f(x)$, and sketch it.

15 A random variable X takes values between 0 and 3, and the cdf is $F(x) = kx^3$ for $0 \leqslant x \leqslant 3$.
What are the values of $F(x)$ when $x < 0$ and when $x > 3$?
By considering $F(3)$, find the value of k.
Calculate the mean and variance of X.

16   Currants are distributed at random in a mixture which is used to make spherical buns of radius a. Given that X is the distance of a currant from the centre of a bun, complete

$P(X \leqslant x) = P(\text{currant lies within a sphere of radius } x)$

$$= \dfrac{\text{Volume of sphere of radius } x}{\text{Volume of bun}}$$

$$= \cdots$$

Hence find the pdf of X, and the mean and standard deviation of the distance of a currant from the centre of a bun.

17   The probability that I have to wait longer than t minutes at a bus-stop is inversely proportional to $(t + 2)^2$. Show that the cdf of the waiting time is

$$F(t) = 1 - \frac{4}{(t + 2)^2} \qquad (\text{for } t \geqslant 0).$$

(i)   Find the probability that I wait between 2 and 3 minutes.
(ii)   Find the mean waiting time.
      (You may find the substitution $t + 2 = u$ helpful in evaluating the integral.)

13.3   Functions of a random variable

If a continuous random variable X has pdf $f(x)$, then we know that

$$E[X] = \int_{\text{all } x} x f(x) \, dx, \quad E[X^2] = \int_{\text{all } x} x^2 f(x) \, dx$$

The expected value of any function of X can be found in a similar way, for example,

$$E[X^3] = \int_{\text{all }x} x^3 f(x)\, dx, \quad E[X^5] = \int_{\text{all }x} x^5 f(x)\, dx, \quad E[\sin X] = \int_{\text{all }x} \sin x\, f(x)\, dx$$

$$E[X^4 - 2X^2] = \int_{\text{all }x} (x^4 - 2x^2) f(x)\, dx, \quad \text{and so on.}$$

In general, we have

$$\boxed{E[g(X)] = \int_{\text{all }x} g(x) f(x)\, dx}$$

Example 1
A random variable X has pdf

$$f(x) = \begin{cases} \dfrac{2}{x^2} & \text{if } 1 \leqslant x \leqslant 2 \\ 0 & \text{otherwise} \end{cases}$$

Find the expected value of $\left(X^5 - \dfrac{2}{X}\right)$.

$$E\left[X^5 - \frac{2}{X}\right] = \int_1^2 \left(x^5 - \frac{2}{x}\right) f(x)\, dx = \int_1^2 \left(x^5 - \frac{2}{x}\right)\left(\frac{2}{x^2}\right) dx$$

$$= \int_1^2 \left(2x^3 - \frac{4}{x^3}\right) dx = \left[\tfrac{1}{2}x^4 + \frac{2}{x^2}\right]_1^2$$

$$= (8 + \tfrac{1}{2}) - (\tfrac{1}{2} + 2)$$

$$= 6$$

Now suppose that X is a random variable whose pdf $f(x)$ is known, and let Y be a random variable defined in terms of X, for example, $Y = 4X^3$.

Then $E[Y]$ and $E[Y^2]$ can be found by first expressing Y and Y^2 in terms of X; and hence the mean and variance of Y may be calculated.

Probabilities concerning Y may be found by first writing the required event in terms of X.

These techniques are illustrated in the following example.

Example 2
A random variable X has pdf

$$f(x) = \begin{cases} \dfrac{2}{x^2} & \text{if } 1 \leqslant x \leqslant 2 \\ 0 & \text{otherwise} \end{cases}$$

and the random variable Y is defined by $Y = 4X^3$.
Find (i)   the mean and variance of Y
     (ii) $P(10 < Y < 20)$
     (iii) the median of Y.

(i) $E[Y] = E[4X^3] = \int_1^2 4x^3 \left(\frac{2}{x^2}\right) dx = \int_1^2 8x \, dx = \left[4x^2\right]_1^2 = 12$

$E[Y^2] = E[16X^6] = \int_1^2 16x^6 \left(\frac{2}{x^2}\right) dx = \int_1^2 32x^4 \, dx = \left[\frac{32}{5}x^5\right]_1^2 = 198.4$

$\text{var}(Y) = 198.4 - (12)^2 = 54.4$

(ii) $P(10 < Y < 20) = P(10 < 4X^3 < 20) = P\left(\left(\frac{10}{4}\right)^{1/3} < X < \left(\frac{20}{4}\right)^{1/3}\right)$

$$= P(1.3572 < X < 1.7100)$$

$$= \int_{1.3572}^{1.7100} \frac{2}{x^2} \, dx = \left[\frac{-2}{x}\right]_{1.3572}^{1.7100}$$

$$= 0.3040$$

(iii) We first find the median m of X.

We require $\frac{1}{2} = P(X < m) = \int_1^m \frac{2}{x^2} \, dx = \left[\frac{-2}{x}\right]_1^m = \frac{-2}{m} + 2,$

so the median of X is $m = \frac{4}{3}$

Then $\frac{1}{2} = P\left(X < \frac{4}{3}\right) = P\left(4X^3 < 4\left(\frac{4}{3}\right)^3\right) = P\left(Y < \frac{256}{27}\right),$

hence the median of Y is $\frac{256}{27} \approx 9.481$

Notes (1) The random variable Y can take values between 4 and 32. However in all the above integrals we are using the pdf *of X*, and so the limits of integration are limits *of X* (which takes values between 1 and 2).
     (2) It is possible to find the pdf of Y, but this is unnecessary; we can always work in terms of X.

Example 3

A semicircular arc, with centre O and radius a, is drawn with AB as diameter (Fig. 13.20). A point Q is taken at random on this arc, such that the angle BOQ = θ has the rectangular distribution between 0 and π, i.e. the pdf of θ is

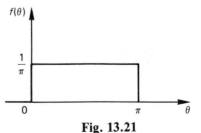

Fig. 13.20

$$f(\theta) = \begin{cases} \dfrac{1}{\pi} & \text{if } 0 \leqslant \theta \leqslant \pi \\ 0 & \text{otherwise} \end{cases}$$

QN is the perpendicular from Q to the diameter AB.
(i)   Calculate the mean and the standard deviation of the length of QN.
(ii) Find the probability that QN is longer than $\frac{1}{2} a$.

Fig. 13.21

Suppose QN has length L. Then L is a random variable taking values between 0 and a, but we cannot assume that L is uniformly distributed (we shall see later that the mean of L is *not* $\frac{1}{2}a$). We know the distribution of θ, so we must first express L in terms of θ; we have

$$L = a \sin \theta$$

(i) $E[L] = E[a \sin \theta] = \displaystyle\int_0^\pi a \sin \theta \, f(\theta) \, d\theta = \int_0^\pi a \sin \theta \left(\frac{1}{\pi}\right) d\theta$

$$= \frac{a}{\pi}\left[-\cos \theta\right]_0^\pi = \frac{2a}{\pi}$$

$E[L^2] = E[a^2 \sin^2 \theta] = \displaystyle\int_0^\pi a^2 \sin^2 \theta f(\theta) \, d\theta = \int_0^\pi a^2 \sin^2 \theta \left(\frac{1}{\pi}\right) d\theta$

$$= \frac{a^2}{\pi}\int_0^\pi \tfrac{1}{2}(1 - \cos 2\theta) \, d\theta = \frac{a^2}{2\pi}\left[\theta - \tfrac{1}{2}\sin 2\theta\right]_0^\pi = \tfrac{1}{2}a^2$$

$$\mathrm{var}(L) = \tfrac{1}{2}a^2 - \left(\frac{2a}{\pi}\right)^2 = \left(\tfrac{1}{2} - \frac{4}{\pi^2}\right)a^2$$

The length of QN has mean $\dfrac{2a}{\pi} \approx 0.6366a$

and standard deviation $\left(\tfrac{1}{2} - \dfrac{4}{\pi^2}\right)^{1/2} a \approx 0.3078a$

(ii) $P(L > \tfrac{1}{2}a) = P(a \sin \theta > \tfrac{1}{2}a) = P(\sin \theta > \tfrac{1}{2}) = P\left(\dfrac{1}{6}\pi < \theta < \dfrac{5}{6}\pi\right)$

$$= \frac{1}{\pi}\left(\frac{5}{6}\pi - \frac{1}{6}\pi\right) \quad \left(\text{since } f(\theta) = \frac{1}{\pi}\right)$$

$$= \tfrac{2}{3}$$

Exercise 13.3 Functions of a random variable

1 The random variable X has pdf $f(x) = 2x$ for $0 \leqslant x \leqslant 1$.
Find the mean of the following random variables

(i) X (ii) X^2 (iii) X^3 (iv) $2X - \sqrt{X}$ (v) $\dfrac{1}{X^2 + 1}$

2 The random variable X has a rectangular distribution with pdf $f(x) = \tfrac{1}{2}$ for $1 \leqslant x \leqslant 3$.
Find

(i) $E[X^4]$ (ii) $E[X(3 - X)]$ (iii) $E\left[\dfrac{1}{X}\right]$ (iv) $E\left[\dfrac{1}{X^2}\right]$

(v) $E[\sin \tfrac{1}{3}\pi X]$

3   The random variable X has pdf $f(x) = \frac{3}{4}(1 - x^2)$ for $-1 \leqslant x \leqslant 1$, and $Y = 2X^2$. Find the mean of Y.
   Find $E[Y^2]$ (i.e. $E[4X^4]$), and hence find the variance of Y.

4   The random variable X has a rectangular distribution with pdf $f(x) = 1$ for $1 \leqslant x \leqslant 2$. Find the mean and variance of $1/X$.

5   The random variable X has an exponential distribution with pdf $f(x) = 2e^{-2x}$ for $x \geqslant 0$. Find the mean and standard deviation of e^{-X}.

6   The radius X of a circle is a random variable having the rectangular distribution on $[0, a]$, i.e. the pdf is $f(x) = 1/a$ for $0 \leqslant x \leqslant a$.
   Find the mean and standard deviation of the area $(A = \pi X^2)$ of the circle.

7   The radius of a sphere is a random variable having pdf $f(x) = kx^2(5 - x)$ for $0 \leqslant x \leqslant 5$. Find
   (i) the value of k (ii) the mean radius   (iii) the mean volume of the sphere.

8   A point P is chosen at random on a stick AB of length 10 cm. The stick is broken at P, and the two pieces AP and PB are used as sides of a rectangle. Find the mean and standard deviation of the area of this rectangle.
   (*Hint*: If the length AP is X cm, then X has the rectangular distribution on $[0, 10]$; PB has length $(10 - X)$ cm, so the area is $X(10 - X)$ cm^2.)

9   A random variable X has pdf $f(x) = 6x(1 - x)$ for $0 \leqslant x \leqslant 1$ and $Y = 3X + 5$. Solve the inequality $6 < 3X + 5 < 7$, and hence find $P(6 < Y < 7)$.

10 The random variable X has pdf $f(x) = 3x^2$ for $0 \leqslant x \leqslant 1$ and $Y = 2\sqrt{X}$.
   Find (i) $P(0.5 < Y < 1.5)$ (ii) $P(Y > 1)$.

11 The random variable X has pdf $f(x) = e^{-x}$ for $x \geqslant 0$, and $Y = 2X^2$.
   Find (i) $P(2 < Y < 18)$ (ii) $P(Y < 4)$

12 The random variable X has a rectangular distribution with pdf $f(x) = \frac{1}{10}$ for $-5 \leqslant x \leqslant 5$, and $Y = X^2$.
   Find $P(Y < 4)$. (Remember that X can take negative values.)

13 The radius of a sphere has the rectangular distribution between 6 cm and 10 cm (i.e. pdf is $f(x) = \frac{1}{4}$ for $6 \leqslant x \leqslant 10$). Find the probability that the surface area is less than 620 cm^2.

14 The speeds of trains passing a particular point are Normally distributed with mean 32 m s^{-1} and standard deviation 5 m s^{-1}. Given that the trains have length 160 m, find the probability that a train takes between 4 and 6 seconds to pass the point.

15 In the triangle ABC, AB = 5 cm, AC = 8 cm, and $B\hat{A}C = \theta$ has the rectangular distribution between 0 and π (i.e. pdf is $f(\theta) = \frac{1}{\pi}$ for $0 \leqslant \theta \leqslant \pi$). Show that the area

of the triangle ABC is $Y = 20 \sin \theta$. Hence find
(i) the mean and standard deviation of the area of the triangle ABC
(ii) the probability that the area is greater than 10 cm^2.

16 AB is a fixed diameter of a circle with radius R. A chord AP is drawn at random
through A such that the angle PAB $= \theta$ has the rectangular distribution between $-\dfrac{\pi}{2}$

and $\dfrac{\pi}{2}$. Show that the length of the chord AP is $L = 2R \cos \theta$. Hence find
(i)   the mean length of the chord
(ii) the probability that the chord is longer than $R\sqrt{2}$
(iii) the median length of the chord.

17 A man walks the distance of 600 m to work each day. The time T which he takes has
the rectangular distribution between 6 and 10 minutes.
(i)   Find the mean, standard deviation and median of the time T.
(ii) Express his speed V m s^{-1} in terms of T.
(iii) Find the mean and standard deviation of the speed V.
(iv) Find the median speed.

13.4   The standard Normal distribution

The standard Normal distribution has pdf

$$\phi(x) = \frac{1}{\sqrt{2\pi}} \, e^{-x^2/2}$$

Let Z be a standard Normal variable.

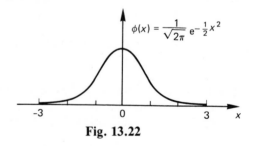

Fig. 13.22

Probabilities such as $P(1 < Z < 2) = \displaystyle\int_1^2 \frac{1}{\sqrt{2\pi}} e^{-x^2/2} \, dx$ cannot be found in the usual way
by integrating, since it is not possible to integrate $e^{-x^2/2}$ in terms of elementary
functions, but numerical integration (for example, Simpson's Rule) can be used.
However some theoretical work is possible.
It can be shown that $\displaystyle\int_{-\infty}^{+\infty} e^{-x^2/2} \, dx = \sqrt{2\pi}$, confirming that $\displaystyle\int_{-\infty}^{+\infty} \phi(x) \, dx = 1$
We shall assume this result.
Clearly $e^{-x^2/2} \rightarrow 0$ as $x \rightarrow \pm \infty$. We shall also need the results that $x e^{-x^2/2}$, $x^2 e^{-x^2/2}$, and
so on, all tend to zero as $x \rightarrow \pm \infty$.

We have

$$\int xe^{-x^2/2} \, dx = -e^{-x^2/2} + C$$

and, integrating by parts,

$$\int x^2 e^{-x^2/2} \, dx = \int x(xe^{-x^2/2}) \, dx = x(-e^{-x^2/2}) - \int (1)(-e^{-x^2/2}) \, dx$$

$$= -xe^{-x^2/2} + \int e^{-x^2/2} \, dx$$

Thus $E[Z] = \displaystyle\int_{-\infty}^{+\infty} x \frac{1}{\sqrt{2\pi}} e^{-x^2/2} \, dx = \frac{1}{\sqrt{2\pi}} \left[-e^{-x^2/2}\right]_{-\infty}^{+\infty} = 0 - 0 = 0$

This is also clear by symmetry.

Also $E[Z^2] = \displaystyle\int_{-\infty}^{+\infty} x^2 \frac{1}{\sqrt{2\pi}} e^{-x^2/2} \, dx = \frac{1}{\sqrt{2\pi}} \left[-xe^{-x^2/2}\right]_{-\infty}^{+\infty} + \frac{1}{\sqrt{2\pi}} \int_{-\infty}^{+\infty} e^{-x^2/2} \, dx$

$$= 0 + 1 = 1$$

and so var $(Z) = 1 - 0^2 = 1$

Hence Z has mean 0 and standard deviation 1.

We may write $Z \sim N(0, 1)$ to mean that Z has the standard Normal distribution.

The cdf is $\Phi(x) = P(Z \leqslant x)$

Since $\Phi(x) = \displaystyle\int \frac{1}{\sqrt{2\pi}} e^{-x^2/2} \, dx$, we cannot express

$\Phi(x)$ in terms of simple functions but we know that

$$\Phi'(x) = \phi(x) = \frac{1}{\sqrt{2\pi}} e^{-x^2/2}$$

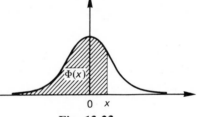

Fig. 13.23

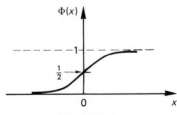

Fig. 13.24

Values of the cdf $\Phi(x)$ are tabulated (for positive values of x) in Normal distribution tables; these values are obtained by numerical methods.

By symmetry, we have $\Phi(-x) = 1 - \Phi(x)$

Probabilities are found as

$$P(Z < a) = \Phi(a)$$
$$P(Z > a) = 1 - \Phi(a)$$
$$P(a < Z < b) = \Phi(b) - \Phi(a)$$

Exercise 13.4 Standard Normal distribution

1   Use Simpson's rule with 8 strips to estimate $\int_0^4 e^{-x^2/2}\, dx$, and hence verify that

$$\int_{-4}^4 e^{-x^2/2}\, dx \approx \sqrt{2\pi}$$

2   If Z has the standard Normal distribution, estimate the probability
$P(1 < Z < 2) = \int_1^2 \frac{1}{\sqrt{2\pi}} e^{-x^2/2}\, dx$ by using Simpson's rule with 2 strips.
Check your answer by using Normal distribution tables.

3   If Z has the standard Normal distribution, explain why
$$P(1.4 < Z < 1.6) \approx \phi(1.5) \times 0.2$$
and hence estimate this probability.
Check your answer by using Normal distribution tables.

4   For the pdf $\phi(x) = \frac{1}{\sqrt{2\pi}} e^{-x^2/2}$, find $\phi'(x)$ and $\phi''(x)$.

Hence show that the points of inflexion on the pdf curve occur at $x = \pm 1$ (i.e. at one standard deviation each side of the mean).

5   Use the series $e^u = 1 + u + \frac{u^2}{2!} + \frac{u^3}{3!} + \cdots$

to expand $e^{-x^2/2}$ as a power series in x as far as the term in x^8. Hence show that

$$\Phi(x) = \tfrac{1}{2} + \frac{1}{\sqrt{2\pi}} \left\{ x - \frac{x^3}{6} + \frac{x^5}{40} - \frac{x^7}{336} + \frac{x^9}{3456} - \cdots \right\}$$

Use this result to calculate $\Phi(0.3)$ correct to 6 decimal places.

6   By integrating by parts, show that

$$\int x^3 e^{-x^2/2}\, dx = -x^2 e^{-x^2/2} - 2e^{-x^2/2} + C$$

$$\int x^4 e^{-x^2/2}\, dx = -x^3 e^{-x^2/2} - 3xe^{-x^2/2} + 3 \int e^{-x^2/2}\, dx$$

Hence find, for a standard Normal variable Z,
(i) $E[Z^3]$ (ii) $E[Z^4]$
(iii) the mean and variance of Z^2.

7   A computer is programmed to generate random values from a standard Normal distribution, but to print out only those values which are positive. Sketch the pdf for the numbers which are printed, and, by considering the area under the curve, explain

why the pdf is

$$f(x) = \begin{cases} \dfrac{2}{\sqrt{2\pi}}\, e^{-x^2/2} & \text{if } x > 0 \\ 0 & \text{if } x \leqslant 0 \end{cases}$$

Calculate the mean and the standard deviation of the printed numbers.

8   The computer program (see Question 7) is now changed so that, while still generating random values from a standard Normal distribution, it only prints out values which are between 1 and 2.
   (i)   What proportion of the numbers generated are printed out?
   (ii) Find the median of the printed numbers.
         (*Hint:* the two shaded areas are equal)

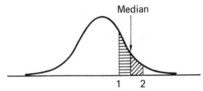

Fig. 13.25

   (iii) Sketch the pdf of the printed numbers.
         Explain why the pdf is *not* just $\dfrac{1}{\sqrt{2\pi}}\, e^{-x^2/2}$ (for $1 \leqslant x \leqslant 2$)

         (*Hint:* consider the area under the curve), and give the equation of the pdf.
   (iv) Calculate the mean of the printed numbers.

13.5   General Normal distributions

Suppose that Z is a standard Normal variable.
Consider the random variable $X = \sigma Z + \mu$
We have $E[X] = \sigma E[Z] + \mu = \mu$
   and $\operatorname{var}(X) = \sigma^2 \operatorname{var}(Z) = \sigma^2$

We say that X has the Normal distribution with mean μ and variance σ^2 (standard deviation σ), and we write

$$X \sim N(\mu, \sigma^2).$$

Then $P(X < a) = P(\sigma Z + \mu < a) = P\left(Z < \dfrac{a - \mu}{\sigma}\right) = \Phi\left(\dfrac{a - \mu}{\sigma}\right)$
This is the usual standardisation process, which we have already used extensively.

The cdf of X is $F(x) = P(X \leqslant x) = \Phi\left(\dfrac{x - \mu}{\sigma}\right)$, and the pdf of X can be found by differentiating $F(x)$.

Writing $F(x) = \Phi\left(\dfrac{x - \mu}{\sigma}\right) = \Phi(u)$ where $u = \dfrac{x - \mu}{\sigma}$,

the pdf of X is

$$f(x) = F'(x) = \Phi'(u) \times \frac{\mathrm{d}u}{\mathrm{d}x} = \Phi'(u) \times \frac{1}{\sigma}$$

Now $\Phi'(u) = \phi(u) = \dfrac{1}{\sqrt{2\pi}}\, e^{-u^2/2}$, so $f(x) = \dfrac{1}{\sigma\sqrt{2\pi}}\, e^{-u^2/2} = \dfrac{1}{\sigma\sqrt{2\pi}}\, e^{-[(x-\mu)/\sigma]^2/2}$

i.e.

$$\boxed{f(x) = \frac{1}{\sigma\sqrt{2\pi}}\, e^{-(x-\mu)^2/2\sigma^2}}$$

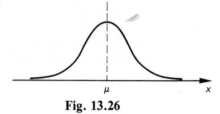

Fig. 13.26

Example 1
The random variable X has a Normal distribution with mean 50 and standard deviation 6, i.e. $X \sim N(50, 36)$.
State the pdf of X.

We have $\mu = 50$, $\sigma = 6$, so the pdf of X is

$$f(x) = \frac{1}{6\sqrt{2\pi}}\, e^{-(x-50)^2/72}$$

Example 2

A random variable X has pdf $f(x) = \dfrac{1}{\sqrt{5\pi}}\, e^{-(x-15)^2/5}$.
State the mean and the standard deviation of X, and find $P(X < 14)$

The pdf is $\dfrac{1}{\sigma\sqrt{2\pi}}\, e^{-(x-\mu)^2/2\sigma^2}$ with $\mu = 15$ and $2\sigma^2 = 5$

i.e. $\sigma = \sqrt{2.5}$

so X has a Normal distribution with mean 15 and standard deviation $\sqrt{2.5} \approx 1.5811$

$$P(X < 14) = P\left(Z < \frac{14 - 15}{1.5811}\right) = P(Z < -0.632) = \Phi(-0.632)$$

$$= 1 - 0.7363$$

$$= 0.2637$$

Truncated Normal distributions

Example 3
The oranges delivered to a supermarket have diameters which are Normally distributed with mean 7.5 cm and standard deviation 0.8 cm. Those with diameters less than 8.0 cm are sold as 'small' oranges, and the remainder are sold as 'large'.

Find (i)   the proportion of oranges which are sold as 'small'
    (ii)   the median diameter of a 'small' orange
    (iii)   the pdf for the diameters of 'small' oranges
    (iv)   an expression for the mean diameter of a 'small' orange.

(i)   P(diameter less than 8.0 cm) $= \mathrm{P}\left(Z < \dfrac{8.0 - 7.5}{0.8}\right)$

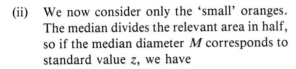

$$= \mathrm{P}(Z < 0.625)$$

$$= 0.7340$$

(ii) We now consider only the 'small' oranges.
The median divides the relevant area in half,
so if the median diameter M corresponds to
standard value z, we have

$$\Phi(z) = \tfrac{1}{2} \times 0.7340 = 0.3670$$
$$z = -0.3398$$

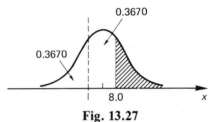

Thus $\dfrac{M - 7.5}{0.8} = -0.3398$, and the median

diameter is $M = 7.23$ cm

(iii) The pdf for the diameters of the 'small' oranges
will have the same shape as the Normal curve for
all oranges, but 'cut off' at 8.0 cm.
This is a 'truncated' Normal distribution. The pdf
for all oranges is

$$\frac{1}{0.8\sqrt{2\pi}} \, e^{-(x-7.5)^2/2 \times 0.8^2}$$

Fig. 13.28

but the area under this curve, when cut off at 8.0 cm, is 0.7340.
To obtain the pdf for the diameters of the 'small' oranges, we must multiply by

$\dfrac{1}{0.7340}$ to make the area one. Thus the pdf is

$$f(x) = \begin{cases} \dfrac{1}{0.7340} \times \dfrac{1}{0.8\sqrt{2\pi}} \, e^{-(x-7.5)^2/2 \times 0.8^2} = 0.6794 \, e^{-(x-7.5)^2/1.28} & \text{if } x < 8.0 \\[2mm] 0 & \text{otherwise} \end{cases}$$

(iv) The mean diameter of the 'small' oranges is

$$\int_{-\infty}^{8.0} x f(x) \, \mathrm{d}x = 0.6794 \int_{-\infty}^{8.0} x \, e^{-(x-7.5)^2/1.28} \, \mathrm{d}x$$

(This integral may be evaluated as follows:

$$\int_{-\infty}^{8.0} xf(x)\,\mathrm{d}x = \int_{-\infty}^{8.0} (x-7.5)f(x)\,\mathrm{d}x + \int_{-\infty}^{8.0} 7.5f(x)\,\mathrm{d}x$$

$$= 0.6794 \int_{-\infty}^{8.0} (x-7.5)\,\mathrm{e}^{-(x-7.5)^2/1.28}\,\mathrm{d}x + 7.5$$

$$\left(\text{since } \int_{-\infty}^{8.0} f(x)\,\mathrm{d}x = 1\right)$$

$$= 0.6794 \times 0.64\left[-\mathrm{e}^{-(x-7.5)^2/1.28}\right]_{-\infty}^{8.0} + 7.5$$

$$\approx 7.14 \text{ cm})$$

Fitting a Normal distribution

We now consider methods for checking whether a given set of data could reasonably have been taken from a Normal distribution. This may be important because several statistical tests (such as t-tests, see p. 330, and tests for sample variance, see p. 325) are based on the assumption that the underlying variable is Normally distributed.

Suppose we are given the following data about the total hours of sunshine during the month of June, measured over a period of 80 years.

Hours of sunshine (nearest hour)	141–150	151–160	161–170	171–180	181–190
Number of years	2	3	11	26	21

Hours of sunshine	191–200	201–210	211–220	221–230
Number of years	10	5	1	1

Method 1 (Expected frequencies)
We first find the mean and standard deviation of the sample,

$$\text{i.e. } \bar{x} = \frac{145.5 \times 2 + 155.5 \times 3 + \cdots}{80} = 180.75$$

and $s = 14.403$

We then calculate the probabilities, and hence the expected frequencies for a Normal distribution having this mean and standard deviation. For example, the class 161–170 hours has class boundaries 160.5 and 170.5, so for a Normal distribution with mean 180.75 and standard deviation 14.403,

$$P(160.5 < X < 170.5) = P\left(\frac{160.5 - 180.75}{14.403} < Z < \frac{170.5 - 180.75}{14.403}\right)$$

$$= P(-1.406 < Z < -0.712)$$

$$= 0.1583$$

and the expected frequency is $0.1583 \times 80 \approx 12.7$.

We obtain the following table.

Hours of sunshine	< 150.5	150.5–160.5	160.5–170.5	170.5–180.5		
Expected frequency	1.4	5.0	12.7	20.4		
Actual frequency	2	3	11	26		

Hours of sunshine	180.5–190.5	190.5–200.5	200.5–210.5	210.5–220.5	> 220.5
Expected frequency	20.6	13.1	5.3	1.3	0.2
Actual frequency	21	10	5	1	1

The expected frequencies are quite close to the actual frequencies, so this sample could reasonably have been taken from a Normal population.
This method is quite lengthy (although it does have the advantage that a χ^2 goodness of fit test could now be applied, see p. 316).
The following graphical methods are often used instead.

Method 2 (Graphical)
We find the cumulative probabilities (i.e. cumulative frequencies divided by the total frequency) for the data. Then, using inverse Normal tables, we find the values (z) in a standard Normal distribution which would give these cumulative probabilities.
For example, for the class 160–170 hours, the cumulative probability is $\frac{16}{80} = 0.2$; if $\Phi(z) = 0.2$, then $z = -0.842$.
We plot these z-values against the upper class boundaries (x)

Upper class boundary x	150.5	160.5	170.5	180.5	190.5	200.5	210.5	220.5	230.5
Cumulative frequency	2	5	16	42	63	73	78	79	80
Cumulative probability	0.025	0.0625	0.2	0.525	0.7875	0.9125	0.975	0.9875	1
z	−1.960	−1.534	−0.842	0.063	0.798	1.357	1.960	2.242	—

For a Normal distribution, the graph of z against x should be a straight line (with equation $z = (x - \mu)/\sigma$, where μ and σ are the mean and the standard deviation).
In this example, the points are fairly close to a straight line (which we have drawn 'by eye'), suggesting that this sample could have come from a Normal distribution.
We can use the line to estimate the mean and standard deviation (see Fig. 13.29).
When $z = 0$, the mean is $\mu \approx 180$.
When $z = 1$ (one standard deviation above the mean), $\mu + \sigma \approx 195$, so the standard deviation is $\sigma \approx 15$.

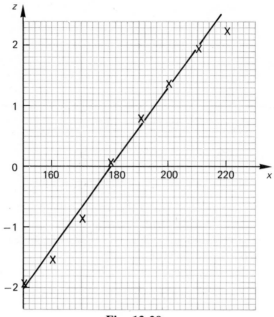

Fig. 13.29

Method 3 (Normal probability paper)
Normal probability paper is a special type of graph paper. One scale is an ordinary linear scale, but the other is marked in probabilities, and is arranged so that the cumulative probability curve for a Normal distribution would be a straight line.

We find the cumulative probabilities, and plot them (on the probability scale) against the upper class boundaries (on the ordinary scale) (Fig. 13.30).

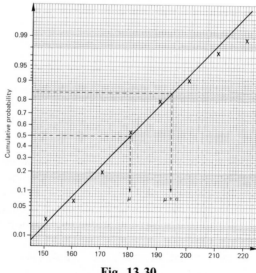

Fig. 13.30

Upper class boundary	150.5	160.5	170.5	180.5	190.5	200.5	210.5	220.5	230.5
Cumulative probability	0.025	0.0625	0.2	0.525	0.7875	0.9125	0.975	0.9875	1

We estimate the mean (when cumulative probability is 0.5) is $\mu \approx 180$.

At one standard deviation above the mean, the cumulative probability is $\Phi(1) = 0.8416$; so $\mu + \sigma \approx 195$, and the standard deviation is $\sigma \approx 15$.

This method is exactly equivalent to Method 2, but is slightly quicker since there is no need to look up the values of z.

Exercise 13.5   Normal distributions

1   Write down the pdf for each of the following Normal distributions:
 (i)   mean 2, standard deviation 3 (ii) mean 0, standard deviation 5
 (iii) mean 57.6, standard deviation 2.4 (iv) mean -10, standard deviation 2
 (v)   N(12, 8) (this means that the *variance* is 8).

2   State the mean and standard deviation of the random variables with the following pdfs:

 (i) $f(x) = \dfrac{1}{4\sqrt{2\pi}}\, e^{-x^2/32}$ (ii) $f(x) = \dfrac{1}{\sqrt{10\pi}}\, e^{-(x-25)^2/10}$

 (iii) $f(x) = \dfrac{1}{\sqrt{\pi}}\, e^{-(x+6)^2}$

3   A random variable X has pdf $f(x) = Ae^{-(x-4)^2/3}$ where A is a constant.
 Find (i)   the mean and variance of X
      (ii) the value of the constant A
      (iii) the probability that X is between 4 and 5.

4   A sugar packing machine produces bags whose weights are distributed Normally with mean 1002 g and standard deviation 5 g. The bags are then weighed individually, and only those weighing more than 1000 g are passed for sale.
 (i)   Find the probability that a bag produced by the machine is passed for sale.
 (ii) Find the median weight of bags which are passed for sale.
 (iii) Sketch the pdf $f(x)$ of the weights of bags which are passed for sale and show that

$$f(x) = \begin{cases} 0.122\, e^{-(x-1002)^2/50} & \text{if } x > 1000 \\ 0 & \text{otherwise} \end{cases}$$

5   The marks in an A level examination have mean 50 and standard deviation 20. Candidates with marks between 42 and 55 are awarded grade C. Assuming that the marks are Normally distributed, sketch the pdf of the marks for candidates awarded grade C, and find its equation.
 Write down as an integral an expression for the mean mark of candidates who are awarded grade C. (Make no attempt to evaluate this integral).

6 Calculate the mean and standard deviation for the following data:

Heights (cm)	150–155	155–160	160–165	165–170	170–175
Frequency	3	6	20	38	47
Heights (cm)	175–180	180–185	185–190	190–195	195–200
Frequency	51	42	26	10	7

Calculate the expected frequencies for a Normal distribution having the same mean and standard deviation.

7 For the data given in Question 6, calculate the cumulative probabilities, and find the values of a standard Normal variable (z) having the same cumulative probabilities.
   Plot the values of z against the upper class boundaries. Hence show that the data could have been taken from a Normal distribution, and estimate the mean and the standard deviation.

8 If Normal probability paper is available, plot the cumulative probabilities against the upper class boundaries for the data in Question 6. Hence show that the data is approximately Normal, and estimate the mean and the standard deviation.

13.6   Random sampling from a distribution

We have seen (on p. 212) how random number tables can be used to select a random sample from a population whose members have been listed and numbered.
   We now demonstrate some ways in which random numbers can be used to simulate taking a random sample from a theoretical distribution.
   In this section we shall use the following random digits, which were obtained on a calculator.

```
33975   65964   91535   30580   45610   93031
25393   57520   99885   81727   01996   61857
53934   09454   15024   57665   76898   15326
54566   79802   49618   07242   82830   44464
24665   40284   25111   72568   57643   94403
34826   65705   45410   61030   78344   77200
44570   60345   32119   76441   48636   55011
46978   78802   17336   73265   91730   58416
98004   70185   00558   05217   90195   02766
02453   12245   20071   44521   64121   88631
```

Example 1
Simulate 15 throws of a fair die.

   We read the digits one at a time, ignoring numbers outside the range 1–6. Using the

first row of the random numbers above, we obtain

$$3, 3, 5, 6, 5, 6, 4, 1, 5, 3, 5, 3, 5, 4, 5.$$

Example 2

Select 10 numbers at random between 1 and 30.

We read the digits in pairs, giving numbers between 00 and 99. We could just ignore numbers outside the range 01–30, but this would result in most pairs being ignored. We shall use the following method:

01–30 : select that number
31–60 : subtract 30
61–90 : subtract 60
00, 91–99 : ignore.

Starting on the second row of random numbers above, reading in pairs gives:

$$25 \quad 39 \quad 35 \quad 75 \quad 20 \quad 99 \quad 88 \quad 58 \quad 17 \quad 27 \quad 01 \quad \dots$$

and the required numbers are

$$25, 9, 5, 15, 20, 28, 28, 17, 27, 1$$

Example 3

Obtain a random sample of 10 values from the Binomial distribution $B(4, \frac{1}{3})$

If $X \sim B(4, \frac{1}{3})$ the probability distribution is

X	0	1	2	3	4
Probability	$\dfrac{16}{81}$	$\dfrac{32}{81}$	$\dfrac{24}{81}$	$\dfrac{8}{81}$	$\dfrac{1}{81}$

If we read random digits in pairs, ignoring numbers outside the range 01–81, we shall have 81 equally likely numbers. We need to allocate 16 of these numbers to the value 0, 32 numbers to the value 1, and so on, to obtain the correct probabilities. We allocate the numbers 01–81 as follows

01–16 : 0
17–48 : 1
49–72 : 2
73–80 : 3
81 : 4
00, 82–99 : ignore.

Starting on the third row of the random numbers above, reading in pairs gives

$$53 \quad 93 \quad 40 \quad 94 \quad 54 \quad 15 \quad 02 \quad 45 \quad 76 \quad 65 \quad 76 \quad 89 \quad 81$$

So a random sample from $B(4, \frac{1}{3})$ is

$$2, 1, 2, 0, 0, 1, 3, 2, 3, 4.$$

Note that $\dfrac{16}{81}, \dfrac{48}{81}, \dfrac{72}{81}, \dots$ are the cumulative probabilities $P(X \leqslant 0), P(X \leqslant 1), P(X \leqslant 2), \dots$

This suggests the systematic approach in the following Example.

Example 4

Obtain a random sample of 8 values from the Poisson distribution with mean 1.4.

If $X \sim$ Poisson (1.4), then

$$
\begin{array}{ll}
P(X = 0) = \quad e^{-1.4} \quad\quad = 0.247 & P(X \leqslant 0) = 0.247 \\
P(X = 1) = \quad e^{-1.4} \times 1.4 \quad = 0.345 & P(X \leqslant 1) = 0.592
\end{array}
$$

$$
P(X = 2) = e^{-1.4} \times \frac{(1.4)^2}{2!} = 0.242 \qquad P(X \leqslant 2) = 0.833
$$

$$
\begin{array}{ll}
P(X = 3) & = 0.113 \qquad P(X \leqslant 3) = 0.946 \\
P(X = 4) & = 0.039 \qquad P(X \leqslant 4) = 0.986 \\
P(X = 5) & = 0.011 \qquad P(X \leqslant 5) = 0.997 \\
P(X = 6) & = 0.003 \qquad P(X \leqslant 6) = 0.999
\end{array}
$$

We read random digits in threes, to obtain numbers between .000 and .999. We ignore .000, and allocate values as follows:

$$
\begin{array}{l}
.001-.247 : 0 \\
.248-.592 : 1 \\
.593-.833 : 2 \\
.834-.946 : 3 \\
.947-.986 : 4 \\
.987-.997 : 5 \\
.998-.999 : 6
\end{array}
$$

Starting on the fourth row of random numbers above, reading in threes gives:

.545   .667   .980   .249   .618   .072   .428   .283

So a random sample from Poisson (1.4) is

1, 2, 4, 1, 2, 0, 1, 1

Note   This method is only approximate; for example, the probability of selecting 1 is $\frac{345}{999} = 0.345\,345\ldots$

whereas the true probability is $e^{-1.4} \times 1.4 = 0.345\,235\ldots$

Also, no values greater than 6 will be selected.

The accuracy could be improved by working to more decimal places.

For a continuous distribution, we use random number tables to obtain numbers between 0 and 1.

We regard these as cumulative probabilities p, and find the corresponding values x, such that x has cumulative probability p, i.e. such that $F(x) = p$ where $F(x)$ is the cumulative distribution function (cdf).

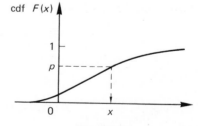

Fig. 13.31

Working to 3 decimal places, we read random digits in threes, ignoring .000, to obtain numbers between 0.001 and 0.999.

We shall use the same numbers as in Example 4,

.545   .667   .980   .249   .618   .072   .428   .283

to obtain random samples of size 8 from the continuous distributions in Examples 5–8.

Example 5

The rectangular distribution between 3 and 7.

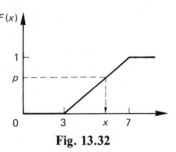

Fig. 13.32

The pdf is $f(x) = \frac{1}{4}$ (for $3 \leqslant x \leqslant 7$) and the cdf is $F(x) = \frac{1}{4}x - \frac{3}{4}$

If $F(x) = p$, then
$$x = 3 + 4p$$

Hence the first value in our random sample is $3 + 4 \times 0.545 = 5.180$ and the complete sample is

5.180,   5.668,   6.920,   3.996,   5.472,   3.288,   4.712,   4.132

Example 6

The distribution with pdf

$$f(x) = \begin{cases} \dfrac{2}{x^2} & \text{if } 1 \leqslant x \leqslant 2 \\ 0 & \text{otherwise} \end{cases}$$

The cdf is $F(x) = -\dfrac{2}{x} + C$,

and $F(1) = 0 \Rightarrow C = 2$,   thus $F(x) = 2 - \dfrac{2}{x}$ (for $1 \leqslant x \leqslant 2$)

If $F(x) = p$, then $2 - \dfrac{2}{x} = p$

giving $x = \dfrac{2}{2 - p}$

Hence the first value in our random sample is $\dfrac{2}{2 - 0.545} = 1.375$, and the complete sample is

1.375,   1.500,   1.961,   1.142,   1.447,   1.037,   1.272,   1.165

Example 7

The standard Normal distribution.

The cdf is $\Phi(x)$. To find values x such that $\Phi(x) = p$, we use the inverse Normal tables (on p. 396).

When $p = 0.545$, we have $x = 0.113$, and so on.
The sample is

$$0.113, \quad 0.432, \quad 2.054, \quad -0.678, \quad 0.300, \quad -1.461, \quad -0.182, \quad -0.574$$

Example 8
The Normal distribution with mean 36 and standard deviation 5.

We first obtain a random sample from the standard Normal distribution, as in Example 7. We then apply the transformation

$$X = 5Z + 36$$

Hence the first value in our random sample is $5 \times 0.113 + 36 = 36.57$, and the complete sample is

$$36.57, \quad 38.16, \quad 46.27, \quad 32.61, \quad 37.50, \quad 28.70, \quad 35.09, \quad 33.13$$

Note   The random samples obtained in Examples 4 to 8 are not independent of each other, since we used the same random numbers to generate each one. If we require independent random samples, we must use a different sequence of random numbers to generate each sample.

Exercise 13.6   Simulated random sampling

Use the random digits given on p. 294, or obtain your own from a calculator, computer, or random number tables.

For each question, list the random digits you are using, and explain your method clearly.

All the random samples you give in this Exercise should be independent of each other.

1   Simulate throwing a coin 20 times.

2   Select 12 numbers at random between 1 and 400.

3   Obtain a random sample of 6 values from the Binomial distribution $B(6, \frac{1}{2})$.

4   Obtain a random sample of 8 values from the Binomial distribution $B(3, \frac{7}{9})$.

5   Obtain a random sample of 8 values from the Poisson distribution with mean 2.

6   Obtain a random sample of 10 values from the following discrete distribution:

X	1	2	3	4	5
Probability	0.08	0.25	0.34	0.20	0.13

7   Obtain a random sample of 8 values from the continuous rectangular distribution between 0 and 3.

8 Obtain a random sample of 5 values from the continuous distribution with pdf

$$f(x) = \begin{cases} 3x^2 & \text{if } 0 \leqslant x \leqslant 1 \\ 0 & \text{otherwise} \end{cases}$$

9 For the exponential distribution with pdf

$$f(x) = \begin{cases} 0.2e^{-0.2x} & \text{if } x \geqslant 0 \\ 0 & \text{if } x < 0 \end{cases}$$

find the cdf $F(x)$,
and express x in terms of p if $F(x) = p$.
  Hence obtain a random sample of 6 values from this distribution.

10 Obtain a random sample of 10 values from the Normal distribution with mean 50 and
standard deviation 20.
Calculate the mean and the standard deviation of your sample.

14
Hypothesis Testing

14.1 Tests for the population mean

Consider a population having mean μ and standard deviation σ. We have seen, in Chapter 10, how to test a *null hypothesis* H_0 asserting that μ takes some particular value, i.e. $H_0 : \mu = \mu_0$ against an *alternative hypothesis* H_1

$$\text{such as } H_1 : \mu \neq \mu_0 \quad \text{(a two-tailed test)}$$
$$\text{or } H_1 : \mu > \mu_0 \quad \text{(a one-tailed test)}.$$

The decision is based on the mean $\bar{X}$ of a random sample from the population.

In this Chapter we shall show how these ideas can be extended to a wide variety of situations.

Suppose that a random sample of size n from the population has sample mean $\bar{X}$. We have $E[\bar{X}] = \mu$ and var $(\bar{X}) = \dfrac{\sigma^2}{n}$

We need to assume that
  (i) the sample mean $\bar{X}$ has an approximately Normal distribution. This will be true if
     EITHER *the population itself has an approximately Normal distribution*
     OR   *the sample size n is reasonably large*
and (ii) the population standard deviation σ is known
     OR   *the sample is sufficiently large (say n > 50), so that it is reasonable to estimate σ by the sample standard deviation s*

Both these conditions are automatically satisfied for large samples; but note that the methods of this section can also be applied to very small samples provided that σ is known and the population is approximately Normally distributed.

The sample mean $\bar{X}$ is approximately Normal, with mean μ and standard deviation $\sigma/\sqrt{n}$, and so $\dfrac{\bar{X} - \mu}{\sigma/\sqrt{n}}$ is (approximately) standard Normal.

We may write $\dfrac{\bar{X} - \mu}{\sigma/\sqrt{n}} \sim N(0, 1)$

Hence

$$\boxed{\begin{array}{c} \text{to test } H_0 : \mu = \mu_0 \\[2mm] \text{we consider } Z = \dfrac{\bar{X} - \mu_0}{\sigma/\sqrt{n}} \\[2mm] \text{If } H_0 \text{ is true, then } Z \sim N(0, 1) \end{array}}$$

A 'rejection region' is formed, consisting of values of Z which favour the alternative hypothesis H_1, and such that, when H_0 is true (and so $Z \sim N(0, 1)$), the probability that Z lies in this region is equal to the *significance level* of the test (usually 5%, unless specified otherwise).

We then calculate the value of the 'test statistic' $Z = \dfrac{\bar{X} - \mu_0}{\sigma/\sqrt{n}}$

for the particular sample given.

If the value lies in the rejection region, we reject H_0;

$$\text{otherwise we accept } H_0$$

Two types of error are possible:

Type 1 error   to reject H_0 when it is actually true.
  The probability of making this error is equal to the significance level of the test.

Type 2 error   to accept H_0 when it is actually false.

Example 1

The contents of a bottle of wine sold by a vineyard have mean μ ml with standard deviation 12 ml. An inspector tests the null hypothesis, $H_0 : \mu = 700$, against the alternative, $H_1 : \mu < 700$, by measuring the mean contents $\bar{X}$ ml of a random sample of 50 bottles, and using the 5% level of significance.

(i) For what values of $\bar{X}$ will H_0 be accepted?

(ii) If, in fact, $\mu = 696$, find the probability that H_0 will be accepted.

(i) Let $Z = \dfrac{\bar{X} - 700}{\sigma/\sqrt{n}} = \dfrac{\bar{X} - 700}{12/\sqrt{50}}$

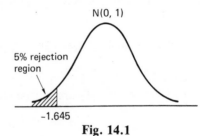

N(0, 1)

5% rejection region

-1.645

Fig. 14.1

If H_0 is true, then $Z \sim N(0, 1)$;
and H_1 is favoured by negative values of Z.
H_0 will be accepted if $Z > -1.645$

$$\text{i.e. if } \frac{\bar{X} - 700}{12/\sqrt{50}} > -1.645$$

$$\text{i.e.} \qquad \bar{X} > 700 - 1.645 \times \frac{12}{\sqrt{50}}$$

$$\text{i.e.} \qquad \bar{X} > 697.21$$

(ii) $\bar{X}$ is (approximately) Normally distributed, with mean μ and standard deviation $\sigma/\sqrt{n}$
  If $\mu = 696$,

  then $\bar{X}$ has mean 696 and standard deviation $\dfrac{12}{\sqrt{50}}$

  Thus

  $P(H_0 \text{ is accepted}) = P(\bar{X} > 697.21)$

  $$= P\left(Z' > \frac{697.21 - 696}{12/\sqrt{50}}\right)$$

  $$= P(Z' > 0.712)$$

  $$= 1 - 0.7617$$

  $$= 0.2383$$

Note If $\mu = 696$, then H_1 is true (the vineyard is selling underfilled bottles); but there is still about a 24% chance that H_0 will be accepted (i.e. the inspector will not detect that the bottles are underfilled). Since we are accepting H_0 when it is, in fact, false, this is an example of a Type 2 error.

Difference in means

Consider two populations,
>the first having mean μ_1 and standard deviation σ_1
>the second having mean μ_2 and standard deviation σ_2

We may wish to decide whether the two population means (μ_1 and μ_2) are equal or different; although we may not have any ideas about the actual values of either of them.

The null hypothesis will assert that the means are equal, i.e. $H_0 : \mu_1 = \mu_2$

The alternative hypothesis may be $H_1 : \mu_1 \neq \mu_2$ (for a two-tailed test),
or, for example, if we have a preconceived idea that the second population has a higher mean,

>the alternative hypothesis is $H_1 : \mu_1 < \mu_2$ (for a one-tailed test).

Suppose that we have two INDEPENDENT *samples, one from each population,* a random sample of size n_1 from the first population having mean $\bar{X}_1$ and a random sample of size n_2 from the second population having mean $\bar{X}_2$.

We have,

$\bar{X}_1$ is approximately Normal with mean μ_1 and standard deviation $\dfrac{\sigma_1}{\sqrt{n_1}}$

$\bar{X}_2$ is approximately Normal with mean μ_2 and standard deviation $\dfrac{\sigma_2}{\sqrt{n_2}}$

Then $E[\bar{X}_1 - \bar{X}_2] = E[\bar{X}_1] - E[\bar{X}_2] = \mu_1 - \mu_2$
and, since $\bar{X}_1$ and $\bar{X}_2$ are independent,

$$\text{var} (\bar{X}_1 - \bar{X}_2) = \text{var} (\bar{X}_1) + \text{var} (\bar{X}_2) = \frac{\sigma_1^2}{n_1} + \frac{\sigma_2^2}{n_2}$$

So $(\bar{X}_1 - \bar{X}_2)$ is approximately Normal,

with mean $(\mu_1 - \mu_2)$ and standard deviation $\sqrt{\dfrac{\sigma_1^2}{n_1} + \dfrac{\sigma_2^2}{n_2}}$,

and thus

$$\frac{(\bar{X}_1 - \bar{X}_2) - (\mu_1 - \mu_2)}{\sqrt{\dfrac{\sigma_1^2}{n_1} + \dfrac{\sigma_2^2}{n_2}}} \sim N(0, 1)$$

Hence

> to test $H_0 : \mu_1 = \mu_2$
>
> we consider $Z = \dfrac{\bar{X}_1 - \bar{X}_2}{\sqrt{\dfrac{\sigma_1^2}{n_1} + \dfrac{\sigma_2^2}{n_2}}}$
>
> If H_0 is true, then $Z \sim N(0, 1)$

Example 2
The mid-day temperatures at two Mediterranean resorts A and B are being compared.
For a random sample of 250 days at resort A, the temperatures had mean 27.2°C and standard deviation 3.9°C.
For a random sample of 180 days at resort B, the temperatures had mean 28.1°C and standard deviation 5.6°C.
At what level of significance does this indicate that the mean mid-day temperatures at these two resorts are different?

Suppose that
    the temperatures at resort A have mean μ_1 and standard deviation σ_1,
and the temperatures at resort B have mean μ_2 and standard deviation σ_2.
We test $H_0 : \mu_1 = \mu_2$
against $H_1 : \mu_1 \neq \mu_2$
If a random sample of n_1 days at resort A has mean temperature $\bar{X}_1$, and
    a random sample of n_2 days at resort B has mean temperature $\bar{X}_2$,
we consider

$$Z = \frac{\bar{X}_1 - \bar{X}_2}{\sqrt{\dfrac{\sigma_1^2}{n_1} + \dfrac{\sigma_2^2}{n_2}}}$$

If H_0 is true, then $Z \sim N(0, 1)$.
We have $n_1 = 250$, $n_2 = 180$, $\bar{X}_1 = 27.2$, $\bar{X}_2 = 28.1$ and since the samples are large, we can estimate $\sigma_1 \approx 3.9$, $\sigma_2 \approx 5.6$ thus

$$Z = \frac{27.2 - 28.1}{\sqrt{\dfrac{3.9^2}{250} + \dfrac{5.6^2}{180}}} = -1.856$$

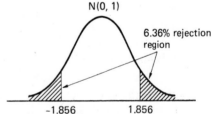

N(0, 1)

6.36% rejection region

−1.856   1.856

Fig. 14.2

If H_0 is true (so that $Z \sim N(0, 1)$)
then $P(Z < -1.856) = 0.0318$.
Since this is a two-tailed test, the value $Z = -1.856$ is significant at a level of $2 \times 3.18\%$, i.e. 6.36%
Since this is greater than 5%, this would not usually be regarded as enough evidence to suggest that the mean temperatures at the two resorts are different.

Note It is important that the two populations are sampled independently. If instead we had selected (say) 200 days at random, and measured the temperatures at the two resorts on these days, then we would have a random sample of size 200 from each population, but the two samples would not be independent (since the same days were used in each case). The result

$$\text{var } (\bar{X}_1 - \bar{X}_2) = \text{var } (\bar{X}_1) + \text{var } (\bar{X}_2) = \frac{\sigma_1^2}{n_1} + \frac{\sigma_2^2}{n_2}$$

would no longer be necessarily true, and the above test would be invalid. In such cases, a paired sample test (see p. 332) should be used.

Exercise 14.1   Tests for population mean

1   A machine produces packages whose weights are Normally distributed with standard deviation 0.25 kg. The mean weight should not be less than 7.5 kg. A random sample of packages is taken, and the machine is stopped if the mean weight of the sample is less than 7.4 kg.
   (i)   Given that a sample of 10 packages is taken, find the significance level of this test. Explain the meaning of this value.
   (ii) For what size of sample would the significance level be (approximately) 5%?

2   When bullets are fired from a rifle, the muzzle velocities are Normally distributed with mean μ metres per second and standard deviation 15 m s^{-1}.
   The null hypothesis $H_0 : \mu = 600$ is to be tested (using the 5% level of significance) against the alternative hypothesis $H_1 : \mu \neq 600$ by measuring the mean velocity $\bar{X}$ for a random sample of bullets.
   (i)   If a random sample of 75 bullets is used, for what values of $\bar{X}$ will H_0 be accepted? Given that, in fact, $\mu = 605$, state the mean and standard deviation of $\bar{X}$, and find the probability that H_0 will be accepted.
   (ii) If a random sample of 150 bullets is used, find the probability that H_0 will be accepted when, in fact, $\mu = 605$.
   Comment on the effect of the sample size on this probability.

3   In a study of the wildlife on two remote islands A and B, fieldmice were trapped and measured. The lengths of the 260 mice trapped on island A had mean 11.8 cm with standard deviation 1.4 cm; and the lengths of the 145 mice trapped on island B had mean 12.1 cm with standard deviation 1.6 cm. Does this suggest that the mice on the two islands have different mean lengths?

4   Some of the plum trees in an orchard were sprayed with insecticide in the hope that this would increase the yield. For a random sample of 80 trees which had been sprayed, the yields x_i kg were measured and $\sum x_i = 1210$, $\sum x_i^2 = 19275$. For a random sample of 80 unsprayed trees, the yields y_i kg were measured and $\sum y_i = 1116$, $\sum y_i^2 = 16850$. Is there sufficient evidence to say that spraying with insecticide increases the mean yield?

5   The ages at which babies learn to walk may be assumed to be Normally distributed with standard deviation 1.8 months. Random samples of babies were taken in two different areas, and the ages (in months) at which they learnt to walk were as follows:

   Area 1 : 14.5, 13.9, 15.2, 14.0, 13.3, 9.8, 11.9, 16.6
   Area 2 : 14.0, 10.0, 10.5, 14.6, 11.4

   Do these figures demonstrate a significant difference between the ages at which the babies learnt to walk in the two areas?

6   Independent random samples of sizes n_1 and n_2 are taken from populations having means μ_1 and μ_2 and standard deviations σ_1 and σ_2 respectively. Given that the sample means are $\bar{X}_1$ and $\bar{X}_2$, state the mean and the standard deviation of $(\bar{X}_1 - \bar{X}_2)$. Assum-

ing that $(\bar{X}_1 - \bar{X}_2)$ is (approximately) Normally distributed, show that

$$(\bar{X}_1 - \bar{X}_2) \pm 1.96 \sqrt{\frac{\sigma_1^2}{n_1} + \frac{\sigma_2^2}{n_2}}$$

are 95% confidence limits for $(\mu_1 - \mu_2)$.

A certain chemical process was repeated 150 times without a catalyst, then 100 times with a catalyst. The reaction times without the catalyst had mean 348 s with standard deviation 16 s; and with the catalyst the reaction times had mean 162 s, with standard deviation 12 s. Find 95% confidence limits for the reduction in mean reaction time caused by the catalyst.

7   It is thought that trees on a south-facing slope will grow taller than those on a north-facing slope. A random sample of 160 trees on south-facing slopes had mean height 28.3 m with standard deviation 4.6 m, and a random sample of 200 trees on north-facing slopes had mean height 26.7 m with standard deviation 3.5 m. Show that this strongly supports the view that the south-facing trees are taller, and give a 99% lower confidence limit for the difference in mean height.

8   A random sample of size n is taken from a population having a Poisson distribution with mean μ. Given that the sample mean is $\bar{X}$, explain why $\dfrac{\bar{X} - \mu}{\sqrt{\mu}/\sqrt{n}}$ has (approximately) the standard Normal distribution.

Over a long period, the daily number of customers visiting a shop has had a Poisson distribution with mean 75. In the 20 days following a big advertising campaign there were a total of 1565 customers. Is there any evidence that the advertising campaign has increased the mean number of customers?

9   On a certain stretch of road, the weekly number of accidents used to have a Poisson distribution with mean 5.8. The layouts of some of the road junctions were altered, and in a period of 52 weeks after these changes the average number of accidents was 6.6 per week. Does this indicate that there has been a change in the accident rate? (Use the 1% level of significance.)

14.2   Tests for proportion

Consider a population in which a proportion θ possess some characteristic (for example, in a population of people, θ might be the proportion of left-handed people; in the population of items produced at a factory, θ might be the proportion of defective items; and so on). We now consider how to test whether θ takes some particular value θ_0; we shall test the null hypothesis $H_0 : \theta = \theta_0$ against a suitable alternative (such as $H_1 : \theta \neq \theta_0$ or $H_1 : \theta > \theta_0$).

Suppose that in a random sample of n members of the population, X possess the characteristic.

Then X has the Binomial distribution $B(n, \theta)$.

We have $E[X] = n\theta$ and $\text{var}(X) = n\theta(1 - \theta)$

We assume that conditions are such that the Binomial distribution $B(n, \theta)$ can be approximated by a Normal distribution.

Then X is approximately Normal, with mean $n\theta$, and standard deviation $\sqrt{n\theta(1-\theta)}$

so $\dfrac{X-n\theta}{\sqrt{n\theta(1-\theta)}} \sim N(0,1)$

Hence

> to test $H_0 : \theta = \theta_0$
>
> we consider $Z = \dfrac{X-n\theta_0}{\sqrt{n\theta_0(1-\theta_0)}}$
>
> If H_0 is true, then $Z \sim N(0,1)$

Example 1

It is known that 12% of people watched the first episode of a new television serial. The following week, in a random sample of 500 people, it was found that 75 had watched the second episode. Does this suggest that the proportion of people watching the second episode was different from that watching the first?

Suppose that the proportion of people watching the second episode was θ; the null hypothesis will say that θ is the same as the proportion watching the first episode, i.e.

we test $H_0 : \theta = 0.12$
against $H_1 : \theta \neq 0.12$

If a random sample of n people is taken, of which X watched the second episode, we consider

$$Z = \frac{X - n \times 0.12}{\sqrt{n \times 0.12 \times 0.88}}$$

If H_0 is true, then $Z \sim N(0,1)$
We have $n = 500$, $X = 75$,
thus

$$Z = \frac{75 - 500 \times 0.12}{\sqrt{500 \times 0.12 \times 0.88}}$$

$$= 2.064$$

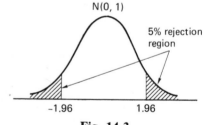

Fig. 14.3

Reject H_0.
This does suggest that the proportion watching the second episode was different from the first.

Continuity correction

We are using the Normal approximation to the Binomial distribution $B(n, \theta)$, so a continuity correction would be appropriate. However, this will only have a small effect on the value of Z, and we shall ignore it.

When calculating a significance level, the use of the continuity correction will give a more accurate value. For example, we shall calculate the significance level corresponding to $X = 75$ in Example 1 above.

If H_0 is true, then $X \sim B(500, 0.12)$,

and

$$P(X \geqslant 75) \approx P(X > 74.5 \text{ in Normal}) = P\left(Z > \frac{74.5 - 500 \times 0.12}{\sqrt{500 \times 0.12 \times 0.88}}\right)$$

$$= P(Z > 1.995)$$

$$= 1 - 0.9770$$

$$= 0.0230$$

Since the test is two-tailed, the result $X = 75$ is significant at a level of $2 \times 2.3\%$, i.e. 4.6%

Difference in proportions

Consider two populations, the first in which a proportion θ_1 possess some characteristic, and the second in which a proportion θ_2 possess some characteristic, for which we wish to test whether the two proportions θ_1 and θ_2 are the same; i.e. we shall test the null hypothesis $H_0 : \theta_1 = \theta_2$

Suppose we have two INDEPENDENT *samples, one from each population*; in a random sample of size n_1 from the first population, X_1 possess the characteristic, in a random sample of size n_2 from the second population, X_2 possess the characteristic.

Then $X_1 \sim B(n_1, \theta_1)$ so X_1 is approximately Normal with mean $n_1\theta_1$ and variance $n_1\theta_1(1 - \theta_1)$ and hence the sample proportion X_1/n_1 is approximately Normal,

with mean $\dfrac{n_1\theta_1}{n_1} = \theta_1$ and variance $\dfrac{n_1\theta_1(1 - \theta_1)}{n_1^2} = \dfrac{\theta_1(1 - \theta_1)}{n_1}$

Similarly, X_2/n_2 is approximately Normal,

with mean θ_2 and variance $\dfrac{\theta_2(1 - \theta_2)}{n_2}$

Then $\left(\dfrac{X_1}{n_1} - \dfrac{X_2}{n_2}\right)$ has mean $E\left[\dfrac{X_1}{n_1} - \dfrac{X_2}{n_2}\right] = E\left[\dfrac{X_1}{n_1}\right] - E\left[\dfrac{X_2}{n_2}\right]$

$$= \theta_1 - \theta_2$$

and (since the samples are independent),

$$\text{var}\left(\frac{X_1}{n_1} - \frac{X_2}{n_2}\right) = \text{var}\left(\frac{X_1}{n_1}\right) + \text{var}\left(\frac{X_2}{n_2}\right)$$

$$= \frac{\theta_1(1 - \theta_1)}{n_1} + \frac{\theta_2(1 - \theta_2)}{n_2}$$

thus $\dfrac{\left(\dfrac{X_1}{n_1} - \dfrac{X_2}{n_2}\right) - (\theta_1 - \theta_2)}{\sqrt{\dfrac{\theta_1(1 - \theta_1)}{n_1} + \dfrac{\theta_2(1 - \theta_2)}{n_2}}} \sim N(0, 1)$

When $\theta_1 = \theta_2$, we have $(\theta_1 - \theta_2) = 0$. We use $p = \dfrac{X_1 + X_2}{n_1 + n_2}$, the proportion in the two

samples combined, to estimate the common value of θ_1 and θ_2.

To test $H_0: \theta_1 = \theta_2$
we consider

$$Z = \frac{\dfrac{X_1}{n_1} - \dfrac{X_2}{n_2}}{\sqrt{\dfrac{p(1-p)}{n_1} + \dfrac{p(1-p)}{n_2}}} \qquad \text{where} \qquad p = \frac{X_1 + X_2}{n_1 + n_2}$$

If H_0 is true, then $Z \sim N(0, 1)$

Example 2

A firm supplies a certain electrical component in two qualities: 'standard' and 'super'. A customer bought 120 'standard' and 80 'super' components, and used all of them in similar applications. After three months, 33 of the 'standard' components and 13 of the 'super' components had failed. Does this provide sufficient evidence (at the 1% level of significance) to say that the 'super' components are better than the 'standard' ones?

Suppose that the probability of failing during the first three months is θ_1 for a 'standard' component, and θ_2 for a 'super' component.

We would expect the 'super' components to have a lower probability of failing, so we shall use a one-tailed test.

We test $H_0: \theta_1 = \theta_2$
against $H_1: \theta_1 > \theta_2$

If X_1 fail out of n_1 'standard' components and X_2 fail out of n_2 'super' components, we consider

$$Z = \frac{\dfrac{X_1}{n_1} - \dfrac{X_2}{n_2}}{\sqrt{\dfrac{p(1-p)}{n_1} + \dfrac{p(1-p)}{n_2}}} \qquad \text{where} \qquad p = \frac{X_1 + X_2}{n_1 + n_2}$$

If H_0 is true, then $Z \sim N(0, 1)$.
H_1 is favoured by positive values of Z.
We have $n_1 = 120$, $n_2 = 80$, $X_1 = 33$, $X_2 = 13$,
so $p = \dfrac{33 + 13}{120 + 80} = 0.23$

Thus

$$Z = \frac{\dfrac{33}{120} - \dfrac{13}{80}}{\sqrt{\dfrac{0.23 \times 0.77}{120} + \dfrac{0.23 \times 0.77}{80}}}$$

$$= 1.852$$

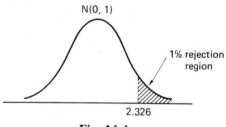

Fig. 14.4

Accept H_0.

There is *not* enough evidence to suggest that the 'super' components are better (at the 1% level of significance).

Exercise 14.2   Tests for proportion

1   A coin was tossed 200 times, and 'heads' appeared 112 times. Is there any evidence that the coin is biased?

2   A die was thrown 120 times, resulting in 30 sixes. At what level of significance does this indicate that the die is biased?
   Find an approximate 95% confidence interval for the probability of scoring six with this die.

3   The organisers of a correspondence course claim that 80% of their pupils complete the course successfully. Someone, who believed that the true proportion was less than 80%, looked at a random sample of 72 pupils and found that 50 of these had been successful. What should he conclude?

4   In the population as a whole, 24% of people wear glasses. In a random sample of 250 university students, it was found that 72 wore glasses. Does this show that the proportion of university students who wear glasses is any different from the proportion in the whole population?

5   One month, of 1000 people interviewed, 376 said that they would vote for the present government; the following month, 500 people were interviewed, of whom 152 said that they would vote for the government. Does this demonstrate a significant change in the proportion of people supporting the government?

6   A new drug is available for the treatment of a certain disease in sheep, and it is hoped that this drug will increase the chances of recovery. In trials, a sample of 120 sheep with the disease were divided at random into two groups of 60. One group was given the traditional treatment, and 37 recovered. The other group was given the new drug, and 48 recovered. Is there sufficient evidence to say, at the 1% level of significance, that the new drug increases the chances of recovery?

7   A manufacturer claims that not more than 10% of his fireworks are defective. A customer buys 20 fireworks, and discovers that 5 of these are defective. Explain why the Normal approximation to the Binomial distribution should not be used in this case.
   Assuming that 10% of fireworks are defective, calculate the probability that a random sample of 20 fireworks will contain 5 or more defective ones.
   Would you accept the hypothesis that the true proportion of defective fireworks is 10%?

8   If the label on a bottle of orange squash states that the volume is '2 litre e' (where e is the EEC mark indicating that the average system is being used), the following 3 conditions must be met:
   (A) The average contents of the bottles must not be less than 2000 ml.
   (B) Not more than 1 in 40 bottles may contain less than 1970 ml.
   (C) No bottles at all may contain less than 1940 ml.
   The contents of a random sample of 500 bottles were measured individually and

were found to be as follows:

Contents (in ml)	1940–1960	1960–1970	1970–1980	1980–1990	1990–2000
Number of bottles	5	12	56	83	162

Contents (in ml)	2000–2010	2010–2020	2020–2030	2030–2040	2040–2060
Number of bottles	88	35	30	13	16

Test (using the 5% level of significance in each case) whether
(i)   condition (A) is being met
(ii) condition (B) is being met.

14.3   The χ^2 test of goodness of fit

χ^2 distributions

If $Z_1, Z_2, \ldots, Z_n$ are independent random variables, each having the standard Normal distribution, then the random variable

$$Y^2 = Z_1^2 + Z_2^2 + \cdots + Z_n^2$$

is said to have the χ_n^2 distribution (the chi-squared distribution with n degrees of freedom). Y^2 is a continuous random variable taking non-negative values, and its pdf is

$$f(x) = \begin{cases} Cx^{n/2-1}\, e^{-x/2} & \text{if } x \geqslant 0 \\ 0 & \text{if } x < 0 \end{cases}$$

where C is a constant, chosen so that $\displaystyle\int_0^\infty f(x)\, dx = 1$.

For each positive integer n, there is a different χ^2 distribution, χ_n^2
For example, when $n = 8$ (Fig. 14.5), the pdf of the χ_8^2 distribution is

$$f(x) = Cx^3\, e^{-x/2}$$

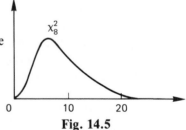

Fig. 14.5

and when $n = 15$ (Fig. 14.6), the pdf of the χ_{15}^2 distribution is

$$f(x) = Cx^{6.5}\, e^{-x/2}$$

Note that χ^2 distributions are positively skew.

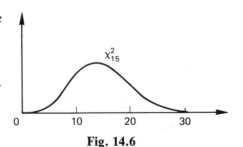

Fig. 14.6

The pdf is difficult to integrate (unless n is small and even). χ^2 tables (see p 398) give the value x which is exceeded with probability $p\%$ by a random variable having the χ^2_n distribution (Fig. 14.7).

i.e. if $Y^2 \sim \chi^2_n$, then $P(Y^2 > x) = \dfrac{p}{100}$

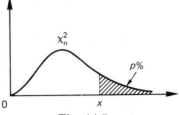

Fig. 14.7

For example,
when $n = 3$ and $p = 5$ (Fig. 14.8), the tables give $x = 7.815$.
If $Y^2 \sim \chi^2_3$, then $P(Y^2 > 7.815) = 0.05$

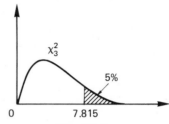

Fig. 14.8

When $n = 7$ and $p = 90$, the tables give $x = 2.833$.
If $Y^2 \sim \chi^2_7$, then $P(Y^2 > 2.833) = 0.90$

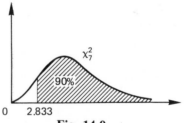

Fig. 14.9

Goodness of fit

On several occasions we have calculated expected frequencies from a theoretical distribution, and compared these with the actual frequencies. We now describe a test for how well the observed frequencies agree with the expected frequencies.

Consider n independent 'trials', where each trial results in one of k possible outcomes. (For example, if a die is thrown, there are 6 possible outcomes: 1, 2, 3, 4, 5, 6.
If a motor vehicle is selected at random, it falls into one of the 8 categories: motorcycle, car, van, bus, coach, lorry, agricultural, others.
A family selected at random can be classified according to the number of children: 'no children', '1 child', '2 children' or 'more than 2 children'; and so on).

The possible outcomes are sometimes called 'classes' or 'cells'. Suppose that the k possible outcomes have probabilities

$$p_1, p_2, ..., p_k, \text{ so that } p_1 + p_2 + \cdots + p_k = 1$$

then for n trials, the 'expected frequencies' are

$$np_1, np_2, ..., np_k$$

In the n trials, let
X_1 be the number of times that the 1st outcome actually occurs,
X_2 be the number of times that the 2nd outcome actually occurs, and so on

The random variables $X_1, X_2, ..., X_k$ are the 'observed frequencies' and we have

$$X_1 + X_2 + \cdots + X_k = n$$

Now X_1 is the number of times that the 1st outcome occurs in n independent trials, so $X_1 \sim B(n, p_1)$. Provided that np_1 is not too small, X_1 will be approximately Normal,

$$\text{and } \frac{X_1 - np_1}{\sqrt{np_1(1 - p_1)}} \sim N(0, 1).$$

Similarly $\dfrac{X_2 - np_2}{\sqrt{np_2(1 - p_2)}} \sim N(0, 1)$, and so on.

Thus
$$\frac{(X_1 - np_1)^2}{np_1(1 - p_1)} + \frac{(X_2 - np_2)^2}{np_2(1 - p_2)} + \cdots + \frac{(X_k - np_k)^2}{np_k(1 - p_k)},$$

being the sum of the squares of k standard Normal variables, might be expected to have the χ_k^2 distribution. However, $X_1, X_2, \cdots, X_k$ are not independent (in particular $X_1 + X_2 + ... + X_k = n$), and so the above argument is not valid. A proper theoretical treatment is difficult; in fact two modifications are needed. The factors $(1 - p_1)$, $(1 - p_2)$ and so on, disappear from the denominators, and the distribution is χ_{k-1}^2 instead of χ_k^2, i.e.

$$\frac{(X_1 - np_1)^2}{np_1} + \frac{(X_2 - np_2)^2}{np_2} + \cdots + \frac{(X_k - np_k)^2}{np_k}$$

has (approximately) the χ_{k-1}^2 distribution.

Now $X_1, X_2, ..., X_k$ are the observed frequencies (O) and $np_1, np_2, ... np_k$ are the expected frequencies (E) so this may be written as $\sum \dfrac{(O - E)^2}{E} \sim \chi_{k-1}^2$

Note $X_1, X_2, ..., X_k$ are whole numbers, so $\displaystyle\sum_{i=1}^{k} \dfrac{(X_i - np_i)^2}{np_i}$

has a discrete distribution, which we approximate by the continuous distribution χ_{k-1}^2. Strictly speaking, a continuity correction should be applied, but this is rather complicated and we shall ignore it.

A null hypothesis H_0, that the probabilities $p_1, p_2, ..., p_k$ have specified values, is tested against the alternative hypothesis H_1 (that the probabilities have some other values), as follows:

Assuming H_0, we calculate the expected frequencies (E); then we consider

$$Y^2 = \sum \frac{(O - E)^2}{E}$$

The test statistic Y^2 is a measure of how much the observed frequencies (O) differ from the expected frequencies (E).

If H_1 is true, we would expect O and E to differ considerably, so $(O - E)^2$ should be large; thus H_1 is favoured by large values of Y^2. We shall reject H_0 if Y^2 is sufficiently large.

If H_0 is true, then $Y^2 \sim \chi_{k-1}^2$
Using χ^2 tables, we choose the rejection region so as to give the required level of significance (usually 5%).

Example 1

A shoe manufacturer makes childrens shoes in five width fittings, C, D, E, F, G, in the following percentages

$$\text{C: 2\%   D: 8\%   E: 30\%   F: 40\%   G: 20\%}$$

A random sample of 500 children was taken, and their width fittings were:

$$\text{C: 12   D: 46   E: 171   F: 178   G: 93}$$

Does this sample suggest that the proportions of the five width fittings are different from those assumed by the manufacturer?

In this example, a 'trial' is to select one child, who is then classified as having width fitting C, D, E, F, or G; this is repeated 500 times.

We test H_0: the proportions are C: 2%   D: 8%   E: 30%   F: 40%   G: 20% against H_1: the proportions are different from those stated above. Assuming H_0, we have

Width fitting	C	D	E	F	G
probability	0.02	0.08	0.3	0.4	0.2
expected frequency (E)	10	40	150	200	100
observed frequency (O)	12	46	171	178	93
$\dfrac{(O-E)^2}{E}$	0.40	0.90	2.94	2.42	0.49

Let $Y^2 = \sum \dfrac{(O-E)^2}{E}$

Since there are 5 classes, if H_0 is true, then $Y^2 \sim \chi_4^2$

We have $Y^2 = 0.40 + 0.90 + 2.94 + 2.42 + 0.49$
$$= 7.15$$

Accept H_0

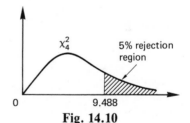

Fig. 14.10

This sample does *not* provide sufficient evidence (at the 5% level of significance) to suggest that the proportions are any different from those assumed by the manufacturer.

Sometimes we wish to test a null hypothesis H_0 which does not, in itself, give sufficient information for us to calculate the probabilities $p_1, p_2, ..., p_k$. Then it is necessary to use the observed frequencies $X_1, X_2, ..., X_k$ to estimate 'parameters' of the population (such as the mean) before we can estimate the probabilities, and hence the expected frequencies.

For example, when fitting a Poisson distribution to some data, we first find the mean of the sample; then we calculate the probabilities for a Poisson distribution having this mean.

In $\sum_{i=1}^{k} \frac{(X_i - np_i)^2}{np_i}$, the probabilities p_i now depend on $X_1, X_2, ..., X_k$ and so p_i are also random variables, making a proper theoretical treatment extremely complicated.

Remarkably, $\sum \frac{(O-E)^2}{E}$ still has (approximately) a χ^2 distribution, but for each time that the data is used to estimate a population parameter, one degree of freedom must be deducted.

The χ^2 test of goodness of fit is applicable to a great variety of situations, but the following conditions should always be checked:

(i)   It must be possible to identify with the general situation: n independent trials, each resulting in one of k possible classes.

(ii)   The classes must account for all possibilities (even if the data does not).

(iii) None of the expected frequencies should be much less than 5; if necessary, combine two (or more) classes into one.

To test a null hypothesis H_0, we calculate expected frequencies, assuming H_0 to be true, then consider

$$Y^2 = \sum \frac{(O-E)^2}{E}$$

If k classes are used to calculate Y^2, and the data has been used to estimate m population parameters, then,

when H_0 is true, $Y^2 \sim \chi^2_{k-1-m}$

Note   The observed frequencies (O) are actual frequencies of occurrence, and must therefore be whole numbers.

The expected frequencies (E) need not be whole numbers.

Example 2

Test (at the 1% level of significance) the fit of a Poisson distribution to the following data.

Number of computers sold in a day	0	1	2	3	4	5	6	7	8	9
Number of days	23	42	35	33	25	18	15	6	0	3

In this example, a 'trial' is to consider one day, which is classified according to the number of computers sold; this is repeated 200 times.

We test H_0: the distribution is Poisson

against H_1: the distribution is not Poisson.

The assumption of H_0, that the distribution is Poisson, is insufficient to calculate the probabilities, since we need to know the mean.

We find the sample mean,

$$\bar{x} = \frac{\sum xf}{\sum f} = \frac{560}{200} = 2.8,$$

and calculate the probabilities, and the expected frequencies, for the distribution Poisson (2.8).

Note that we must include an extra class 'more than 9 sales', to account for all possibilities (and to ensure that the probabilities add up to one).

Number of sales	0	1	2	3	4	5
Probability from Poisson (2.8)	0.0608	0.1703	0.2384	0.2225	0.1557	0.0872
Expected frequency	12.2	34.1	47.7	44.5	31.1	17.4

Number of sales	6	7	8	9	> 9
Probability from Poisson (2.8)	0.0407	0.0163	0.0057	0.0018	0.0006
Expected frequency	8.1	3.3	1.1	0.4	0.1

The last four classes have expected frequencies less than 5; we combine these into a single class '7 or more sales'.

Number of sales	0	1	2	3	4	5	6	$\geqslant 7$
Expected frequency (E)	12.2	34.1	47.7	44.5	31.1	17.4	8.1	4.9
Observed frequency (O)	23	42	35	33	25	18	15	9
$\dfrac{(O-E)^2}{E}$	9.56	1.83	3.38	2.97	1.20	0.02	5.88	3.43

Let $Y^2 = \sum \dfrac{(O-E)^2}{E}$.

We now have 8 classes; and the data was used to estimate the mean; hence the number of degrees of freedom is $8 - 1 - 1 = 6$

If H_0 is true, then $Y^2 \sim \chi^2_6$

We have $Y^2 = 9.56 + 1.83 + \cdots + 3.43$
$\qquad\quad = 28.27$

Reject H_0

There is very strong evidence that the daily number of sales does *not* have a Poisson distribution.

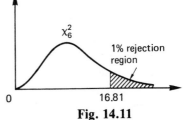

Fig. 14.11

Example 3

Examine whether the following sample could reasonably have come from a Normal distribution.

Hours of sunshine in June (nearest hr)	141–150	151–160	161–170	171–180	181–190
Number of years	2	3	11	26	21

Hours of sunshine in June (nearest hr)	191–200	201–210	211–220	221–230	
Number of years	10	5	1	1	

In this example, a 'trial' is to consider one year, which is classified according to the hours of sunshine during June; this is repeated 80 times.

We test H_0: the sample is taken from a Normal distribution

against H_1: the sample does not come from a Normal distribution.

This data was considered in Section 13.5, see p. 290.

We found the mean $\bar{x} = 180.75$ and standard deviation $s = 14.403$ of the sample, and calculated the expected frequencies for a Normal distribution having this mean and standard deviation, as follows.

Hours of sunshine	< 150.5	150.5 –160.5	160.5 –170.5	170.5 –180.5	180.5 –190.5	190.5 –200.5	200.5 –210.5	210.5 –220.5	> 220.5
Expected frequency (E)	1.4 6.4	5.0	12.7	20.4	20.6	13.1	5.3 6.8	1.3	0.2
Observed frequency (O)	5		11	26	21	10	7		
$\dfrac{(O-E)^2}{E}$	0.31		0.23	1.54	0.01	0.73	0.01		

Let $Y^2 = \sum \dfrac{(O-E)^2}{E}$.

We have used 6 classes in the calculation of $\sum \dfrac{(O-E)^2}{E}$; and the data was used to estimate 2 parameters (the mean and the standard deviation); hence the number of degrees of freedom is $6 - 1 - 2 = 3$

If H_0 is true, then $Y^2 \sim \chi^2_3$

We have $Y^2 = 0.31 + 0.23 + \cdots + 0.01$

$\qquad = 2.82$

Accept H_0.

The Normal distribution fits the data satisfactorily.

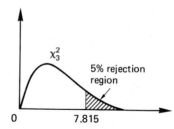

Fig. 14.12

Contingency tables

Consider a population which can be classified in two different ways (for example, a person can be classified by eye colour: brown or blue, and also by hair colour: dark, fair or ginger). The question which then arises is whether the two classifications are related in some way (for example, do people with blue eyes tend to have fair hair?), or whether they are independent.

In such cases, the null hypothesis H_0 always states that there is no association between the two classifications (i.e. they are independent of each other), because this is a precise statement which enables us to calculate expected frequencies. If instead we assumed that there *was* some association, we could not make any detailed calculations unless we knew precisely *how* the two classifications were related.

Example 4

For a random sample of 160 members of the armed forces, the branch of service (Navy, Army or Air-Force) and the type of the most recent school attended (State or Independent) are noted. The results are given in a 'contingency table'

	State	Independent	Totals
Navy	27	24	51
Army	52	18	70
Air Force	27	12	39
Totals	106	54	160

(where, for example, 52 members of the sample are in the Army and had attended a State School).

Calculate the expected frequencies on the assumption that there is no association between the branch of service and the type of school attended.

If the two classifications are independent, then, for example,

$$P(\text{Navy and State school}) = P(\text{Navy}) \times P(\text{State school})$$

Since 51 members of the sample are in the Navy, we estimate $P(\text{Navy}) = \dfrac{51}{160}$

Since 106 members had attended a State school, we estimate $P(\text{State school}) = \dfrac{106}{160}$

thus $P(\text{Navy and State school}) = \dfrac{51}{160} \times \dfrac{106}{160}$

and the expected frequency for 'Navy and State school' is

$$160 \times \frac{51}{160} \times \frac{106}{160} = 33.8$$

Similarly, the expected frequency for 'Navy and Independent school' is

$$160 \times \frac{51}{160} \times \frac{54}{160} = 17.2,$$

and so on.

We obtain the following expected frequencies (E)

E	State	Independent
Navy	33.8	17.2
Army	46.4	23.6
Air Force	25.8	13.2

We can now test the null hypothesis (that there is no association) by carrying out a χ^2 goodness-of-fit test, comparing these expected frequencies (E) with the observed frequencies (O) given in the original contingency table.

In this example, the 'trial' is to select one person, who then fits into one of the six compartments in the contingency table; this is repeated 160 times.

In calculating the expected frequencies, we needed to use the data to estimate

$$P(\text{Navy}) = \frac{51}{160}, \ P(\text{Army}) = \frac{70}{160} \text{ and } P(\text{State school}) = \frac{106}{160}$$

The remaining two probabilities are then estimated as

$$P(\text{Air Force}) = 1 - P(\text{Navy}) - P(\text{Army})$$
$$\text{and} \quad P(\text{Independent school}) = 1 - P(\text{State school}),$$

without further use of the data.

Hence the data has been used to estimate 3 parameters.

This was a 3×2 contingency table, with 6 classes, so the number of degrees of freedom is $6 - 1 - 3 = 2$

In general, for an $r \times s$ contingency table (i.e. r rows and s columns), there are rs classes.

The data is used to estimate $(r - 1)$ probabilities for one classification (given by the rows), and $(s - 1)$ probabilities for the other classification (given by the columns).

Hence the number of degrees of freedom is

$$rs - 1 - (r - 1) - (s - 1)$$
$$= rs - r - s + 1$$
$$= (r - 1)(s - 1)$$

For an $r \times s$ contingency table, to test H_0: there is no association, we consider

$$Y^2 = \sum \frac{(O - E)^2}{E}$$

If H_0 is true, then $Y^2 \sim \chi^2_{(r-1)(s-1)}$

Example 5

Does the data given in the contingency table in Example 4 provide any evidence of an association between the branch of the armed forces and the type of school attended?

We test H_0: there is no association
against H_1: there is some association
The observed frequencies (O) are given in the original contingency table.
Assuming H_0, the expected frequencies are calculated as in Example 4.

For each class we now calculate $\dfrac{(O-E)^2}{E}$; for the class

'Navy and State school', $\dfrac{(O-E)^2}{E} = \dfrac{(27-33.8)^2}{33.8} = 1.37$ and so on.

O	State	Indep	
Navy	27	24	51
Army	52	18	70
Air Force	27	12	39
	106	54	

E	State	Indep
Navy	33.8	17.2
Army	46.4	23.6
Air Force	25.8	13.2

$\dfrac{(O-E)^2}{E}$	State	Indep
Navy	1.37	2.69
Army	0.68	1.33
Air Force	0.06	0.11

$$\text{Let } Y^2 = \sum \frac{(O-E)^2}{E}$$

Since this is a 3×2 contingency table, the number of degrees of freedom is $2 \times 1 = 2$
If H_0 is true, then $Y^2 \sim \chi_2^2$

We have $Y^2 = 1.37 + 2.69 + \cdots + 0.11$
$\qquad\quad = 6.24$
Reject H_0.

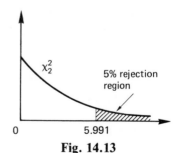

Fig. 14.13

There is some evidence (just significant at the 5% level) of association between branch
of service and type of school.
The highest value of $(O-E)^2/E$, 2.69, occurs in the 'Navy and Independent school'
class; comparing the observed and expected frequencies ($O = 24$, $E = 17.2$), it appears that
in the Navy there are more people from Independent schools than might be expected.

Exercise 14.3 χ^2 distributions and goodness of fit

1 Using χ^2 tables
   (i) Given $Y^2 \sim \chi_5^2$, find the value of a if $P(Y^2 > a) = 0.05$
   (ii) Given $Y^2 \sim \chi_{10}^2$, find the value of b if $P(Y^2 < b) = 0.01$
   (iii) Given $Y^2 \sim \chi_8^2$, find $P(Y^2 > 20.09)$
   (iv) Given $Y^2 \sim \chi_{30}^2$, find $P(18.49 < Y^2 < 50.89)$
   (v) Given $Y^2 \sim \chi_n^2$, and $P(Y^2 > 10.64) = 0.1$, find the value of n.
   (vi) Given $Y^2 \sim \chi_2^2$, find the value of c if $P(Y^2 > c) = 0.005$
   (vii) Given $Y^2 \sim \chi_{50}^2$, find the value of d if $P(Y^2 < d) = 0.05$

2 Given that the pdf of the χ_n^2 distribution is $f(x) = Cx^{n/2-1}\,e^{-x/2}$ (for $x \geqslant 0$), find $f'(x)$, and deduce that the mode is $(n-2)$ provided that $n \geqslant 3$.
   Sketch the pdfs for χ_3^2, χ_4^2 and χ_5^2, paying particular attention to the gradient at $x = 0$, and to the mode.
   Also sketch the pdfs for χ_1^2 and χ_2^2. State the mode in each case.

3 Given that Z is a standard Normal variable, use Normal tables to find
   (i) $P(Z^2 < 2.706)$
   (ii) the value of a if $P(Z^2 > a) = 0.01$
   Check your results using χ^2 tables (Z^2 has the χ_1^2 distribution).

4 The random variable Y^2 has the χ_2^2 distribution.
   Find, by integration,
   (i) the constant C in the pdf of Y^2
   (ii) $P(Y^2 > 6)$
   (iii) the value of a if $P(Y^2 > a) = 0.9$
   (iv) the median of Y^2.
   Check your answers to (ii) and (iii) using χ^2 tables.

5 The random variable Y^2 has the χ_4^2 distribution.
   Find the constant C in the pdf of Y^2, and show that the cdf of Y^2 is

   $$F(x) = 1 - \tfrac{1}{2}(x+2)e^{-x/2} \text{ (for } x \geqslant 0)$$

   Hence find $P(Y^2 < 1)$ and $P(Y^2 > 8)$.

6 The random variable Y^2 has the χ_n^2 distribution.
   Using $Y^2 = Z_1^2 + Z_2^2 + \cdots + Z_n^2$, together with the results

   $$E[Z^2] = 1, \qquad \text{var}\,(Z^2) = 2 \qquad \text{(see Exercise 13.4, question 6),}$$

   show that Y^2 has mean n and variance $2n$.
   Explain why Y^2 should have an approximately Normal distribution when n is large.
   Given $n = 30$, use the Normal approximation to find a and b such that $P(Y^2 < a) = 0.05$ and $P(Y^2 > b) = 0.05$.
   Compare your values with the true values (given in χ^2 tables).

7   In fact the Normal approximation to χ_n^2 (see Question 6) is not satisfactory unless n is very large. A better approximation is given by:

If $Y^2 \sim \chi_n^2$, then $\sqrt{2Y^2}$ is approximately Normal with mean $\sqrt{2n-1}$ and variance 1

Using this approximation,

(i) if $Y^2 \sim \chi_{30}^2$, find a and b such that $P(Y^2 < a) = 0.05$ and $P(Y^2 > b) = 0.05$

(ii) if $Y^2 \sim \chi_{320}^2$, find c such that $P(Y^2 > c) = 0.01$.

8   A die was thrown 100 times, with the following results:

Number showing uppermost	1	2	3	4	5	6
Number of throws	24	10	18	9	13	26

Does this suggest that the die might be biased?

9   According to theory, genetic types A, B, C, D should occur in the offspring of a certain population in the ratio $1:2:2:1$

In a random sample of 150 offspring, there were

19 Type A,   66 Type B,   42 Type C,   and   23 Type D.

Is this consistent with the theory?

10   The frequencies of digits in the first 800 decimal places of the number $\pi = 3.14159\ldots$ are as follows

Digit	0	1	2	3	4	5	6	7	8	9
Decimal places 1–400	39	43	44	39	47	39	42	24	44	39
Decimal places 401–800	35	49	39	40	33	34	35	51	32	52

Test the hypothesis that all digits are equally likely

(i) using the first 400 decimal places

(ii) using the first 800 decimal places.

11   For a sample of 50 families of six children, the numbers of girls in the families were as follows

Number of girls	0	1	2	3	4	5	6
Number of families	2	6	11	19	9	3	0

Assuming that girls and boys are equally likely, calculate the expected frequencies, and test the goodness of fit.

12 Fit a Binomial distribution to the following data, and test the goodness of fit.

Number of rainy days in a week	0	1	2	3	4	5	6	7
Number of weeks	85	118	113	29	8	4	0	3

13 Test whether the following sample is consistent with the claim that it is taken from a Poisson distribution with mean 2.5.

Number of minor faults detected	0	1	2	3	4	5	6	7	8	9
Number of cars	10	20	29	31	22	16	6	5	0	1

14 Test the fit of a Poisson distribution to the following data.

Number of accidents	0	1	2	3	4	5	6	7	8
Number of days	61	115	94	65	24	4	0	1	1

15 Test the fit of a Normal distribution to the data given in Exercise 14.2, question 8 (the contents of 500 bottles of orange squash).

16 A random sample of 100 people were classified by eye colour and hair colour, as follows.

		hair colour		
		fair	dark	ginger
eye colour	blue	15	10	3
	brown	15	41	16

Is there any evidence of association between eye colour and hair colour?

17 For a random sample of 300 car owners, the age of the owner and the type of car were noted, as follows.

		Small car	Family car	Luxury car
age of owner	Under 25	38	28	4
	25–40	23	62	20
	Over 40	31	70	24

Analyse this data for evidence of association between age-group and type of car owned.

18 Does the following contingency table provide any evidence of association between the size of a house and the type of fuel used for heating?

Heating fuel

		Solid fuel	Gas	Oil	Electricity
size of house	2 bedrooms	94	144	30	72
	3 bedrooms	107	220	28	70
	4 bedrooms	18	43	5	11
	5 bedrooms	2	1	5	0

19 Two insecticides, type A and type B, were tested as follows: when controlled doses of type A were administered to a sample of 90 cockroaches, 55 of them died; and when similar doses of type B were administered to a different sample of 60 cockroaches, 28 of them died.
   (i) Write this data in the form of a 2 × 2 contingency table, using the categories 'type A' or 'type B', and 'died' or 'survived'. Use the χ^2 test to test for association between the type of insecticide and its effectiveness.
   (ii) By considering the proportions of cockroaches killed for type A and for type B, use the difference in proportions test to test the null hypothesis (H_0) that the true proportions killed by the two types are equal.
       By writing this test in the form: Reject H_0 if $Z^2 > (1.96)^2$ and calculating the value of Z^2, verify that this test is exactly equivalent to that in (i).

20 For a 2 × 2 contingency table, a continuity correction (called Yates' correction) can be applied: the magnitude of each difference $(O - E)$ is reduced by 0.5 before squaring it, so the test statistic is calculated as

$$Y^2 = \sum \frac{\{|O - E| - 0.5\}^2}{E}$$

Using this correction, calculate Y^2 for the 2 × 2 contingency table in Question 19(i).

14.4   The distribution of the sample variance

Suppose $\{X_1, X_2, ..., X_n\}$ is a random sample of size n from a population having mean μ and variance σ^2
   Then $E[X_i] = \mu$ and var $(X_i) = E[(X_i - \mu)^2] = \sigma^2$
   The sample mean $\bar{X} = \frac{1}{n}(X_1 + X_2 + \cdots + X_n)$

has mean $E[\bar{X}] = \mu$ and variance var $(\bar{X}) = E[(\bar{X} - \mu)^2] = \frac{\sigma^2}{n}$

We now consider the sample variance

$$S^2 = \frac{\sum\limits_{i=1}^{n} (X_i - \bar{X})^2}{n} = \frac{\sum\limits_{i=1}^{n} X_i^2}{n} - \bar{X}^2$$

This is of course a random variable (varying from sample to sample); we shall now derive its mean $E[S^2]$

We have $\sum\limits_{i=1}^{n} (X_i - \mu)^2 = \sum\limits_{i=1}^{n} \{(X_i - \bar{X}) + (\bar{X} - \mu)\}^2$

$$= \sum\limits_{i=1}^{n} \{(X_i - \bar{X})^2 + 2(X_i - \bar{X})(\bar{X} - \mu) + (\bar{X} - \mu)^2\}$$

$$= \sum\limits_{i=1}^{n} (X_i - \bar{X})^2 + 2(\bar{X} - \mu) \sum\limits_{i=1}^{n} (X_i - \bar{X}) + n(\bar{X} - \mu)^2$$

Now $\sum\limits_{i=1}^{n} (X_i - \bar{X})^2 = nS^2$, and $\sum\limits_{i=1}^{n} (X_i - \bar{X}) = \sum\limits_{i=1}^{n} X_i - n\bar{X} = 0$,

and thus $\sum\limits_{i=1}^{n} (X_i - \mu)^2 = nS^2 + n(\bar{X} - \mu)^2$ $\qquad\qquad$ (1)

Taking expectations,

$$\sum\limits_{i=1}^{n} E[(X_i - \mu)^2] = nE[S^2] + nE[(\bar{X} - \mu)^2]$$

i.e. $\qquad\qquad n\sigma^2 = nE[S^2] + n\left(\frac{\sigma^2}{n}\right)$

$$(n - 1)\sigma^2 = nE[S^2]$$

Hence

$$\boxed{E[S^2] = \frac{(n - 1)}{n} \sigma^2}$$

Note   The average value of S^2, taken over all possible samples of size n, is *less* than the population variance σ^2. (This does *not* mean that S^2 is always less than σ^2; but the *average* value is less). We say that the sample variance S^2 gives a 'biased' estimate of the population variance σ^2. We can correct this bias by multiplying the sample variance by $\dfrac{n}{(n - 1)}$; we define the

$$\textit{corrected sample variance } S_*^2 = \frac{n}{(n - 1)} S^2$$

$$= \frac{\sum\limits_{i=1}^{n}(X_i - \bar{X})^2}{(n - 1)}$$

which is often used to estimate the population variance σ^2.

On a calculator with statistics functions, the value of S_* can be found directly, by entering the values of $X_1, X_2, \ldots, X_n$ and pressing the appropriate key (usually labelled σ_{n-1}, or s).

We have shown that the sample variance S^2 always has mean

$$E[S^2] = \frac{(n-1)}{n}\sigma^2.$$

However, the distribution of S^2 depends on the distribution of the population from which the sample is taken. We shall only consider one special case.

For the remainder of this Section,

we assume that the random sample is taken from a NORMAL population.

Then $\dfrac{X_i - \mu}{\sigma} \sim N(0, 1)$, and $\bar{X}$ is also Normally distributed, so

$$\frac{\bar{X} - \mu}{\sigma/\sqrt{n}} \sim N(0, 1)$$

We assume the result that, for samples from a Normal population, the sample mean $\bar{X}$ and the sample variance S^2 are independent random variables.

Also we note that if Y_1^2 and Y_2^2 are independent random variables, with $Y_1^2 \sim \chi_m^2$ and $Y_2^2 \sim \chi_n^2$, then $Y_1^2 + Y_2^2 \sim \chi_{m+n}^2$ (since Y_1^2 is the sum of the squares of m independent standard Normal variables, and Y_2^2 is the sum of the squares of n such variables; hence $Y_1^2 + Y_2^2$ is the sum of the squares of $(m + n)$ such variables).

Dividing equation (1) on p. 324 by σ^2, we have

$$\sum_{i=1}^{n} \left(\frac{X_i - \mu}{\sigma}\right)^2 = \frac{nS^2}{\sigma^2} + \left(\frac{\bar{X} - \mu}{\sigma/\sqrt{n}}\right)^2$$

Now

$$\sum_{i=1}^{n} \left(\frac{X_i - \mu}{\sigma}\right)^2$$

is the sum of the squares of n independent standard Normal variables, so

$$\sum_{i=1}^{n} \left(\frac{X_i - \mu}{\sigma}\right)^2 \sim \chi_n^2$$

Also

$$\left(\frac{\bar{X} - \mu}{\sigma/\sqrt{n}}\right)^2 \sim \chi_1^2,$$

since it is the square of a standard Normal variable.
Since S^2 and $\bar{X}$ are independent, it follows that

$$\boxed{\dfrac{nS^2}{\sigma^2} \sim \chi_{n-1}^2}$$

A null hypothesis that σ^2 takes some particular value can therefore be tested as follows:

> To test $H_0 : \sigma^2 = \sigma_0^2$
>
> we consider $Y^2 = \dfrac{nS^2}{\sigma_0^2}$
>
> If H_0 is true, then $Y^2 \sim \chi_{n-1}^2$

Example 1

Over a long period, the lengths of glass rods produced at a factory have been Normally distributed with standard deviation 4.2 mm. The process of manufacture was then modified with the aim of reducing the standard deviation. After the modification, the lengths of a random sample of 20 rods were measured, and found to have standard deviation 3.5 mm.

Does this sample demonstrate that the standard deviation has been reduced?

Suppose that, after the modification, the lengths of the rods are Normally distributed with standard deviation σ.

We test $H_0 : \sigma = 4.2$
against $H_1 : \sigma < 4.2$

If a random sample of 20 rods has standard deviation S,

we consider $Y^2 = \dfrac{20S^2}{4.2^2}$

H_1 is favoured by small values of Y^2.
If H_0 is true, then $Y^2 \sim \chi_{19}^2$
We have $S = 3.5$, so

$$Y^2 = \frac{20 \times 3.5^2}{4.2^2}$$

$$= 13.89$$

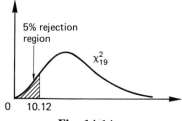

Fig. 14.14

Accept H_0.
This sample does *not* provide sufficient evidence to say that the standard deviation has been reduced.

Example 2

A random sample of 8 ropes broke at the following tensions (in newtons)
8419, 8147, 8094, 8586, 8531, 8197, 8396, 7895.
Assuming that the breaking tensions are Normally distributed, give a 95% confidence interval for the standard deviation of the breaking tensions of all the ropes.

Suppose that the breaking tensions are Normally distributed with standard deviation σ newtons.
If a random sample of 8 rods has standard deviation S, then

$$\frac{8S^2}{\sigma^2} \sim \chi_7^2$$

For 95% of samples,

$$1.690 < \frac{8S^2}{\sigma^2} < 16.01$$

Rearranging,

$$\frac{8S^2}{16.01} < \sigma^2 < \frac{8S^2}{1.690}$$

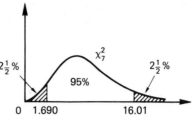

Fig. 14.15

We now calculate the variance S^2 of the given sample. Direct use of the formula

$$S^2 = \frac{\sum x^2}{n} - \left(\frac{\sum x}{n}\right)^2$$

may lead to rounding errors, since the mean is large compared with the standard deviation. So we shall work with

$$u = x - 8000: 419, 147, 94, 586, 531, 197, 396, -105.$$

Then $\sum u = 2265$, $\sum u^2 = 1038013$

$$\text{and } S_x^2 = S_u^2 = \frac{1\,038\,013}{8} - \left(\frac{2265}{8}\right)^2$$

$$= 49\,592$$

Hence the 95% confidence interval is

$$\frac{8 \times 49\,592}{16.01} < \sigma^2 < \frac{8 \times 49\,592}{1.690}$$

i.e. $24780 < \sigma^2 < 234754$

i.e. $157 < \sigma < 485$

Exercise 14.4   Distribution of the sample variance

1   Random samples of size 10 are taken from a population which is Normally distributed with standard deviation 7.2.
Give two values between which the sample standard deviation will lie for 95% of samples.

2   Random samples of size 25 are taken from a population which is Normally distributed with standard deviation 36.0.
Give two values between which the sample standard deviation will lie for 98% of samples.

3   Random samples of size 15 are taken from a population which is Normally distributed with standard deviation 10.0.
Find the value which the sample standard deviation will exceed for only 5% of samples.

4 The heights of the pine trees in one forest are Normally distributed with standard deviation 2.5 m. For a random sample of 16 pine trees in another forest, the heights had standard deviation 3.2 m. Does this suggest that the standard deviation of the heights of the trees in the second forest is different from that in the first?

5 A machine is packing sugar into bags, and the standard deviation of the weights of the bags should not be more than 8 g.
A random sample of 11 bags were weighed individually as follows:

1003.3, 999.0, 1007.7, 1000.1, 995.0, 980.2, 986.1, 1017.4, 1013.3, 986.6, 990.3 g.

Is there any evidence that the standard deviation is greater than 8 g? State any assumptions you have made.

6 A woman has found that her journey times to work have been Normally distributed with standard deviation 6 minutes. She tries a new route, and the times x_i minutes for 20 journeys by the new route satisfy

$$\sum x_i = 965, \sum x_i^2 = 46875$$

Has changing the route altered the standard deviation of the journey times?

7 The following is a random sample of 5 values from a Normal population:

3.54, 4.17, 3.90, 4.30, 4.56.

Find 95% confidence limits for the standard deviation of the population.

8 The heights of people may be assumed to be Normally distributed; and for a random sample of 30 people, their heights had standard deviation 11.4 cm. Find 98% confidence limits for the standard deviation of the heights of all people.

9 Given $Y^2 \sim \chi^2_{499}$, find numbers a and b such that

$$P(Y^2 < a) = 0.025 \text{ and } P(Y^2 > b) = 0.025$$

(Use the result that if $Y^2 \sim \chi^2_n$ then $\sqrt{2Y^2}$ is approximately Normal with mean $\sqrt{2n-1}$ and variance 1).
  A random sample of 500 values from a Normal population has sample standard deviation 26.8
(i) Test the hypothesis that the population standard deviation is 25.0
(ii) Give 95% confidence limits for the population standard deviation.

10 A population is Normally distributed with variance σ^2, and a random sample of size n from this population has sample variance S^2.
(i) State the expected value of S^2, and show that

$$\text{var} (S^2) = \frac{2(n-1)\sigma^4}{n^2}$$

(You may assume that if $Y^2 \sim \chi^2_m$, then var $(Y^2) = 2m$; see Exercise 14.3 Question 6.)

(ii) Given that n is large, show that the sample standard deviation S is approximately Normally distributed with mean $\sigma\sqrt{(2n-3)/2n}$ and standard deviation $\sigma/\sqrt{2n}$. (Use the Normal approximation to $\sqrt{2Y^2}$ given in Question 9).

14.5   *t* distributions

If Z and Y^2 are independent random variables, with $Z \sim N(0, 1)$ and $Y^2 \sim \chi_n^2$, then the random variable

$$T = \frac{Z\sqrt{n}}{\sqrt{Y^2}}$$

is said to have the t_n distribution (Student's t-distribution with n degrees of freedom). T is a continuous random variable, which can take any value. Its pdf is

$$f(x) = C\left(1 + \frac{x^2}{n}\right)^{-(n+1)/2},$$

where C is a constant chosen so that $\displaystyle\int_{-\infty}^{\infty} f(x)\, dx = 1$

The distribution is symmetrical about $x = 0$, and the pdf looks rather like that of a Normal distribution. For each positive integer n, there is a different t distribution.

The diagram shows the pdf of the t_3 distribution, with the standard Normal curve shown dotted for comparison.
Note that the t distribution is more 'spread out' than the standard Normal distribution.

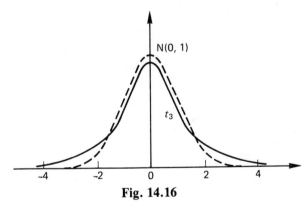

Fig. 14.16

As n increases, the pdf of the t_n distribution becomes closer and closer to that of the standard Normal distribution.
The pdf of the t_n distribution is difficult to work with, so again we use tables. The t tables on p. 399 give *two-tailed* percentage points; i.e. the table gives the value x which is exceeded with probability $\frac{1}{2}p\%$ by a random variable having the t_n distribution (Fig. 14.17).

If $T \sim t_n$, $P(T > x) = \dfrac{\frac{1}{2}p}{100}$

Then by symmetry, $P(T < -x) = \dfrac{\frac{1}{2}p}{100}$

and hence $\quad P(|T| > x) = \dfrac{p}{100}$

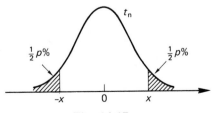

Fig. 14.17

For example, when $n = 4$ and $p = 5$, the tables give $x = 2.776$ (Fig. 14.18).
(Note that the corresponding value for a standard Normal distribution would be 1.96.)

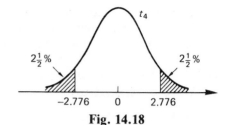

Fig. 14.18

If we wish to cut 5% off one end of the distribution, we must look under $p = 10$.
For example, when $n = 12$ and $p = 10$, the tables give $x = 1.782$ (Fig. 14.19).

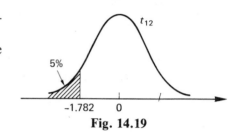

Fig. 14.19

Application to samples

Up to now, hypothesis tests for the population mean μ have been based on the result

$$\frac{\bar{X} - \mu}{\sigma/\sqrt{n}} \sim N(0, 1)$$

and it was assumed that the population standard deviation σ is known.

In practice, it is unlikely that σ will be known; so usually we have to estimate σ from the sample. This causes no difficulty for large samples, but for small samples (of size less than about 50), the sample standard deviation S might differ considerably from σ, and so using S instead of σ could introduce a large error. We now show how this problem can be overcome by using a t distribution, but we can proceed only if the following condition is satisfied:

we must assume that the population is NORMALLY *distributed.*

Consider a Normal population having mean μ and variance σ^2.
Suppose that a random sample of size n from this population has sample mean $\bar{X}$ and sample variance S^2.
Then $\bar{X}$ and S^2 are independent random variables,

$$\frac{\bar{X} - \mu}{\sigma/\sqrt{n}} \sim N(0, 1) \qquad \text{and} \qquad \frac{nS^2}{\sigma^2} \sim \chi^2_{n-1}$$

Hence

$$\frac{\left(\dfrac{\bar{X} - \mu}{\sigma/\sqrt{n}}\right)\sqrt{n-1}}{\sqrt{(nS^2)/\sigma^2}} \sim t_{n-1}$$

i.e.

$$\boxed{\frac{\bar{X} - \mu}{S/\sqrt{n-1}} \sim t_{n-1}}$$

Thus, when only the *sample* standard deviation S is available, we should use

$$\frac{\bar{X}-\mu}{S/\sqrt{n-1}} \quad \text{instead of} \quad \frac{\bar{X}-\mu}{\sigma/\sqrt{n}},$$

and the t_{n-1} distribution instead of the standard Normal distribution.

With these modifications, we carry out hypothesis tests and form confidence intervals, in the same way as before.

Note   If the 'corrected' sample standard deviation $S_* = S\sqrt{\dfrac{n}{n-1}}$

is used, we have

$$\frac{\bar{X}-\mu}{S_*/\sqrt{n}} \sim t_{n-1}$$

which corresponds more closely with $\dfrac{\bar{X}-\mu}{\sigma/\sqrt{n}} \sim N(0,1)$

Hypothesis testing for μ

To test $H_0 : \mu = \mu_0$

we consider $T = \dfrac{\bar{X}-\mu_0}{S/\sqrt{n-1}}$

If H_0 is true, then $T \sim t_{n-1}$

Example 1

A packing machine is supposed to produce bags of sweets with mean weight 225 g. A random sample of 10 bags had the following weights:

$$238, 223, 226, 244, 218, 233, 240, 230, 222, 235 \text{ g.}$$

Does this sample suggest that the mean weight is different from 225 g? (You may assume that the weights are Normally distributed.)

Suppose that the weights are Normally distributed with mean μ.

We test $H_0 : \mu = 225$

against $H_1 : \mu \neq 225$

If a random sample of 10 bags has mean weight $\bar{X}$ and standard deviation S, we consider

$$T = \frac{\bar{X}-225}{S/\sqrt{9}}.$$

If H_0 is true, then $T \sim t_9$ (Fig. 14.20).

For the given sample,

we have $\bar{X} = 230.9, S = 8.117$,

and $\quad T = \dfrac{230.9 - 225}{8.117/\sqrt{9}}$

$\qquad = 2.181$

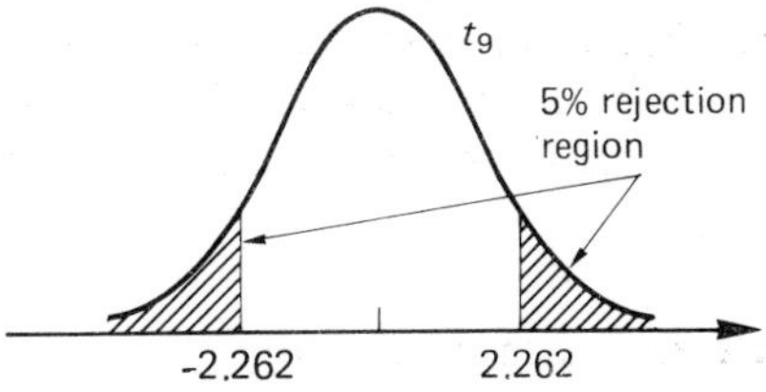

Fig. 14.20

Accept H_0.

This sample does *not* suggest that the mean weight is different from 225 g.

Confidence intervals for μ

Example 2

For a certain train journey, the journey times may be assumed to be Normally distributed. For a random sample of 8 journeys, the times had mean 46.2 minutes with standard deviation 2.3 minutes. Give 99% confidence limits for the mean journey time.

Suppose that the journey times are Normally distributed with mean μ. If a random sample of 8 journeys has mean time $\bar{X}$ and standard deviation S then

$$\frac{\bar{X}-\mu}{S/\sqrt{7}} \sim t_7$$

For 99% of samples,

$$-3.499 < \frac{\bar{X}-\mu}{S/\sqrt{7}} < 3.499 \qquad \text{(Fig. 14.21)}.$$

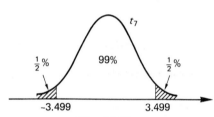

Fig. 14.21

Rearranging,

$$\bar{X} - 3.499 \times \frac{S}{\sqrt{7}} < \mu < \bar{X} + 3.499 \times \frac{S}{\sqrt{7}}$$

99% confidence limits for μ are $\bar{X} \pm 3.499 \times \dfrac{S}{\sqrt{7}}$

$$= 46.2 \pm 3.499 \times \frac{2.3}{\sqrt{7}}$$

$$= 46.2 \pm 3.04$$

$$= 43.16, 49.24 \text{ minutes}.$$

Notes
(1) If the population standard deviation *is* known, then we shall use the result

$$\frac{\bar{X}-\mu}{\sigma/\sqrt{n}} \sim N(0, 1),$$

however small the sample.
(2) For a large sample (of size greater than about 50), there is very little difference between the *t* distribution and the standard Normal distribution; for convenience, we usually use the Normal distribution even when σ is unknown (as in Section 14.1).

Paired sample *t* test

Suppose that we have samples from two populations, and we wish to test whether the population means, μ_1 and μ_2 could be equal.

We first consider the case where each value in one sample pairs off naturally with

a value in the other sample; then the two samples are necessarily of the same size, say n.

We subtract the corresponding values, to obtain a set of n differences $D_1, D_2, ..., D_n$.

These differences come from a population having mean $(\mu_1 - \mu_2)$; but we do not know the standard deviation of the population of the differences (we could not find it even if we knew the standard deviations of the individual populations, since these samples are not independent). We can use a t distribution provided that the following condition is satisfied:

we must assume that the differences are NORMALLY *distributed*.

Then if the differences $D_1, D_2, ..., D_n$ have mean $\bar{D}$, and standard deviation S,

$$\text{we have } \frac{\bar{D} - (\mu_1 - \mu_2)}{S/\sqrt{n-1}} \sim t_{n-1}$$

Hence

to test $H_0 : \mu_1 = \mu_2$

$$\text{we consider } T = \frac{\bar{D}}{S/\sqrt{n-1}}$$

If H_0 is true, then $T \sim t_{n-1}$

Example 3
A company claims that its petrol (Brand X) is better than that of a rival company (Brand Y). A random sample of 7 cars was taken; each car was driven as far as possible on one gallon of petrol Brand X, then on one gallon of petrol Brand Y. The distances were as follows:

Car	A	B	C	D	E	F	G
Miles on Brand X	43.8	22.8	15.3	35.5	9.7	30.3	28.2
Miles on Brand Y	37.1	24.0	14.6	27.9	8.0	31.1	23.2

(i) Do these results provide any evidence to support the company's claim?
(ii) Provide a 95% confidence interval for the difference in mean mpg (miles per gallon) between the two brands.

The differences, (distance on Brand X) – (distance on Brand Y), are
for car A, $43.8 - 37.1 = 6.7$ miles
for car B, $22.8 - 24.0 = -1.2$ miles, and so on, i.e.

Car	A	B	C	D	E	F	G
Difference	6.7	-1.2	0.7	7.6	1.7	-0.8	5.0

These seven differences have mean $\bar{D} = 2.814$
and standard deviation $S = 3.331$
Suppose that the mean mpg for Brand X is μ_1
and the mean mpg for Brand Y is μ_2

Then the differences have mean $(\mu_1 - \mu_2)$. We assume that the differences are Normally distributed.

If a random sample of 7 differences has mean $\bar{D}$ and standard deviation S, then

$$\frac{\bar{D} - (\mu_1 - \mu_2)}{S/\sqrt{6}} \sim t_6$$

(i) We test $H_0 : \mu_1 = \mu_2$
  against $H_1 : \mu_1 > \mu_2$

Let $T = \dfrac{\bar{D}}{S/\sqrt{6}}$

If H_0 is true, then $T \sim t_6$
We have $\bar{D} = 2.814$, $S = 3.331$,
giving

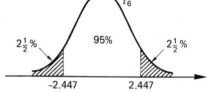

t_6

5% rejection region

1.943

Fig. 14.22

$$T = \frac{2.814}{3.331/\sqrt{6}}$$

$$= 2.069$$

Reject H_0.
There is some evidence (just significant at the 5% level) that Brand X is better.

(ii) We have $\dfrac{\bar{D} - (\mu_1 - \mu_2)}{S/\sqrt{6}} \sim t_6$

For 95% of samples,

$$-2.447 < \frac{\bar{D} - (\mu_1 - \mu_2)}{S/\sqrt{6}} < 2.447$$

Rearranging,

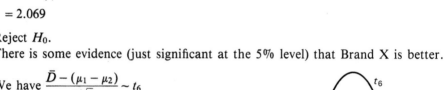

t_6

$2\frac{1}{2}\%$ 95% $2\frac{1}{2}\%$

-2.447   2.447

Fig. 14.23

$$\bar{D} - 2.447 \times \frac{S}{\sqrt{6}} < (\mu_1 - \mu_2) < \bar{D} + 2.447 \times \frac{S}{\sqrt{6}}$$

Substituting $\bar{D} = 2.814$, $S = 3.331$,
a 95% confidence interval for $(\mu_1 - \mu_2)$ is

$$-0.51 < (\mu_1 - \mu_2) < 6.14$$

Two-sample *t* test

Now suppose that we have two samples which do not pair off naturally, but which are independent. It is no longer necessary that the two samples have the same size.
  Consider two populations,
the first having mean μ_1 and (unknown) standard deviation σ,
the second having mean μ_2 and standard deviation σ.
Note that

We must assume that both populations are NORMALLY *distributed and that their standard deviations (although unknown) are equal.*
Suppose that we have two INDEPENDENT *samples, one from each population,*

the first of size n_1, having sample mean $\bar{X}_1$ and sample standard deviation S_1,
the second of size n_2, having sample mean $\bar{X}_2$ and sample standard deviation S_2

Then, as shown in Section 14.1, $\dfrac{(\bar{X}_1 - \bar{X}_2) - (\mu_1 - \mu_2)}{\sqrt{\sigma^2/n_1 + \sigma^2/n_2}} \sim N(0, 1)$

Also $\dfrac{n_1 S_1^2}{\sigma^2} \sim \chi^2_{n_1-1}$ and $\dfrac{n_2 S_2^2}{\sigma^2} \sim \chi^2_{n_2-1}$,

thus $\dfrac{n_1 S_1^2}{\sigma^2} + \dfrac{n_2 S_2^2}{\sigma^2} \sim \chi^2_{n_1+n_2-2}$

and so

$$\frac{\left(\dfrac{(\bar{X}_1 - \bar{X}_2) - (\mu_1 - \mu_2)}{\sqrt{\sigma^2/n_1 + \sigma^2/n_2}} \right) \sqrt{n_1 + n_2 - 2}}{\sqrt{n_1 S_1^2/\sigma^2 + n_2 S_2^2/\sigma^2}} \sim t_{n_1+n_2-2}$$

i.e. $\dfrac{(\bar{X}_1 - \bar{X}_2) - (\mu_1 - \mu_2)}{S\sqrt{1/n_1 + 1/n_2}} \sim t_{n_1+n_2-2}$ where $S^2 = \dfrac{n_1 S_1^2 + n_2 S_2^2}{(n_1 + n_2 - 2)}$

This compares with $\dfrac{(\bar{X}_1 - \bar{X}_2) - (\mu_1 - \mu_2)}{\sigma\sqrt{1/n_1 + 1/n_2}} \sim N(0, 1)$

Hence

to test $H_0 : \mu_1 = \mu_2$

we consider $T = \dfrac{\bar{X}_1 - \bar{X}_2}{S\sqrt{\dfrac{1}{n_1} + \dfrac{1}{n_2}}}$

where $S^2 = \dfrac{n_1 S_1^2 + n_2 S_2^2}{(n_1 + n_2 - 2)}$

If H_0 is true, then $T \sim t_{n_1+n_2-2}$

Example 4
To test the effectiveness of two sleeping pills, 5 patients were given Drug A, and 8 patients were given Drug B.
The times of sleep were as follows:
Hours of sleep (Drug A): 5.2, 9.8, 8.4, 7.1, 3.4.
Hours of sleep (Drug B): 10.1, 7.5, 2.1, 12.0, 11.7, 9.3, 14.4, 8.0.
   Do these figures demonstrate a significant difference between Drug A and Drug B?

Suppose that the sleeping times for Drug A are Normally distributed

with mean μ_1 and standard deviation σ,

and that the sleeping times for Drug B are Normally distributed

with mean μ_2 and standard deviation σ.

Note that it is necessary to assume that both populations are Normally distributed, with the same standard deviation. We must also assume that the two samples of patients were independent random samples.

We test $H_0 : \mu_1 = \mu_2$

against $H_1 : \mu_1 \neq \mu_2$

If the sleeping times for a random sample of 5 patients given Drug A have mean $\bar{X}_1$ and standard deviation S_1, and those for a random sample of 8 patients given Drug B have mean $\bar{X}_2$ and standard deviation S_2, we consider

$$T = \frac{\bar{X}_1 - \bar{X}_2}{S\sqrt{\frac{1}{5} + \frac{1}{8}}} \qquad \text{where} \qquad S^2 = \frac{5S_1^2 + 8S_2^2}{11}$$

If H_0 is true, then $T \sim t_{11}$ (Fig. 14.24).

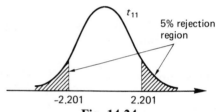

Fig. 14.24

For the first sample (Drug A), we have $\bar{X}_1 = 6.78, S_1 = 2.2702$
For the second sample (Drug B), we have $\bar{X}_2 = 9.3875, S_2 = 3.4715$
Thus

$$S^2 = \frac{5 \times 2.2702^2 + 8 \times 3.4715^2}{11}$$

$$S = 3.3327$$

and

$$T = \frac{6.78 - 9.3875}{3.3327\sqrt{\frac{1}{5} + \frac{1}{8}}} = -1.372$$

Accept H_0.
There is no evidence of any difference between the two drugs.

Note If we have two independent samples, and the population standard deviations σ_1 and σ_2 are *known;*
or if both samples are reasonably large;
then we would use instead the result

$$\frac{(\bar{X}_1 - \bar{X}_2) - (\mu_1 - \mu_2)}{\sqrt{\frac{\sigma_1^2}{n_1} + \frac{\sigma_2^2}{n_2}}} \sim N(0, 1)$$

(as shown in Section 14.1).

Exercise 14.5   *t* distributions

1   Using *t* tables,
   (i)   Given $T \sim t_4$, find the value of a if $P(|T| > a) = 0.05$
   (ii)   Given $T \sim t_8$, find the value of b if $P(T > b) = 0.01$
   (iii)   Given $T \sim t_{20}$, find the value of c if $P(T < c) = 0.05$
   (iv)   Given $T \sim t_7$, find $P(1.895 < T < 3.499)$
   (v)   Given $T \sim t_2$, find $P(-2.92 < T < 2.92)$
   (vi)   Given $T \sim t_n$, and $P(T > 2.65) = 0.01$, find the value of n.
   (vii)   Given $T \sim t_5$, find the value of d if $P(-d < T < d) = 0.95$ where $d > 0$.
   (viii) Given $T \sim t_1$, find the value of e if $P(T < e) = 0.995$

2   Given that $T \sim t_1$, write down the pdf of T, and find the value of the constant.

   Show that the cdf of T is $F(x) = \frac{1}{2} + \frac{1}{\pi} \tan^{-1} x$

   Hence find (i) $P(-12.7 < T < 12.7)$
               (ii) the value of a if $P(T > a) = 0.01$.
   Check your answers using *t* tables.

3   A random variable T has cdf $F(x) = \frac{1}{2} + \dfrac{x}{2\sqrt{2 + x^2}}$

   Find the pdf $f(x)$, and hence verify that T has the t_2 distribution.
   Find (i) $P(T < 4)$
         (ii) the value of a if $P(T > a) = 0.1$
   Draw an accurate graph of the pdf between $x = -5$ and $x = +5$

   Plot the pdf of the standard Normal distribution $\left(\phi(x) = \dfrac{1}{\sqrt{2\pi}} e^{-x^2/2} \right)$

   using the same axes, and compare the two distributions t_2 and $N(0, 1)$.

4   The diameters of the ball-bearings produced by a machine are Normally distributed, and the mean diameter should be 12.00 mm. A random sample of 8 ball-bearings were found to have diameters:

   11.89, 11.76, 11.98, 12.44, 12.70, 12.45, 13.76, 12.79 mm.

   Is there sufficient evidence to say that the mean diameter is incorrect?

5   The lengths of the jumps of a long jumper have been Normally distributed with mean 6.46 m. After some special coaching, he did six jumps with lengths:

   6.77, 6.37, 6.48, 6.59, 6.46, 6.64 m

   Is there any evidence that the coaching has improved his performance?

6   A girl noted the time she awoke each morning, and discovered that the times (measured in minutes past 7 o'clock) were approximately Normally distributed with mean 25.

For 12 days whilst she was on holiday, her waking times were:

7.24, 7.35, 7.30, 7.37, 7.42, 7.32, 7.35, 7.33, 7.17, 7.42, 7.18, 7.25

Is her mean waking time while on holiday significantly different from usual?

7 For a random sample of 15 bags of flour (said to contain 1.5 kg) taken from the shelf of a supermarket, the weights of these 15 bags had mean 1.490 kg and standard deviation 0.014 kg.

 Does this provide evidence, at the 1% level of significance, that the mean weight is less than 1.5 kg?

8 Ten years ago, the average speed of cars on a certain stretch of road was 93 km h^{-1}. Recently, a random sample of 21 cars had speeds x_i km h^{-1}, where $\sum x_i = 1800$, and $\sum x_i^2 = 159\,660$.

 Is there any evidence that the average speed has changed?

9 The lengths of a certain type of snake may be assumed to be Normally distributed. The lengths of a random sample of 4 snakes were measured as 2.72, 2.68, 1.89, 3.23 m.

 Find 95% confidence limits for the mean length of this type of snake.

10 For a certain type of cable, the breaking tensions are Normally distributed. A random sample of 16 cables were tested, and their breaking tensions x_i kN were found to satisfy $\sum x_i = 268.8$, $\sum x_i^2 = 4746.88$

 Find the 99% lower confidence limit for the mean breaking tension of this type of cable.

11 The weights of the apples from a certain tree are known to be Normally distributed with standard deviation 22 g.

 A random sample of 6 apples had weights:

$$150, 148, 109, 175, 139, 145 \text{ g}.$$

Find 95% confidence limits for the mean weight of apples from this tree.

12 For two seaside resorts A and B, the temperatures were measured at the same time on each of seven days, as follows.

Day	1	2	3	4	5	6	7
Temp at resort A ($^\circ$C)	24.7	18.5	25.6	27.3	22.2	28.2	31.2
Temp at resort B ($^\circ$C)	17.9	20.0	26.7	21.8	19.7	26.3	24.4

 Is there any evidence that the mean temperatures at the two resorts are different? State any assumptions which you need to make.

13 Eight specimens of metal were cut in half. One half of each specimen was given a special treatment designed to increase its resistance to corrosion; the other half was left untreated. All sixteen pieces were then placed in the same harsh environment, and the times taken for them to corrode away were as follows.

Specimen		A	B	C	D	E	F	G	H
Time to corrode (hours)	Treated half	75	60	31	58	40	80	72	97
	Untreated half	64	35	30	53	33	84	62	80

Does this show that the treatment is effective?

14 The gestation periods for random samples of two species of monkeys were as follows.
Species A: 208, 217, 216, 219, 211, 203, 212, 207, 207 days.
Species B: 201, 209, 209, 195, 219, 206, 208 days.
   Does this indicate that the two species have different gestation periods? (You may assume that the gestation periods are Normally distributed).

15 Twenty slimmers went on a diet, and after one month their weight losses had mean 5.12 kg and standard deviation 1.92 kg. At the same time, another twelve slimmers went on a similar diet, and also followed a vigorous exercise programme; their weight losses had mean 6.15 kg and standard deviation 1.75 kg.
   Does this show that exercise increases the mean weight loss?

16 For a random sample of 10 employees of a certain firm, their journey times to and from work on a particular day were as follows.

Employee	A	B	C	D	E	F	G	H	I	J
Journey time to work (min)	25	68	47	19	7	58	71	35	60	21
Journey time from work (min)	33	64	45	28	7	62	80	45	63	27

Does this indicate that the mean journey times to and from work are different?

17 For a random sample of 9 students on a certain course, their marks in a mid-term examination and in the end-of-term examination were as follows.

Student	A	B	C	D	E	F	G	H	I
Mark in mid-term exam	23	60	71	15	43	80	38	26	64
Mark in end-of-term exam	40	74	91	30	55	79	45	51	87

Assuming that the differences of the two marks for each student are approximately Normally distributed, find a 95% confidence interval for the difference between the mean marks of the two examinations.

18 To test the effect of a fertiliser, ten similar plants were chosen. Five plants were treated with the fertiliser, and five were left untreated. Some time later, the growth of each plant was measured as follows:

With fertiliser: 85, 102, 82, 107, 75 mm
Without fertiliser: 65, 60, 63, 45, 53 mm

Find 90% confidence limits for the mean increase in growth of this type of plant due to the fertiliser. State any assumptions which you need to make.

19 For a random sample of 150 people, each person was given two intelligence tests A and B. Their scores in test A had mean 58.6 and standard deviation 18.3, and their scores in test B had mean 60.8 and standard deviation 16.2. For each person, the difference

$$\text{(mark in test A)} - \text{(mark in test B)}$$

was calculated, and these differences had mean -2.2 and standard deviation 8.5.
   Is there any evidence that people perform differently in the two tests?

20 A random sample of 12 values, $x_1, x_2, \ldots, x_{12}$ is taken from a Normal distribution, and $\sum x_i = 243$, $\sum x_i^2 = 5226$.
Find 95% confidence limits for
(i) the population mean
(ii) the population standard deviation.

14.6   Non-parametric tests

When carrying out hypothesis tests using small samples, it has been necessary to assume that the population is Normally distributed (or at least approximately so). This is particularly important when a t distribution is used.
   If the assumption of Normality is not a reasonable one, we need to devise other tests. We now describe a few *non-parametric* or *distribution-free* tests for which no assumptions about the distribution of the population are necessary.

The sign test

This is a test for the *median* of a population.
   Suppose we wish to test the hypothesis that the median has a particular value m_0, and we have a random sample of size n.
   We attribute the sign $+$ to each value in the sample which is greater than m_0; and the sign $-$ to each value less than m_0.
   If the median really *is* m_0, then the probability that a value is greater than m_0 will be $\frac{1}{2}$; so the number of $+$ signs will have the Binomial distribution $B(n, \frac{1}{2})$

> To test H_0: median is m_0
> we consider $U =$ number of $+$ signs
> If H_0 is true, then $U \sim B(n, \frac{1}{2})$

We can use the Normal approximation for $B(n, \frac{1}{2})$ if appropriate (say for $n \geqslant 8$)

Example 1

For a certain type of light bulb, the median lifetime is 520 hours. A new long-life version is introduced, and a random sample of 12 long-life bulbs had the following lifetimes (in hours):

324, 816, 552, 1570, 512, 640, 1242, 602, 758, 410, 645, 1857.

Is there any evidence that the long-life bulbs have a higher median life-time?

Suppose that the long-life bulbs have median life-time m.

We test $H_0 : m = 520$

against $H_1 : m > 520$

Attributing signs (+ if greater than 520, − if less than 520) to the sample values, we obtain,

$$- \quad + \quad + \quad + \quad - \quad + \quad + \quad + \quad + \quad - \quad + \quad +$$

Let U be the number of + signs.

H_1 is favoured by large values of U.

If H_0 is true, then $U \sim B(12, \frac{1}{2})$,

which we can approximate by a Normal distribution having mean $12 \times \frac{1}{2} = 6$

and standard deviation $\sqrt{12 \times \frac{1}{2} \times \frac{1}{2}} = 1.732$

We have $U = 9$.

To find the significance level of the result, we calculate $P(U \geqslant 9)$ when H_0 is true;

then $P(U \geqslant 9) \approx P(U > 8.5 \text{ in Normal}) = P\left(Z > \dfrac{8.5 - 6}{1.732}\right)$

$$= P(Z > 1.443)$$

$$= 1 - 0.9255$$

$$= 0.0745$$

This is greater than 5%; so at the 5% level of significance we accept H_0. There is no evidence that the long-life bulbs have a higher median lifetime.

Notes (1) When we use a sign test, a lot of information is ignored; for example the values 552 and 1857 were treated in the same way; each was given the sign +.

(2) If any values in the sample were equal to 520, we would give them sign 0. Such values are then disregarded and the value of n is reduced accordingly.

Sign test for paired samples

Suppose that we have two random samples, and we wish to test the null hypothesis H_0 that the two samples come from populations having identical distributions.

(Note that there is no need to specify what that distribution might be.)

We first consider the case when the two samples are paired, and each has size n. We attribute signs

+ if the value in the first sample is greater than the corresponding value in the second sample

− if the value in the first sample is less than the corresponding value in the second sample.

If H_0 is true, then for each pair of values, the probability that the first sample value

is greater will be $\frac{1}{2}$, so the number of + signs will have the Binomial distribution $B(n, \frac{1}{2})$

> To test H_0: populations have identical distributions
> we consider
>
> $$U = \text{number of} + \text{signs}$$
>
> If H_0 is true, then $U \sim B(n, \frac{1}{2})$

Example 2

For a random sample of 10 people, their pulse rates (in beats per minute) were measured just before, and just after, a meal, as follows:

Person	A	B	C	D	E	F	G	H	I	J
Pulse rate (before meal)	82	63	85	77	83	86	74	79	58	88
Pulse rate (after meal)	78	63	82	71	73	81	74	80	58	86

Is there a significant difference between the pulse rates before and after the meal?

We test H_0: pulse rates before and after the meal are identically distributed against H_1: pulse rates before and after the meal are different.

Attributing signs, we obtain

A	B	C	D	E	F	G	H	I	J
+	0	+	+	+	+	0	−	0	+

There are 7 non-zero signs.
Let U be the number of + signs.
If H_0 is true, then $U \sim B(7, \frac{1}{2})$
We have $U = 6$;

when H_0 is true, $P(U \geqslant 6) = P(U = 6) + P(U = 7)$

$$= 7(\tfrac{1}{2})^7 + (\tfrac{1}{2})^7$$

$$= \frac{1}{16}$$

$$= 0.0625$$

Since this is a two-tailed test, the value $U = 6$ would be significant at a level of $2 \times 6.25\%$, i.e. 12.5%
Hence at the 5% level of significance, we
Accept H_0
These results do not demonstrate any significant difference.
Note   The Normal approximation to $B(7, \frac{1}{2})$ gives

$$P(U \geqslant 6) \approx P\left(Z > \frac{5.5 - 3.5}{\sqrt{1.75}}\right) = 0.0653,$$

so even with $n = 7$, the Normal approximation seems very satisfactory.

Rank sum test

Now suppose that we have two independent samples, of sizes n_1 and n_2. We now use a completely different method.

Considering all $(n_1 + n_2)$ values as a single set of numbers, we rank them.

In this situation it is usual to assign rank 1 to the *lowest* value, rank 2 to the next lowest, and so on.

(If two or more values are equal, we assign to each the average rank for the tied places.)

Let R_1 be the sum of the ranks in the first sample.

If the two samples come from identical populations, the ranks

$$1, 2, 3, ..., (n_1 + n_2)$$

will be randomly divided between the two samples.

The total of the ranks is $\frac{1}{2}(n_1 + n_2)(n_1 + n_2 + 1)$, and since the first sample contains n_1 out of the $(n_1 + n_2)$ values, we would expect R_1 to be about

$$\frac{n_1}{n_1 + n_2} \times \tfrac{1}{2}(n_1 + n_2)(n_1 + n_2 + 1) = \tfrac{1}{2}n_1(n_1 + n_2 + 1)$$

It can be shown that $E[R_1] = \tfrac{1}{2}n_1(n_1 + n_2 + 1)$

and $\operatorname{var}(R_1) = \tfrac{1}{12} n_1 n_2 (n_1 + n_2 + 1)$

and if n_1 and n_2 are not too small (say provided that $n_1 \geqslant 8$ and $n_2 \geqslant 8$), then R_1 is approximately Normally distributed.

Hence

---

to test H_0: populations have identical distributions
we consider

$$Z = \frac{R_1 - \tfrac{1}{2}n_1(n_1 + n_2 + 1)}{\sqrt{\tfrac{1}{12} n_1 n_2 (n_1 + n_2 + 1)}}$$

If H_0 is true, then $Z \sim N(0, 1)$

Example 3

From a large number of children who had taken part in a charity walk, random samples of 8 girls and 12 boys were selected.

The amounts of money they had collected were as follows:

Girls: £16.40, £8.50, £27.00, £13.30, £17.65, £39.25, £20.00, £13.90

Boys: £12.60, £9.75, £7.50, £18.10, £27.00, £11.40
£3.00, £12.60, £15.00, £5.25, £6.70, £14.00

Is there any evidence that the amounts collected by girls and boys are different?

We test H_0: there is no difference
against H_1: there is a difference.

We rank the 20 amounts

Girls: 14, 5, 18.5, 10, 15, 20, 17, 11

Boys: 8.5, 6, 4, 16, 18.5, 7, 1, 8.5, 13, 2, 3, 12

Let R_1 be the sum of the girls ranks.
Consider

$$Z = \frac{R_1 - \frac{1}{2} \times 8 \times 21}{\sqrt{\frac{1}{12} \times 8 \times 12 \times 21}}$$

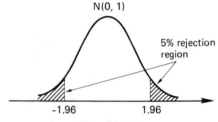

If H_0 is true, then $Z \sim N(0, 1)$

We have $R_1 = 14 + 5 + \cdots + 11 = 110.5$

and $Z = \dfrac{110.5 - 84}{\sqrt{168}}$

$= 2.045$

Fig. 14.25

Reject H_0.
There is some evidence that girls and boys collect differing amounts. Since R_1 is greater than the expected value (84), this suggests that the girls ranks are higher; i.e. that girls collect more money than boys.

Note   When the Normal approximation cannot be used (i.e. when one of the sample sizes is much less than 8), we must consider the distribution of R_1 from first principles. If H_0 is true, then the ranks of the first sample are equally likely to be any set of n_1 numbers chosen from $1, 2, 3, \ldots, (n_1 + n_2)$.

Exercise 14.6 Non-parametric tests

1   The ages (in years) of 15 cars brought to a scrapyard were as follows:

11, 26, 13, 7, 11, 8, 18, 16, 9, 11, 22, 11, 14, 12, 21.

Test the hypothesis that the median age at which cars are scrapped is 10 years.

2   The median time for a certain train journey was 144 minutes. After faster trains were introduced, seven journeys took the following times:

138, 140, 138, 139, 152, 138, 142 minutes.

At what level of significance does this indicate that the median journey time has decreased?

3   It is claimed that the median weekly wage for a certain type of employee is £165. In a random sample of these employees, 103 earned less than £165, 15 earned £165, and 82 earned more than £165.

Is this consistent with the claim that the median wage is £165?

4 Identical programs were run on two computers, and the run-times (in seconds) were as follows.

Program	A	B	C	D	E	F	G	H	I	J	K	L	M
Computer 1	15	37	11	78	45	50	25	28	8	21	56	18	30
Computer 2	14	30	13	65	44	41	32	24	6	18	40	17	26

Using the sign test, determine whether there is any evidence that the two computers differ in their run-times.

5 To test whether it is hotter at the bottom of a hill than at the top, the temperatures were measured at the bottom and at the top at the same time on 10 different days, as follows.

Day	1	2	3	4	5	6	7	8	9	10
Temp at bottom ($^\circ$C)	21	24	28	22	17	15	20	25	16	20
Temp at top ($^\circ$C)	17	26	24	22	17	11	22	21	12	16

(i) Use the sign test to decide whether there is any evidence that the bottom of the hill is hotter than the top.
(ii) Use a t test to decide whether the mean temperature at the bottom is greater than that at the top. What assumption is necessary for this to be a valid test? Does this seem a reasonable assumption in this case?

6 For random samples of 12 town houses and 10 country houses, the amounts spent on repairs during the last five years were:

Town houses: £540 £290 £80 £250 £4900 £300
£330 £1350 £750 £180 £350 £560

Country houses: £120 £730 £1750 £310 £800
£710 £1700 £2800 £260 £9500

Use the rank sum test to decide whether there are any differences between the amounts spent on town and country houses.

7 Nine 'standard' car batteries and nine 'long-life' batteries were found to have the following life-times (in months)
'Standard' batteries: 38, 24, 44, 22, 35, 41, 22, 29, 46
'Long-life' batteries: 36, 52, 47, 28, 45, 61, 49, 41, 54
Use the rank sum test to decide whether the 'long-life' batteries do last longer than the 'standard' ones.

8 When the birthweights of 20 boys and 25 girls were ranked, the total of the boys' ranks was 375. Does this suggest that the birthweights of boys and girls are different?

9 Three girls and eight boys took part in a shooting competition, and their scores were as follows:

Girls: 85, 66, 74
Boys: 91, 75, 98, 90, 83, 89, 95, 88

Rank these 11 scores (with 1 for the lowest score, and so on), and let R denote the sum of the girls' ranks. Show that $R = 8$.
List all the possible sets of girls' ranks for which $R \leqslant 8$ and deduce the probability that $R \leqslant 8$ when girls and boys are assumed to perform equally (so that the girls' ranks are equally likely to be any set of 3 numbers chosen from $1, 2, 3, ..., 11$).
   At what level of significance do the above scores indicate that girls and boys perform differently?

10 Twelve swimmers were timed over the same course in the morning and in the afternoon of the same day, as follows.

Swimmer	A	B	C	D	E	F
Time for morning swim (s)	123.2	117.2	128.3	142.5	130.0	120.4
Time for afternoon swim (s)	122.0	117.4	127.6	140.7	128.7	120.3

Swimmer	G	H	I	J	K	L
Time for morning swim (s)	126.8	136.3	119.2	125.4	147.6	116.9
Time for afternoon swim (s)	127.7	133.9	117.7	126.0	146.0	116.5

(i) Allocate signs to each swimmer in the usual way (+ if the morning time is longer, − if the morning time is shorter), and use the sign test to examine any difference between the morning and afternoon times.

(ii) The test can be improved to take into account the size of the differences, as follows: Calculate the differences between the morning and afternoon times for each swimmer, and, ignoring the signs, rank these differences in order of increasing size (1 for the smallest difference, and so on). Let R be the sum of the ranks for the swimmers given + signs in (i). Calculate R for the given data.
   You are given that, for paired samples of size n taken from the same population, R is approximately Normally distributed with mean $\frac{1}{4}n(n + 1)$ and variance $\frac{1}{24}n(n + 1)(2n + 1)$.
   Use this result to test whether the morning and afternoon swimming times are different.

11 A certain volcano is believed to erupt every 80 years, on average; but successive eruptions occurred in the years

1751, 1769, 1799, 1827, 1887, 1977 and 1982

(i)   Assuming that the time intervals between eruptions are Normally distributed, use the t distribution to test the hypothesis that the mean is 80 years.

(ii) Now assume instead that the time intervals have an exponential distribution with pdf $f(x) = \lambda e^{-\lambda x}$ (for $x \geqslant 0$).

You are given that the mean of this distribution is $\dfrac{1}{\lambda}$;

also, if a random sample of size n from this distribution has mean $\bar{X}$, then $2n\lambda\bar{X}$ has the χ^2_{2n} distribution.

Use these results to test the hypothesis that the mean is 80 years.

(iii) Making no assumptions about the distribution of the time intervals, use the sign test to test the hypothesis that the median time interval is 80 years.

Answers to Exercises

Exercise 1.1 (page 7)

1

Goals	0	1	2	3	4	5	6	7
No of matches	2	9	11	11	5	2	1	1

2 (a) $-\frac{1}{2} \to \frac{1}{2}$
$\frac{1}{2} \to 1\frac{1}{2}$
$1\frac{1}{2} \to 2\frac{1}{2}$

(b) $65\frac{1}{2} \to 70\frac{1}{2}$
$70\frac{1}{2} \to 75\frac{1}{2}$
$75\frac{1}{2} \to 80\frac{1}{2}$

(c) $-\frac{1}{2} \to 4\frac{1}{2}$
$4\frac{1}{2} \to 5\frac{1}{2}$
$5\frac{1}{2} \to 6\frac{1}{2}$

(d) $2\frac{3}{4} \to 5\frac{3}{4}$
$5\frac{3}{4} \to 6\frac{1}{4}$
$6\frac{1}{4} \to 6\frac{3}{4}$

(e) $149.5 \to 159.5$
$159.5 \to 169.5$
$169.5 \to 179.5$

(f) $4.95 \to 5.45$
$5.45 \to 5.95$
$5.95 \to 6.45$

(g) $0 \to 27.5$
$27.5 \to 52.5$
$52.5 \to 77.5$

(h) $11 \to 14$
$14 \to 17$
$17 \to 20$

(i) $-\frac{1}{2} \to 10\frac{1}{2}$
$10\frac{1}{2} \to 20\frac{1}{2}$
$20\frac{1}{2} \to 30\frac{1}{2}$

(j) $37\frac{1}{2} \to 39\frac{1}{2}$
$39\frac{1}{2} \to 41\frac{1}{2}$
$41\frac{1}{2} \to 43\frac{1}{2}$

(k) $109.5 \to 119.5$
$119.5 \to 129.5$
$129.5 \to 139.5$

(l) $0.2 \to 0.4$
$0.4 \to 0.6$
$0.6 \to 0.8$

(m) $9.5 \to 19.5$
$19.5 \to 29.5$
$29.5 \to 39.5$

(n) $-0.105 \to -0.045$
$-0.045 \to +0.045$
$+0.045 \to +0.105$

3 Frequency densities 0.67, 0.9, 1.5, 2.2, 3.6, 2.8, 2.0, 1.6, 0.8
No of candidates = (area between 42.5 and 54.5) = $(3 \times 2.2) + 18 + (4 \times 2.8)$
= 35.8

4 Frequency densities 0.6, 0.8, 1.15, 1.65, 0.9.
Probability = $\dfrac{\text{area between 112 and 128}}{\text{total area}} = \dfrac{13 \times 1.15 + 3 \times 1.65}{125} = \dfrac{19.9}{125} = 0.1592.$

5 (i) Probability densities 0.021, 0.022, 0.020, 0.010, 0.003
(ii) Probability densities 0.008, 0.005, 0.012, 0.019, 0.022, 0.015, 0.004
Difficulties direct from the tables: Ages are grouped differently; villages have different numbers of people.
Comparing the histograms: 1st village has mainly young population; the 2nd village has mainly an elderly one.

Exercise 1.2 (page 12)

1 Points

Temperature	11.5	13.5	15.5	17.5	19.5	21.5	23.5	25.5	27.5	29.5
cf	3	5	18	45	76	101	119	139	148	150

(i) when temp = 19.0, cf ≈ 68, i.e. 68 days
(ii) when temp = 22.0, cf ≈ 105 } hence 80 days
when temp = 16.0, cf ≈ 25
(iii) when cf = 125, temp ≈ 24.1°C.

2

Population	4950	9950	19950	29950	39950	59950	79950	99950	
cf	28	63	102	130	150	173	191	200	220

(i) when cf = 110, median ≈ 23000
(ii) when cf = 143, 65th percentile ≈ 36000
(iii) when pop = 45000, cf ≈ 156, percentage = $\dfrac{64}{220} \times 100 = 29.1\%$
(iv) when pop = 36000, cf ≈ 142 } probability = $\dfrac{71}{220} = 0.323$
when pop = 12000, cf ≈ 71
(v) when cf = 187, population ≈ 76000

3

Age	10	20	30	40	50	60	70	80	
cum probability	0.129	0.292	0.433	0.571	0.683	0.8	0.9	0.971	1.0

(i) when cp = 0.5, median ≈ 35 years
(ii) when cp = 0.2, 20th percentile ≈ 14.5 years
(iii) when age = 65, cp ≈ 0.85 } proportion ≈ 0.59
when age = 18, cp ≈ 0.26

4

Marks	30.5	40.5	45.5	50.5	55.5	60.5	70.5	100.5
cf	51	89	115	154	191	235	303	350

(i) when cf = 175, median = $50.5 + \dfrac{21}{37} \times 5 = 53.34$
(ii) when mark = 64.5, cf = $235 + \dfrac{4}{10} \times 68 = 262.2$ } 78.6 candidates
when mark = 54.5, cf = $154 + \dfrac{4}{5} \times 37 = 183.6$

5 (iii) when cf = 140, pass mark $= 45.5 + \dfrac{25}{39} \times 5 = 48.71$.

Capacity	1000	1600	2800
cf Export	64	190	252
cf Home	126	438	541
cf Total	190	628	793

(i) 1295 c.c.   (ii) 1278 c.c.   (iii) 1283 c.c.

6

Salary	£3148	£3721	£4449	£5202	£6036
cum probability	0.1	0.25	0.5	0.75	0.9

The median is £4449
(i) when cp = 0.35, 35th percentile is £4012
(ii) when salary is £5500, cp = 0.8036
i.e. 19.64% earn more than £5500.

Exercise 1.3   (page 18)

1 (i) mode ≈ 17, median (dividing the area in half) ≈ 29 minutes

(ii)

Time	0–10	10–20	20–30	30–50	50–70	70–100
Frequency	7	20	15	20	10	9

Mean $= \dfrac{5 \times 7 + 15 \times 20 + 25 \times 15 + 40 \times 20 + 60 \times 10 + 85 \times 9}{81} \approx 35.5$ minutes

2 (i) median ≈ 21.4 years

(ii)

Age	10–14	14–16	16–17	17–18	18–19	19–20
Frequency	5	10	11	22	22	10

Age	20–21	21–22	22–23	23–24	24–30
Frequency	11	21	44	34	10

(iii) Frequency densities 1.25, 5, 11, 22, 22, 10, 11, 21, 44, 34, 1.67
Mean = 20.57 years; mode = 22.7 years

3 Midpoints are 5.5, 15.5, etc; mean = 28 marks
This is only an estimate since we do not know the actual marks: we have assumed that those in the class 1–10 have average mark 5.5, etc.
(i) for greatest possible value take the marks as 10, 20, 30, ...,
i.e. all values 4.5 more than before; so greatest possible mean = 28 + 4.5 = 32.5.
(ii) for least possible value take the marks as 1, 11, 21, ...,
i.e. all values 4.5 less than before; so least possible mean = 28 − 4.5 = 23.5.

4 He should have used mid-points 15, 25, 35, 45, which will give mean $= 19.88 + 5$
$= 24.88$ kg.

5 (a) (i) mean $= \dfrac{10 \times 60 + 30 \times 55 + 50 \times 45 + 68 \times 40}{200} = \dfrac{4500 + 2720}{200} = 36.1$ years

(ii) mean $= \dfrac{4500 + 78 \times 40}{200} = 38.1$ years

(b) mean $= \dfrac{4500 + (x \times 40)}{200} = 36.7$, giving $x = \dfrac{7340 - 4500}{40} = 71$ years.

6 $\bar{u} = \dfrac{\sum uf}{\sum f} = \dfrac{41}{110} = 0.3727$

(i) $\bar{x} = \bar{u} + 53 = 53.3727$
(iii) $\bar{z} = \bar{u} \times 3 + 46 = 47.1182$
(ii) $\bar{y} = 53.3727$
(iv) $\bar{w} = \bar{u} \times 10 + 95.5 = 99.2273$.

7 Midpoints 14.5, 24.5, ...,   mean = 48.11

8 Midpoints 35.28, 35.33, ..., mean = 35.40

9 Midpoints 17.5, 20.5, ...,   mean = 30.80 years.

Exercise 2.1   (page 25)

1 (i) range is 10   (ii) mean is 7, deviations 1, −5, 0, −4, 3, 1, 1, −2, 5
(iii) mean deviation $= \dfrac{22}{9} = 2.44$.

2 (i) range is 60 cm   (ii) frequencies are 14 + 42 + 24 = 80
lower quartile = 22.86 cm, upper quartile = 45 cm, interquartile range = 22.14 cm
(iii) half of them, i.e. 40 trees.

3 uq = 22.9, lq = 18.1; iq range ≈ 4.8 years

4 uq = 22.8, lq = 16.9; iq range ≈ 5.9°C

5 uq ≈ 56, lq ≈ 17; iq range ≈ 39 years.

6 median ≈ £4700; uq ≈ £5850; lq ≈ £3800; iq range ≈ £2050

7

Score	264.5	268.5	272.5	276.5	280.5	284.5	288.5
cf	1	1	3	4	9	19	35

Score	292.5	296.5	300.5	304.5	308.5	312.5	316.5
cf	59	82	110	124	127	127	128

when cf = 96, uq = $296.5 + \frac{14}{28} \times 4 = 298.5$

when cf = 32, lq = $284.5 + \frac{13}{16} \times 4 = 287.75$

iq range ≈ 10.75.

8

Current	12.95	13.45	13.95	14.45	14.95	15.45	15.95	16.45	16.95	
cf	3	18	54	98	155	204	225	232	234	243

when cf = 182.25, uq = $14.95 + \frac{27.25}{49} \times 0.5 = 15.23$

when cf = 60.75, lq = $13.95 + \frac{6.75}{44} \times 0.5 = 14.03$ iq range ≈ 1.20 A.

9

xf	$x - \bar{x}$	$|x - \bar{x}| f$
0	−2.4	7.2
5	−1.4	7.0
16	−0.4	3.2
21	0.6	4.2
20	1.6	8.0
10	2.6	5.2
72		34.8

$\bar{x} = \frac{72}{30} = 2.4$, mean deviation $= \frac{34.8}{30} = 1.16$.

10

uf	$u - \bar{u}$	$|u - \bar{u}| f$
−34	−2.4	40.8
−44	−1.4	61.6
0	−0.4	41.6
86	0.6	51.6
70	1.6	56
42	2.6	36.4
120		288

$\bar{u} = 0.4$, mean deviation $= \frac{288}{300} = 0.96$

for lengths, $\bar{x} = 0.4 \times 0.05 + 35.38 = 35.40$ mm

mean deviation $= 0.96 \times 0.05 = 0.048$ mm.

Exercise 2.2 (page 33)

1 $\bar{x} = 1$, standard deviation $= \sqrt{48/40} \approx 1.095$

2 (i) $\bar{x} = 8.25$

(ii) Deviations are −5.25, 2.75, −1.25, −3.25, 6.75, 3.75, 0.75, −4.25, standard deviation $= \sqrt{\frac{125.5}{8}} = 3.961$

(iii) $\sum x^2 = 670$; standard deviation $= \sqrt{\frac{670}{8} - 8.25^2} = 3.961$.

3 mean = 3.4, standard deviation = 1.29 kg
(i) between 2.11 and 4.69 kg: 7 weights
(ii) between 0.82 and 5.98 kg: 9 weights.

4 mean = 19.71, standard deviation 7.26 minutes; longer than 26.97 minutes there are 2 journeys.

5 mean = £0.80, standard deviation = £11.16.

6 mean = 28.83, standard deviation = 13.44.

7

No of spades	0	1	2	3	4	5	6	7
No of hands	1	3	14	10	7	6	6	2

mean = 3.48, standard deviation = 1.70.

8 Mid-interval values: 1, 3, 6, 10.5, 19.5, 39, 78.

mean = 30.622, standard deviation = 28.19 weeks

Frequency densities: 11.9, 7.75, 6.3, 3.62, 1.77, 1.08
1 standard deviation each side: 2.4, 58.8
Number between 2.4 and 58.8 is approximately $14 + 31 + 30 + 47 + 46 + 7 = 175$ i.e. 70%.

9 A: mean 84.64, standard deviation 3.90
B: mean 80.00, standard deviation 14.48
C: mean 73.00, standard deviation 0.65

Method C is the best. A consistently low value can be allowed for. Although Method B gives the correct mean value, the high standard deviation makes it very unreliable.

10 Mean = 3, standard deviation = 0.4472; $\bar{x} \pm 4s = 1.21, 4.79$; 4 values i.e. 5% lie outside these limits. Usually over 99% of values are less than 3 standard deviations from the mean.

Excerise 2.3 (page 37)

1 (i) 100.5, 28.87   (iii) 102, 57.74
(iv) 500.6, 288.7
(v) 4.505, 0.2887.

2 u: 0.3727, 1.249, w: 116.3727, 1.249, x: 116.3727, 1.249
y: 73.8635, 6.245 z: 301.954, 24.98,

3 Mid-values: 5, 15, 25, ... mean = 46.1 years, standard deviation = 19.2 years
Number between 7.7 and 84.5 is approximately
$1 + 13 + 24 + 35 + 40 + 36 + 22 + 16 + 4 = 191$ people.

4 Mid-values: 35.28, 35.33,...., mean = 35.40 mm, standard deviation = 0.0589 mm.

5 Mid-values: 12.5, 17.5,...., mean = 30.02 g, standard deviation = 7.872 g
true mean = 27.02 g, standard deviation = 7.872 g.

6 (i) lq = 2.375, uq = 6.364; siqr = 1.994
(ii) $\bar{x} = 5$, mean deviation $= \dfrac{144}{60} = 2.4$
(iii) standard deviation = 3.507.

7 Mean = 4.867 m, variance = 0.632 m^2
(i) 48.67 km, 63.2 km^2
(ii) 34.867 cm, 0.632 cm^2
(iii) 0.973 m, 0.0253 m^2
(iv) 6486.7 mm, 6322 mm^2

8 mean = 24.69 m s^{-1} = 88.875 km h^{-1}, variance = 42.46 m^2 s^{-2} = 550.3 km^2 h^{-2}.

Exercise 2.4 (page 40)

1 (i) 0.75, 0.583, 1.769, 1.714, −0.5 Since most of these standard scores are positive, her performance is well above average for her class. French is best; Physics is worst.
(ii) Standard scores: −1.75 and −1.80. The English mark is better.

2 Original marks ($\bar{x} = 42.3, s = 15.8$): 64   29   44   92   10
Standardised marks ($\bar{x} = 0, s = 1$): 1.373   −0.842   0.108   3.146   −2.044
Scaled marks ($\bar{x} = 50, s = 20$): 77.5   33.2   52.2   112.9   9.1

3 $\bar{x} = 12, s = 7.616$; z: −1.182, 0.263, 0.394, −0.394, 1.838, −0.919
$\sum z = 0, \sum z^2 = 6.000$: the z-scores have mean 0 and standard deviation 1. Rescaled values ($\bar{x} = 40, s = 15$): 22.3, 43.9, 45.9, 34.1, 67.6, 26.2.

4 Standardised values: (a) 1.6   (b) 2.8   (c) −4.2   (d) −2.0
We would suspect the coin of bias in cases (b) and (c), since values which are 2.8 and 4.2 standard deviations from the mean are very unlikely.

5 (i) By spreading out his marks much more than the other two judges.
(ii) and (iii)

	1st judge. $\bar{x}=61, s=9.96$	2nd judge $\bar{x}=78, s=3.03$	3rd judge $\bar{x}=55.8, s=23.34$	Total	Position
A	−1.10	1.32	0.61	0.83	1st
B	1.10	−0.99	−0.68	−0.57	4th
C	0.60	0.33	−0.16	0.77	2nd
D	0.70	0.66	−1.32	0.04	3rd
E	−1.31	−1.32	1.55	−1.08	5th

This method gives equal weights to the opinions of all 3 judges.

Exercise 2.5 (page 42)

1 $\bar{x} = 5.3, s = 2.216$
New $\sum x = 53 + 12 + 8 = 73, \sum x^2 = 330 + 12^2 + 8^2 = 538$: $\bar{x} = 6.083, s = 2.798$.

2 $\sum w = 20 \times 28 = 560, \sum w^2 = 20(28^2 + 8^2) = 16960$.
For 21 children $\sum w = 560 + 35 = 595, \sum w^2 = 18185$; $\bar{w} = 28.33, s = 7.948$ kg.

3 Top set: $\sum x = 1188, \sum x^2 = 94\,704$
2nd set: $\sum x = 590, \sum x^2 = 36\,202.4$
Combined: $\sum x = 1778, \sum x^2 = 130\,906.4$
$\bar{x} = 71.12,$ $s = 13.35$.

4

	$n = 8$	$\sum x =$	$\sum x^2 =$
Management	8	2080,	545 800
Office staff	32	4640,	677 408
Factory	110	18 700,	3 278 000
All	150	25 420,	4 501 208

$\bar{x} = £169.47,\ s = £35.90.$

5 For 8 children: $\sum x = 113.6,\ \sum x^2 = 1\,624.64$
 For 7 remaining: $\sum x = 98.6,\ \sum x^2 = 1399.64$
 $\bar{x} = 14.086,\ s = 1.241$ years.

6 All: $n = 250,\ \sum x = 43\,500,\ \sum x^2 = 7\,605\,000$
 Men: $n = 100,\ \sum x = 17\,800,\ \sum x^2 = 3\,178\,400$
 Women: $n = 150,\ \sum x = 25\,700,\ \sum x^2 = 4\,426\,600$
 $\bar{x} = 171.33,\ s = 12.47$ cm.

7 Mean = 31, standard deviation = 0.79
 The value 27 is more than 5 standard deviations from the mean ($z = -5.06$) – hence surprise!
 For the other 15 rock samples, $\sum x = 469,\ \sum x^2 = 14657$;

 standard deviation $= \sqrt{\dfrac{14657}{15} - \left(\dfrac{469}{15}\right)^2} = \sqrt{-0.47}$. Impossible.

Exercise 3.1 (page 47)

1 (i) (a)

X	2	3	4	5	6	7	8	9	10	11	12
Probability	$\frac{1}{36}$	$\frac{2}{36}$	$\frac{3}{36}$	$\frac{4}{36}$	$\frac{5}{36}$	$\frac{6}{36}$	$\frac{5}{36}$	$\frac{4}{36}$	$\frac{3}{36}$	$\frac{2}{36}$	$\frac{1}{36}$

(b) mode is 7;   (c) $P(X > 2) = \frac{35}{36}$.

(ii) (a)

X	0	1	2
Probability	$\frac{25}{36}$	$\frac{10}{36}$	$\frac{1}{36}$

(b) mode is 0   (c) $P(X > 2) = 0$

(iii) (a)

X	1	2	3	4	5	6
Probability	$\frac{1}{36}$	$\frac{3}{36}$	$\frac{5}{36}$	$\frac{7}{36}$	$\frac{9}{36}$	$\frac{11}{36}$

(b) mode is 6;   (c) $P(X > 2) = \frac{8}{9}$.

(iv) (a)

X	0	1	2	3	4	5
Probability	$\frac{6}{36}$	$\frac{10}{36}$	$\frac{8}{36}$	$\frac{6}{36}$	$\frac{4}{36}$	$\frac{2}{36}$

(b) mode is 1   (c) $P(X > 2) = \frac{1}{3}$.

2 (i)

X	0	1	2	3	4
Probability	$\frac{1}{16}$	$\frac{4}{16}$	$\frac{6}{16}$	$\frac{4}{16}$	$\frac{1}{16}$

X	0	1	2	3	4
Frequency	2	8	12	8	2

(ii)

Y	1	2	3	4
Probability	$\frac{2}{16}$	$\frac{8}{16}$	$\frac{4}{16}$	$\frac{2}{16}$

3 (i)

X	10	20	50
Probability	$\frac{5}{10}$	$\frac{2}{10}$	$\frac{3}{10}$

(ii)

X	20	30	40	60	70	100
Probability	$\frac{20}{90}$	$\frac{20}{90}$	$\frac{30}{90}$	$\frac{2}{90}$	$\frac{12}{90}$	$\frac{6}{90}$

$P(25 < X < 65) = \dfrac{26}{45}$.

4 (i)

X	0	1	2	3
Probability	$\frac{2}{56}$	$\frac{15}{56}$	$\frac{27}{56}$	$\frac{12}{56}$

(ii)

Y	15	18	21	24
Probability	$\frac{12}{56}$	$\frac{27}{56}$	$\frac{15}{56}$	$\frac{2}{56}$

5

X	0	1	3
Probability	$\frac{2}{6}$	$\frac{3}{6}$	$\frac{1}{6}$

$E[X] = 1, E[X^2] = 2, \text{var}(X) = 1.$

6

X	1	2	3	4	5
Probability	$\frac{1}{5}$	$\frac{1}{5}$	$\frac{1}{5}$	$\frac{1}{5}$	$\frac{1}{5}$

$E[X] = 3, E[X^2] = 11, \text{var}(X) = 2$

$P(X \text{ within } \sigma \text{ of } \mu) = P(1.586 < X < 4.414) = \frac{3}{5}.$

7

X	1	2	3	4	5
Probability	$\frac{5}{15}$	$\frac{4}{15}$	$\frac{3}{15}$	$\frac{2}{15}$	$\frac{1}{15}$

$E[X] = \frac{7}{3}.$

8

X	-10	0	30
Probability	$\frac{4}{8}$	$\frac{3}{8}$	$\frac{1}{8}$

$E[X] = -\frac{5}{4}, E[X^2] = \frac{325}{2}, \text{var}(X) = \frac{2575}{16} \approx 160.94.$

If they play 50 times, A should expect to lose $\frac{5}{4} \times 50 = 62.5$ pence

9 (i)

X	0	1	2	3
Probability	$\frac{4}{16}$	$\frac{4}{16}$	$\frac{4}{16}$	$\frac{4}{16}$

Y	0	1	2	3
Probability	$\frac{4}{16}$	$\frac{4}{16}$	$\frac{4}{16}$	$\frac{4}{16}$

$X+Y$	0	1	2	3	4	5	6
Probability	$\frac{1}{16}$	$\frac{2}{16}$	$\frac{3}{16}$	$\frac{4}{16}$	$\frac{3}{16}$	$\frac{2}{16}$	$\frac{1}{16}$

XY	0	1	2	3	4	6	9
Probability	$\frac{7}{16}$	$\frac{1}{16}$	$\frac{2}{16}$	$\frac{2}{16}$	$\frac{1}{16}$	$\frac{2}{16}$	$\frac{1}{16}$

$E[X] = 1.5, E[Y] = 1.5, E[X+Y] = 3 = E[X] + E[Y]$
$$E[XY] = 2.25 = E[X]E[Y].$$

5

X	1	2	3	4	...
Probability	$\frac{1}{3}$	$\frac{2}{3} \times \frac{1}{3}$	$\frac{2}{3} \times \frac{2}{3} \times \frac{1}{3}$	$\frac{2}{3} \times \frac{2}{3} \times \frac{2}{3} \times \frac{1}{3}$	...
	$\frac{1}{3}$	$\frac{2}{9}$	$\frac{4}{27}$	$\frac{8}{81}$	...

The probabilities form an infinite G.P. with first term $\frac{1}{3}$ and common ratio $\frac{2}{3}$, hence the sum of probabilities

$$\frac{1}{3} + \frac{2}{9} + \frac{4}{27} + \frac{8}{81} + \cdots$$
$$= \frac{\frac{1}{3}}{1 - \frac{2}{3}}$$
$$= 1.$$

For a sample of 243 people

X	1	2	3	$\geqslant 4$
Frequency	81	54	36	72

Exercise 3.2 (page 55)

1 (i) $E[X] = 7$ (ii) $E[X] = \frac{1}{3}$ (iii) $E[X] = \frac{161}{36}$ (iv) $E[X] = \frac{35}{18}.$

2 (i) $E[X] = 2, E[X^2] = 5, \text{var}(X) = 5 - 2^2 = 1, \sigma = 1$
(ii) $E[Y] = 2.375, E[Y^2] = 6.375, \text{var}(Y) = 0.7344; \sigma = 0.857.$

3 (i) $E[X] = 24, E[X^2] = 880, \text{var}(X) = 304$
(ii) $E[X] = 48, E[X^2] = \frac{25600}{9}, \text{var}(X) = 540.44$
$P(X > \mu + \sigma) = P(X > 71.25) = \frac{1}{15}.$

4 (i) $E[X] = \frac{15}{8}, E[X^2] = \frac{33}{8}, \text{var}(X) = \frac{39}{64}$
(ii) $E[Y] = \frac{147}{8}, E[Y^2] = \frac{2745}{8}, \text{var}(Y) = \frac{351}{64}$
(iii) If there are X red discs, then there are $(3 - X)$ green ones; hence the total number of points is $Y = 5X + 8(3 - X) = 24 - 3X.$
If $Y = 24 - 3X$, then $E[Y] = 24 - 3E[X] = 24 - 3 \times \frac{15}{8} = \frac{147}{8}$
and $\text{var}(Y) = 3^2 \text{var}(X) = 9 \times \frac{39}{64} = \frac{351}{64}.$

7 B(10, 0.9); P(X < 8) = 1 − P(X = 8) − P(X = 9) − P(X = 10)
= 1 − 0.1937 − 0.3874 − 0.3487 = 0.0702.

8 B(5, 0.2); P(X ⩾ 3) = P(X = 3) + P(X = 4) + P(X = 5)
= 0.0512 + 0.0064 + 0.00032 = 0.05792.

9 If X is the number of forward steps, X is B(8, ½)

X	8	7	6	5	4	3	2	1	0
Finishing position (metres forward)	8	6	4	2	0	−2	−4	−6	−8

(i) P(at start) = P(X = 4) = $\frac{70}{256}$

(ii) P(more than 4 m from the start) = P(X = 8, 7, 1 or 0) = $\frac{1+8+8+1}{256} = \frac{18}{256}$.

10 Number of correct answers is B$\left(12, \frac{1}{5}\right)$

(i) P(13 marks) = P(X = 5) = 0.0532
(ii) P(less than 0 marks) = P(X = 0, 1 or 2)
= 0.0687 + 0.2062 + 0.2835 = 0.5583.

Exercise 3.4 (page 64)

1 (i) B$\left(15, \frac{1}{6}\right)$ (ii) No, since the 5 cards are not drawn independently (e.g. there cannot be 5 aces)   (iii) B$\left(8, \frac{4}{17}\right)$

(iv) Possibly B$(7, \frac{1}{3})$, but the assumption of independence (i.e. that the probability of a rainy day is unaffected by the weather on the preceding days) is dubious here.
(v) No; not independent   (vi) B(50, 0.01).

2 (i)

X	0	1	2	3
Probability	$\frac{27}{64}$	$\frac{27}{64}$	$\frac{9}{64}$	$\frac{1}{64}$

$E[X] = \frac{3}{4}$, $E[X^2] = \frac{72}{64}$, var $(X) = \frac{9}{16}$

Modes 0, 1

(ii)

X	0	1	2	3
Probability	$\frac{3}{12}$	$\frac{3}{12}$	$\frac{3}{12}$	$\frac{3}{12}$

Y	0	1	2	3
Probability	$\frac{3}{12}$	$\frac{3}{12}$	$\frac{3}{12}$	$\frac{3}{12}$

X + Y	1	2	3	4	5
Probability	$\frac{2}{12}$	$\frac{2}{12}$	$\frac{4}{12}$	$\frac{2}{12}$	$\frac{2}{12}$

XY	0	2	3	6
Probability	$\frac{6}{12}$	$\frac{2}{12}$	$\frac{2}{12}$	$\frac{2}{12}$

E[X] = 1.5, E[Y] = 1.5, E[X + Y] = 3 = E[X] + E[Y]
$$E[XY] = \frac{11}{6} \neq E[X]E[Y].$$

Exercise 3.3 (page 60)

1

X	0	1	2	3	4
Probability	$\frac{1}{16}$	$\frac{4}{16}$	$\frac{6}{16}$	$\frac{4}{16}$	$\frac{1}{16}$
Expected frequency	4	16	24	16	4

2

X	0	1	2	3	4	5	6
Probability	$\frac{729}{4096}$	$\frac{1458}{4096}$	$\frac{1215}{4096}$	$\frac{540}{4096}$	$\frac{135}{4096}$	$\frac{18}{4096}$	$\frac{1}{4096}$

P(X < 3) = $\frac{3402}{4096}$.

3 B$\left(5, \frac{1}{6}\right)$ (i) P(X = 3) = $\frac{250}{7776}$
(ii) P(X ⩾ 3) = $\frac{250}{7776} + \frac{25}{7776} + \frac{1}{7776} = \frac{276}{7776}$.

4 B(6, 0.1)   (i) P(X = 0) = 0.5314   (ii) P(X = 1) = 0.3543
(iii) 1 − P(X = 0) = 0.4686.

5 B(12, 0.65); P(X = 5) = 792 × (0.65)⁵ × (0.35)⁷ = 0.0591.

6 Number of damaged fruit in the sample is B(20, 0.03)
(i) P(Class I) = P(X = 0) = (0.97)²⁰ = 0.5438
(ii) P(Class II) = P(X = 1) + P(X = 2) = 20 × 0.03 × (0.97)¹⁹
+ 190 × (0.03)² × (0.97)¹⁸
= 0.3364 + 0.0988 = 0.4352
(iii) P(Class III) = 1 − 0.5438 − 0.4352 = 0.0210.

(ii)

X	0	1	2	3	4	5
Probability	0.00243	0.02835	0.1323	0.3087	0.36015	0.16807

Mode 4, $E[X] = 3.5$, $E[X^2] = 13.3$, var $(X) = 1.05$.

3 X is $B(12, 0.1)$, $\mu = 1.2$, $\sigma = 1.039$
$P(\mu - \sigma < X < \mu + \sigma) = P(X = 1 \text{ or } 2) = 0.3766 + 0.2301 = 0.6067$.

4 X is $B(150, \frac{1}{6})$, $\mu = 25$, $\sigma = 4.564$.

5 X is $B(16, 0.05)$; $P(X = 0) = 0.4401$, $P(X \leqslant 0) = 0.4401$
$P(X = 1) = 0.3706$, $P(X \leqslant 1) = 0.8108$
$P(X = 2) = 0.1463$, $P(X \leqslant 2) = 0.9571$
$P(X = 3) = 0.0359$, $P(X \leqslant 3) = 0.9930$

Hence at least 3 plugs should be available.

6 (i) P(none) $= (0.8)^n$ (ii) P(at least one) $= 1 - (0.8)^n$
If $1 - (0.8)^n > 0.95$ then $n > 13.43$, so at least 14 pots should be left.

7 (i) $B(8, p)$ has mean $8p$; the sample has mean $\bar{x} = 4.8$; so $8p = 4.8$ and hence $p = 0.6$.
(ii) Variance of sample is $s^2 = 3.62$
Variance of $B(8, 0.6)$ is $8 \times 0.6 \times 0.4 = 1.92$

(iii)

X	0	1	2	3	4
Probabilities from B(8, 0.6)	0.001	0.008	0.041	0.124	0.232
Expected frequencies	0.1	0.8	4.1	12.4	23.2

X	5	6	7	8
Probabilities from B(8, 0.6)	0.279	0.209	0.090	0.017
Expected frequencies	27.9	20.9	9.0	1.7

(iv) The probability p that a question is answered correctly must be the same for every question – unlikely, since some questions will probably be easier than the others. Responses to the 8 questions must be independent – this is also unlikely, since for example if a recruit has answered the first 6 questions correctly he has probably been paying close attention and it is more likely that he will also answer the last 2 correctly; conversely a recruit who has got the first 6 wrong has probably been asleep, and he will get the last two wrong as well.

Exercise 4.1 (page 70)

1 (i) $h = \frac{1}{4}$ (ii) $P(2.5 < X < 5.0) = \frac{1}{4}(5.0 - 2.5) = 0.625$
(iii) $P(X > 4.5) = 0.375$; expected number $= 75 \times 0.375 = 28.125$
(iv) area $= 0.95$, time $= 2 + \dfrac{0.95}{0.25} = 5.8$ minutes.

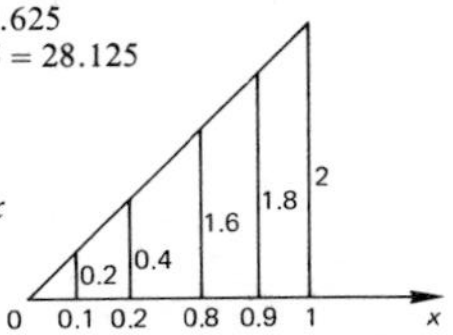

2 (i) mode is 1 m (ii) height $= 2$, hence $f(x) = 2x$
(iii) (a) $P(X < 0.2) = 0.04$ (b) $P(X > 0.8) = 0.36$
(iv) $\frac{1}{2} \times M \times 2M = \frac{1}{2}$, so $M = \dfrac{1}{\sqrt{2}} = 0.707$ m
(v) $P(0.1 < X < 0.9) = 0.8$; expected number $= 800 \times 0.8 = 640$ trees
(vi) (a) $0.36 \times 0.36 = 0.1296$ (b) $0.36 \times 0.04 \times 2 = 0.0288$

3 (i) $h = 5$ (ii) $P(-0.1 < X < 0.1) = 1 - 2 \times \frac{1}{2} \times 0.1 \times 2.5 = 0.75$
(iii) measured total weight is 28.3 kg; we require an error between -0.05 and 0.15.
$P(-0.05 < X < 0.15) = 1 - \frac{1}{2} \times 0.15 \times 3.75 - \frac{1}{2} \times 0.05 \times 1.25 = 0.6875$.

4 (i)

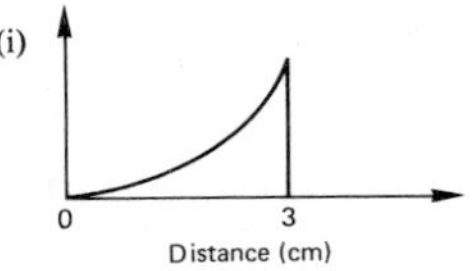

$$\text{Area} = \int_0^3 f(x)\,dx = \int_0^3 kx^2\,dx = \left[\frac{1}{3}kx^3\right]_0^3 = 9k$$

and hence $k = \dfrac{1}{9}$

(ii) $P(X > 2) = \displaystyle\int_2^3 \frac{1}{9}x^2\,dx = \left[\frac{1}{27}x^3\right]_2^3 = \frac{19}{27}$

(iii) For median M, $\dfrac{1}{2} = \displaystyle\int_0^M \frac{1}{9}x^2\,dx = \frac{1}{27}M^3$, so median $= 2.381$ cm

similarly for uq, $\dfrac{1}{27}Q^3 = \dfrac{3}{4}$, uq $= 2.726$; lq $= 1.890$;
iq range $= 0.836$ cm

(iv) $P(X < 1) = \displaystyle\int_0^1 \frac{1}{9}x^2\,dx = \frac{1}{27}$: expected number $= 250 \times \dfrac{1}{27} = 9.26$ currants.

5

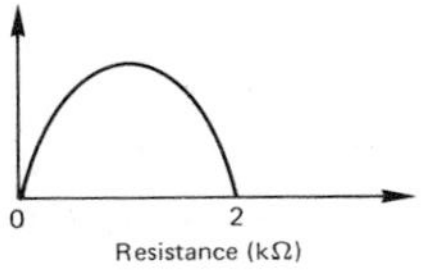

$$\text{Area} = \int_0^2 \tfrac{3}{4}x(2 - x)\,dx = \tfrac{3}{4}[x^2 - \tfrac{1}{3}x^3]_0^2 = \tfrac{3}{4}(4 - \tfrac{8}{3}) = 1$$

(i) $P(X > 1.5) = \displaystyle\int_{1.5}^2 \tfrac{3}{4}x(2 - x)\,dx = \tfrac{3}{4}[x^2 - \tfrac{1}{3}x^3]_{1.5}^2 = 0.15625$

(ii) $P(0.9 < X < 1.1) = \tfrac{3}{4}[x^2 - \tfrac{1}{3}x^3]_{0.9}^{1.1} = 0.1495$.

356

6 (i)

Area $= \int_0^4 kx^2(4-x)\,dx = k[\tfrac{1}{3}x^3 - \tfrac{1}{4}x^4]_0^4 = \tfrac{64}{3}k$

Hence $k = \tfrac{3}{64}$

(ii) for the mode, $f'(x) = 0$,

$\tfrac{3}{64}(8x - 3x^2) = 0$; hence mode $= \tfrac{8}{3}$ months

(iii) $P(X < 1) = \int_0^1 \tfrac{3}{64}x^2(4-x)\,dx = \tfrac{3}{64}[\tfrac{4}{3}x^3 - \tfrac{1}{4}x^4]_0^1 = \tfrac{13}{256}$

7 (i)

Area $= \int_0^{2\pi} \tfrac{1}{2\pi}(1 - \cos x)\,dx$

$= \tfrac{1}{2\pi}[x - \sin x]_0^{2\pi} = 1$

(ii) the mode is π (i.e. due South)

(iii) $P\left(\tfrac{5\pi}{6} < X < \tfrac{7\pi}{6}\right)$

$= \int_{5\pi/6}^{7\pi/6} \tfrac{1}{2\pi}(1 - \cos x)\,dx$

$= \tfrac{1}{2\pi}[x - \sin x]_{5\pi/6}^{7\pi/6} = \tfrac{1}{6} + \tfrac{1}{2\pi}$

$= 0.326.$

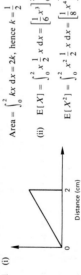

Exercise 4.2 (page 76)

1 (i)

Area $= \int_0^2 kx\,dx = 2k$, hence $k = \tfrac{1}{2}$

(ii) $E[X] = \int_0^2 x \cdot \tfrac{1}{2}x\,dx = [\tfrac{1}{6}x^3]_0^2 = \tfrac{4}{3}$

$E[X^2] = \int_0^2 x^2 \cdot \tfrac{1}{2}x\,dx = [\tfrac{1}{8}x^4]_0^2 = 2$

$\text{var}(X) = 2 - \left(\tfrac{4}{3}\right)^2 = \tfrac{2}{9}$

(iii) $P\left(\tfrac{4}{3} - \sqrt{\tfrac{2}{9}} < X < \tfrac{4}{3} + \sqrt{\tfrac{2}{9}}\right) = P(0.862 < X < 1.805)$

$= [\tfrac{1}{4}x^2]_{0.862}^{1.805} = 0.629.$

2 (i) $1 < X < 4$

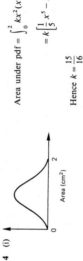

$f(x) = \tfrac{1}{3}$ for $1 \leqslant X \leqslant 4$

(ii) $E[X] = \int_1^4 x\,\tfrac{1}{3}\,dx = [\tfrac{1}{6}x^2]_1^4 = 2.5$ kN

$E[X^2] = \int_1^4 x^2\,\tfrac{1}{3}\,dx = [\tfrac{1}{9}x^3]_1^4 = 7$

$\text{var}(X) = 7 - 2.5^2 = 0.75$

$\sigma = 0.866$ kN.

3 (i)

Area $= \int_2^3 k(x^2 - 1)\,dx = k[\tfrac{1}{3}x^3 - x]_2^3 = \tfrac{16}{3}k$

Hence $k = \tfrac{3}{16}$

(ii) $E[X] = \int_2^3 x \cdot \tfrac{3}{16}(x^2-1)\,dx = \tfrac{3}{16}[\tfrac{1}{4}x^4 - \tfrac{1}{2}x^2]_2^3 = 2.578$ kg

$E[X^2] = \int_2^3 x^2 \cdot \tfrac{3}{16}(x^2-1)\,dx = \tfrac{3}{16}[\tfrac{1}{5}x^5 - \tfrac{1}{3}x^3]_2^3 = 6.725$

$\text{var}(X) = 6.725 - 2.578^2 = 0.0783$ kg^2.

4 (i)

Area under pdf $= \int_0^2 kx^2(x-2)^2\,dx$

$= k[\tfrac{1}{5}x^5 - x^4 + \tfrac{4}{3}x^3]_0^2 = \tfrac{16}{15}k$

Hence $k = \tfrac{15}{16}$

(ii) $E[X] = \int_0^2 x \cdot \tfrac{15}{16}x^2(x-2)^2\,dx$

$= \tfrac{15}{16}[\tfrac{1}{6}x^6 - \tfrac{4}{5}x^5 + x^4]_0^2$

$= 1$ cm^2

$E[X^2] = \int_0^2 x^2 \cdot \tfrac{15}{16}x^2(x-2)^2\,dx = \tfrac{8}{7}$; $\text{var}(X) = \tfrac{8}{7} - 1^2 = \tfrac{1}{7}$ cm^4.

5 (i) $a = 7.5$ minutes

(ii) (a) $f(x) = 0.16$ (b) $f(x) = 0.48 - 0.064\,x$

(iii) $E[X] = \int_0^5 x(0.16)\,dx + \int_5^{7.5} x(0.48 - 0.064x)\,dx = 2 + \tfrac{7}{6} = 3\tfrac{1}{6}$ minutes

(iv) For median M, $0.16 \times M = 0.5$; so median $= 3.125$ minutes.

Exercise 5.1 (page 81)

1 (a) 0.1841   (b) 0.0322   (c) 0.4068   (d) 0.9108.

2 (i) 0.9010   (ii) 0.2743   (iii) 0.7257   (iv) 0.0060   (v) 0.0584
  (vi) 0.5947   (vii) 0.9316   (viii) 0.3126.

3 $a = 0.8416$, $b = -1.645$, $c = -1.751$, $d = 2.326$
  $\Phi(e) = \Phi(-0.5) + 0.6 = 0.9085$; $e = 1.332$.

4 (i) $a = 1.282$ (ii) $b = -2.054$ (iii) $c = -0.3853$ (iv) $d = 1.036$
  (v) $e = 2.326$ (vi) $f = 0.9154$
  (vii) for uq, $\Phi(z) = 0.75$, so uq $= 0.6745$; lq $= -0.6745$; iqr $= 1.349$
  (viii) $\Phi(z) = 0.88$; 88th percentile is 1.175.

Exercise 5.2 (page 86)

1 (i) $\mu = 60$, $\sigma = 6$ kg   (ii) $\mu = 2625$, $\sigma = 50$ ohms.

2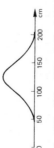

3 (i) $P(Z > 300) = P(X > 1.25) = 0.1056$
  (ii) $P(X < 175) = P(Z < -1.875) = 0.0303$
  (iii) $P(260 < X < 270) = P(0.25 < Z < 0.5) = 0.0928$
  (iv) $P(235 < X < 265) = P(-0.375 < Z < 0.375) = 0.2924$.

4 (i) $P(X > 4.8) = P(Z > 2.0) = 0.0228$; expected number is 136.8
  (ii) $P(X < 4.7) = P(Z < -1.333) = 0.0913$; expected number is 547.8.

5 (i) $P(T < 100) = P(Z < -1.111) = 0.1333$, or 13.33%
  (ii) T corresponds to z where $\Phi(z) = 0.99$; $z = 2.326$ and $T = 161.9\,^{\circ}$C.

6 (i) $z = -0.2533$, i.e. speed 82.45 km h^{-1}
  (ii) $z = \pm 1.960$; so $\mu = 58.56$, $v = 113.44$ km h^{-1}
  (iii) uq has $z = 0.6745$; uq $= 95.44$, lq $= 76.56$, siqr $= 9.44$ km h^{-1}.

7 (a) $P(X > 6) = P(Z > 1.5) = 0.0668$
  (b) (i) $(0.0668)^2 = 0.0045$
      (ii) $1 - P(\text{neither will last}) = 1 - (0.9332)^2 = 0.1291$.

8 For one man, $P(X > 170) = P(Z > -0.6) = 0.7257$
  For five men, the number taller than 170 cm is B(5, 0.7257)
  (i) P(all 5 are taller than 170 cm) $= (0.7257)^5 = 0.2013$

(ii) P(exactly 3 are taller than 170 cm) $= \binom{5}{3} \times (0.7257)^3 \times (0.2743)^2 = 0.2876$.

9 $z = -2.054$; $\dfrac{3-\mu}{0.08} = -2.054$, so $\mu = 3.164$ kg
  $P(3.1 < X < 3.2) = P(-0.804 < Z < 0.446) = 0.4614$.

10 (i) $\dfrac{15.0 - 14.5}{\sigma} = 0.5244$, so $\sigma = 0.9535$ A.
   (ii) $P(X < 13.0) = P(Z < -1.573) = 0.0578$, i.e. 5.78%.

11 (i) 18 and 22 mm^3 correspond to $z = \pm 1.645$
     Thus $\dfrac{22 - 20}{\sigma} = 1.645$, so $\sigma = 1.216$ mm^3
   (ii) $z = \pm 2.576$, i.e. between 16.87 and 23.13 mm^3.

12 (i) $\dfrac{20-\mu}{\sigma} = -0.3853$ and $\dfrac{30-\mu}{\sigma} = 1.405$ giving $\mu = 22.15$ minutes and
     $\sigma = 5.59$ minutes
   (ii) $P(X > 25) = P(Z > 0.510) = 0.3050$
   (iii) $z = 2.054$; she should allow 33.6 minutes.

13 (i) $\dfrac{18-\mu}{\sigma} = -1.762$ and $\dfrac{24-\mu}{\sigma} = 0.9463$, giving $\mu = 21.90$ m, $\sigma = 2.22$ m
   (ii) $P(16 < X < 20) = P(-2.665 < Z < -0.859) = 0.1913$: expected number 122.4.

14 $\bar{x} = 53.69$, $s = 9.61$. For a Normal distribution
   $P(60.5 < X < 65.5) = P(0.709 < Z < 1.229) = 0.1297$; expected frequency 31.1
   $P(35.5 < X < 40.5) = P(-1.893 < Z < -1.372) = 0.0558$; expected frequency 13.4.

15

Selling price	Probability		
£25	$P(119.8 < X < 120.2) = 0.5403$		
£20	$P(119.5 < X < 119.8) + P(120.2 < X < 120.5)$		
	=	0.1069 +	0.2898 = 0.3967
£2	$P(X < 119.5)$ +	$P(X > 120.5)$	
	=	0.0082 +	0.0548 = 0.0630

Mean selling price $= 25 \times 0.5403 + 20 \times 0.3967 + 2 \times 0.0630 = \pounds 21.57$.

Exercise 5.3 (page 92)

1 (i) $\mu \pm 3\sigma = 0.3, 11.7$; good approximation
  (ii) $\mu \pm 3\sigma = 1.8, 18.2$; good approximation
  (iii) $\mu \pm 3\sigma = 42.8, 56.2$; not a good approximation
  (iv) $\mu \pm 3\sigma = -2.2, 6.2$; not a good approximation
  (v) $\mu \pm 3\sigma = 9.9, 38.1$; very good approximation.

358

2 (a) $\mu = 5, \sigma = 2$
(b) (i) $300 \times (0.2)^2 \times (0.8)^{23} = 0.0708$
(ii) $P(1.5 < X < 2.5) = P(-1.75 < Z < -1.25) = 0.0655$
(c) $P(X > 6.5) = P(Z > 0.75) = 0.2266.$

3 $\mu = 100, \sigma = 7.071$; $P(89.5 < X < 110.5) = P(-1.485 < Z < 1.485) = 0.8626.$

4 (i) Number of sixes is $B\left(120, \frac{1}{6}\right) \sim N(20, 4.082^2)$
$P(X < 14.5) = P(Z < -1.347) = 0.0890$
(ii) Number of sixes is $B\left(120, \frac{1}{10}\right) \sim N(12, 3.286^2)$
$P(X < 14.5) = P(Z < 0.761) = 0.7767.$

5 Number of correct answers is $B(45, \frac{1}{4}) \sim N(11.25, 2.905^2)$
(i) $P(X > 15.5) = P(Z > 1.463) = 0.0717$
(ii) $P(X < 9.5) = P(Z < -0.602) = 0.2737$
(iii) $P(7.5 < X < 12.5) = P(-1.291 < Z < 0.430) = 0.5681.$

6 (i) Number of people under 30 is $B(100, 0.46) \sim N(46, 4.984^2)$
$P(X > 50.5) = P(Z > 0.903) = 0.1833$
(ii) Number of people under 30 is $B(225, 0.46) \sim N(103.5, 7.476^2)$
$P(X > 112.5) = P(Z > 1.204) = 0.1144.$

7 Number of secretaries requested is $B\left(80, \frac{1}{5}\right) \sim N(16, 3.578^2)$
If k secretaries are available
$$P(X < k + \tfrac{1}{2}) = 0.99 \Rightarrow \frac{(k + 0.5) - 16}{3.578} = 2.326 \Rightarrow k = 23.8$$
Hence 24 secretaries are needed. $P(X < 11.5) = P(Z < -1.258) = 0.1041.$

8 Number of defectives is $B(25000, 0.02) \sim N(500, 22.14^2)$
With probability 0.95, the number of defectives is between $\mu \pm 1.96\,\sigma$, i.e. between 457 and 543.

Exercise 6.1 (page 99)

1 (a) Negative correlation; a good relationship but non-linear
(b) No correlation; x and y appear to be independent.
(c) Almost perfect positive correlation; linear relationship
(d) Fair positive correlation; could be linear
(e) No correlation; but there is a strong non-linear relationship
(f) There seem to be two distinct groups (men and women?). Overall the pattern shows negative correlation, but this is misleading.
(g) The x-scale is badly chosen (it should be more spread out) and so it is difficult to see what the pattern is.

2 (a) Fairly good negative correlation: could be linear
(b) The relationship is non-linear.
(c) x and y appear to be independent.

3 $\sum x = 28, \sum y = 29, \sum xy = 150$; $\text{cov}\,(x, y) = \frac{150}{8} - \left(\frac{28}{8}\right)\left(\frac{29}{8}\right) = 6.0625.$

4 $\sum u = 31, \sum v = 33, \sum uv = 169$; $\text{cov}\,(u, v) = \frac{169}{5} - \left(\frac{31}{5}\right)\left(\frac{33}{5}\right) = -7.12$
(i) $\text{cov}\,(x, y) = -7.12$ (ii) $\text{cov}\,(s, t) = -71.2$
(iii) $\text{cov}\,(p, q) = -0.356$ (iv) $\text{cov}\,(w, z) = -71.2.$

Exercise 6.2 (page 106)

1 (i) Points lie close to a straight line with positive gradient
(ii) Points lie fairly close to a straight line with negative gradient.
(iii) Points lie exactly on a straight line with positive gradient
(iv) There is no *linear* relationship between the variables (although there could be a non-linear relationship).
(v) An impossible value; there must be a mistake.

2 $\sum x = 11, \sum y = 9, \sum x^2 = 31, \sum y^2 = 31, \sum xy = 11,$
$s_x = \sqrt{\frac{31}{6} - \left(\frac{11}{6}\right)^2} = 1.344, s_y = \sqrt{\frac{19}{6} - \left(\frac{9}{6}\right)^2} = 0.957$
$\text{cov}\,(x, y) = \frac{11}{6} - \left(\frac{11}{6}\right)\left(\frac{9}{6}\right) = -0.917$, giving $r = -0.7125.$

3 $\sum x = 461, \sum y = 578, \sum x^2 = 26591, \sum y^2 = 36000, \sum xy = 29474$
$s_x = 23.106, s_y = 16.098, \text{cov}\,(x, y) = 282.82; r = 0.760.$

4

$u = \dfrac{x - 360}{5}$	−5	−2	−1	0	2	4	8
$v = \dfrac{y - 18.70}{0.01}$	8	6	5	2	3	2	0

$\sum u = 6, \sum v = 26, \sum u^2 = 114, \sum v^2 = 142, \sum uv = -43$
$s_u = 3.943, s_t = 2.548, \text{cov}\,(u, v) = -9.327; r = -0.928$
$\text{cov}\,(x, y) = 5 \times 0.01 \times (-9.327) = -0.466.$

5. $s_x = 1.982, s_y = 7.881$, cov $(x, y) = -1.069$: $r = -0.068$

We deduce that there is no linear relationship between x and y; the scatter diagram reveals a strong non-linear relationship.

6. $s_x = 0.4, s_y = 3$, cov $(x, y) = 0.72$; $r = 0.6$.

7. $\sum x = n\bar{x}, \sum y = n\bar{y}, \sum x^2 = n(s_x^2 + \bar{x}^2), \sum y^2 = n(s_y^2 + \bar{y}^2)$

$\sum xy = n(\text{cov}(x, y) + \bar{x}\bar{y}); \text{cov}(x, y) = rs_xs_y = 0.6 \times 2.6 \times 0.8 = 1.248$

$\sum xy = 50(1.248 + 15 \times 4) = 3062.4$.

8. For first 20, $\sum x = 240, \sum y = 440, \sum x^2 = 3005, \sum y^2 = 10000, \sum xy = 5300$

For other 10, $\sum x = 300, \sum y = 540, \sum x^2 = 9090, \sum y^2 = 29200, \sum xy = 16176$

For all 30, $\sum x = 540, \sum y = 980, \sum x^2 = 12095, \sum y^2 = 39200, \sum xy = 21476$

$s_x = 8.898, s_y = 15.48$, cov $(x, y) = 127.87$; $r = 0.929$.

9. $\sum x = 67.2, \sum y = 144, \sum x^2 = 393.6, \sum y^2 = 1776, \sum xy = 833.472$

With the two new points

$\sum x = 88.7, \sum y = 149, \sum x^2 = 625.85, \sum y^2 = 1789, \sum xy = 887.972$

$s_x = 2.136, s_y = 3.810$, cov $(x, y) = -4.004$; $r = -0.492$.

Exercise 6.3 (page 109)

1 (i) mid-interval values $\dfrac{5.5 + 15.5}{2} = 10.5$, etc   (ii) $v = \dfrac{y - 29.5}{20}$

(iii) 13 applicants scored between 6 and 15 in the initiative test *and* between 0 and 19 in the general knowledge test.

(iv) $\sum u^2 f = 101, \sum vf = 2, \sum uvf = 38$

$s_u = \sqrt{\dfrac{101}{96} - \left(\dfrac{33}{96}\right)^2} = 0.9664, \quad s_v = \sqrt{\dfrac{60}{96} - \left(\dfrac{2}{96}\right)^2} = 0.7903$

cov $(u, v) = \dfrac{38}{96} - \left(\dfrac{33}{96}\right)\left(\dfrac{2}{96}\right) = 0.3887; r = 0.509$

(v) cov $(x, y) = 10 \times 20 \times 0.3887 = 77.73$.

2

u\v	-1	0	1	2	f	uf	u²f
-2					6	-12	24
-1					21	-21	21
0					14	0	0
1					9	9	9
2					2	4	8
3					2	6	18
f	11	9	22	12	54	-14	80
vf	-11	0	22	24			$\sum uvf = -7$
v²f	11	0	22	48	81		

$s_u = 1.189, s_v = 1.039$, cov $(u, v) = 0.0384, r = 0.031$.

Exercise 6.4 (page 115)

1 Ranks: 10 11 7 8 5 4 2 1 3 6 9 12
            8 10 1 7 5 11 9 12 2 3 6 4

$\sum d^2 = 344; \quad r_s = 1 - \dfrac{6 \times 344}{12 \times 143} = -0.20$.

2 Ranks: 3.5 2 1 3.5 5 6 7 8 9 10
            4.5 1.5 6 4.5 1.5 3 7 9 8 10

$\sum d^2 = 50.5; r_s = 0.69$.

3 Ranks: 3 8 2 6 11 10 5 1 9 4 7
            1.5 4 4 6.5 8.5 10.5 4 1.5 8.5 6.5 10.5

$\sum d^2 = 49; r_s = 0.78$.

4 $\sum d^2 = 40; r_s = -1$

Whenever x increases, y decreases

r will be negative, but it is not necessarily -1 because the relationship may not be linear.

cov $(x, y) = -107.2, s_x = 31.158, s_y = 7.616, r = -0.45$.

5

Finalist	A	B	C	D	E	F
Ranks: First judge	3	4	2	6	1	5
Second judge	6	3	4	5	1	2
Third judge	1	3	4.5	6	2	4.5

360

(i) $\sum d^2 = 24, r_s = 0.31$
(ii) $\sum d^2 = 12.5, r_s = 0.64$
(iii) $\sum d^2 = 33.5, r_s = 0.04$
(iv) first and third
(v) second and third.

6 (i) $r_s = 0.21$
(ii) $r_s = -0.58$
(iii) Between job effectiveness and appearance

qualifications	(a) $r_s = 0.05$
experience	(b) $r_s = -0.29$
personality	(c) $r_s = 0.29$
alertness	(d) $r_s = -0.42$
	(e) $r_s = 0.76$

Alertness correlates best.
(iv) $r_s = 0.24$
(v) b and d ignored since they correlate negatively with effectiveness; a, c and e correlate positively, c more than a and e most of all; hence suggested weighting $a + 2c + 3e$. $r_s = 0.89$.

7 Ranks: u 9   7.5   7.5   5   5   5   2.5   2.5   1
        v 7   9   2.5   8   5.5   4   2.5   1   5.5
(i) cov $(u, v) = 2.8889, s_u = 2.5166, s_v = 2.5604, r_s = 0.448343$

(ii) $\sum d^2 = 64, r_s = 1 - \dfrac{6 \times 64}{9 \times 80} = 0.466667$

These are not equal because the formula in (ii) assumes that no ranks are tied.

(iii) $r_s = 1 - \dfrac{6 \times 64}{9 \times 80 - 24} = 0.44276.$

8 $r_s = 1 - \dfrac{6 \times 50.5}{10 \times 99 - 9} = 0.69$

Correcting for tied ranks makes no difference (to 2 decimal places).

Exercise 7.1 (page 126)

1 $\sum x = 26, \sum x^2 = 128, \sum y = 17, \sum xy = 85, s_x^2 = 2.556,$ cov $(x, y) = 1.889$
line of regression is $y - 2.833 = 0.739(x - 4.333)$
$$y = 0.739x - 0.370.$$

2 $\sum u = 3, \sum u^2 = 39, \sum v = -5, \sum uv = -46$
$\bar u = 0.6, \bar x = 0.6 \times 0.2 + 17.0 = 17.12; \bar v = -1, \bar y = -1 \times 25 + 675 = 650$
$s_u^2 = 7.44, s_x^2 = 7.44 \times 0.2^2 = 0.2976;$ cov $(u, v) = -8.6,$ cov $(x, y) = -8.6 \times 0.2 \times 25$
$$= -43$$

$v + 1 = -1.1559(u - 0.6),$ $y - 650 = -144.49(x - 17.12)$
$v = -1.156u - 0.306$ $y = 3123.7 - 144.5x$

$\left[\text{Alternatively } \dfrac{y - 675}{25} = -1.1559 \left(\dfrac{x - 17.0}{0.2}\right) - 0.3065, \text{ etc}\right].$

3 $\sum x = 37.4, \sum x^2 = 168.78, \sum y = 105.5, \sum xy = 456.7,$
$\bar x = 3.74, \bar y = 10.55, s_x^2 = 2.8904,$ cov $(x, y) = 6.213$
The line of regression is $y - 10.55 = 2.1495(x - 3.74)$
$$y = 2.1495x + 2.5108$$

When $x = 3.2, y = 9.39$ m.

4 Scatter diagram shows that the relationship could be linear.
Working with $u = \dfrac{x}{12}, v = y - 200,$ we have

u	0	1	2	3	4
v	-18	-3	23	64	95

$\sum u = 10, \sum u^2 = 30, \sum v = 161, \sum uv = 615, \bar u = 2, \bar v = 32.2$
$s_u^2 = 2,$ cov $(u, v) = 58.6; \bar x = 24, \bar y = 232.2, s_x^2 = 288,$ cov$(x, y) = 703.2.$
The line of regression is $y - 232.2 = 2.4417(x - 24)$
$$y = 2.4417x + 173.6$$

(i) when $x = 42, y = 276.1$. This is reasonably reliable since $x = 42$ lies within the given range of values (0-48).
(ii) when $x = -12, y = 144.3$. We have extrapolated outside the range of the given values, so this value is not so reliable.
(iii) when $x = 156, y = 554.5$. This is way outside the given points, and so is thoroughly unreliable.

(Note   The actual values of the RPi were   (i) 276.5,   (ii) 157.1
                      (iii) who knows?)

5 $s_x^2 = 1.2489,$ cov $(x, y) = -14.572$. The line of regression is:
$y - 41.333 = -11.668(x - 1.967)$
$$y = 64.281 - 11.668x$$

When $x = 1.4, y = 48$. The scatter diagram shows that the relationship is non-linear. A better estimate when $x = 1.4$ is $y = 28$.

361

Exercise 7.2 (page 134)

1 $\sum x = 24.2, \sum x^2 = 97.38, \sum y = 50.8, \sum y^2 = 354.16, \sum xy = 100.38$
$s_x^2 = 3.5899, \quad s_y^2 = 7.4914, \quad \text{cov}(x,y) = -4.024$
l.o.r. of y on x is $y - 5.644 = -1.1209(x - 2.689)$, i.e. $y = 8.658 - 1.1209x$
l.o.r. of x on y is $x - 2.689 = -0.5371(y - 5.644)$, i.e. $x = 5.721 - 0.537y$
(i) when $x = 1.6, y = 6.86$ i.e. 6860 tourists
(ii) when $y = 3.2, x = 4.00$ cm
(iii) [graph: y on x, x on y]

2 $\bar{x} = 65, s_x = 15, \bar{y} = 200, s_y = 8, \text{cov}(x,y) = 117.75$
$r = 0.98125$; this shows that the points lie close to a straight line, and so linear regression should give good results
(i) l.o.r. is $y - 200 = 0.5233(x - 65)$, i.e. $y = 0.5233x + 165.98$
when $x = 78, y = 206.8$ ml; $x = 78$ is close to the measured values (which have $\bar{x} = 65, s_x = 15$) so this should be very reliable.
(ii) l.o.r. is $x - 65 = 1.8398(y - 200)$, i.e. $x = 1.8398y - 302.97$
when $y = 260, x = 175.4$°C; $y = 260$ is well outside the measured values, (which have $\bar{y} = 200, s_y = 8$), so this estimate is *not* reliable — the liquid may well have boiled!

3 $\text{cov}(x,y) = 0.6 \times 18 \times 12 = 129.6$
l.o.r. of y on x is $y = 0.4x + 38.8$; l.o.r. of x on y is $x = 0.9y - 4.2$
(i) when $y = 71, x = 59.7$
(ii) when $x = 25, y = 48.8$

4 (i) $\bar{y} = 21.6875 - 0.3125\bar{x} = 17$°C
(ii) $\frac{\text{cov}(x,y)}{s_x^2} = -0.3125 \Rightarrow \text{cov}(x,y) = -20$
(iii) $r = \frac{\text{cov}(x,y)}{s_x s_y}$
(iv) $\bar{x} = 28.6 - 0.8\bar{y}$
(v) when $y = 21, x = 11.8$ km h^{-1}

5 (i) when $y = 78, x = 180.8$ cm   (ii) when $x = 180.8$ cm
(iii) lines meet at $\bar{x} = 172$ cm, $\bar{y} = 68$ kg
(iv) $r^2 = 0.56 \times 0.88$, so $r = 0.702$
(v) $\frac{\text{cov}(x,y)}{s_x^2} = 0.56, \frac{\text{cov}(x,y)}{s_y^2} = 0.88 \Rightarrow \frac{s_x^2}{s_y^2} = \frac{0.56}{0.88}$
$\Rightarrow \frac{s_y}{s_x} = 0.798.$

---

6 Working with $u = \dfrac{x - 165}{10}, \quad v = \dfrac{y - 165}{10}$ we have,

u \ v	-1	0	1	2	f	uf	u²f
-1	3	2	1	0	6	-6	6
0	5	13	8	6	32	0	0
1	9	16	24	10	59	59	59
2	2	4	10	12	28	56	112
f	19	35	43	28	125	109	177
vf	-19	0	43	56	80		

$\sum uvf = 101$

$\bar{u} = 0.872, \bar{v} = 0.64, s_u^2 = 0.6556, \text{cov}(u,v) = 0.2499$
$\bar{x} = 173.72, \bar{y} = 171.4, s_x^2 = 65.56, \text{cov}(x,y) = 24.99$
l.o.r. is $y = 0.3812x + 105.18$; when $x = 182, y = 174.6$ cm.

7 For first 8, $\sum x = 14.3, \sum x^2 = 32.31, \sum y = 258, \sum xy = 542.1$
$\bar{x} = 1.7875, \bar{y} = 32.25, s_x^2 = 0.8436, \text{cov}(x,y) = 10.1156.$
l.o.r. is $y = 11.99x + 10.82$
For all 9, $\sum x = 19.3, \sum x^2 = 57.31, \sum y = 278, \sum xy = 642.1$
$\bar{x} = 2.144, \bar{y} = 30.889, s_x^2 = 1.769, \text{cov}(x,y) = 5.1049.$
l.o.r. is $y = 2.886x + 24.70$
Introducing the ninth point greatly affects the line of regression which then becomes quite useless!

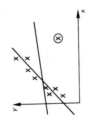

8

x	21	22	23	24	25	26
Average yield w	4.60	5.22	5.46	5.98	6.48	6.80

$\bar{x} = 23.5, s_x^2 = 2.9167, \bar{w} = 5.7567, \text{cov}(x,w) = 1.275$

l.o.r. of w on x is $w = 0.437x - 4.516$
thus l.o.r. of y on x is $y = 0.437x - 4.516$
when $x = 23.7, y = 5.84$ kg.

6 (i) $\bar{x} = 3$, $s_{\bar{x}}^2 = 2.8$

(ii) $\bar{y} = 2\bar{x} - 2 = 4$, $\dfrac{\operatorname{cov}(x,y)}{s_x^2} = 2 \Rightarrow \operatorname{cov}(x,y) = 5.6$

$\sum y = 5 \times 4 = 20$, $\sum xy = 5(5.6 + 3 \times 4) = 88$

(iii) If the missing values are a, b we have:

$a + b + 8 = 20$ and $6a + 3b + 22 = 88$

giving $a = 10$, $b = 2$.

7 $\sum x = 36$, $\sum x^2 = 204$, so $\bar{x} = 4.5$, $s_x^2 = 5.25$

$\bar{y} = 13.6 - 0.8 \times 4.5 = 10$, $\sum y = 80$; $\dfrac{\operatorname{cov}(x,y)}{5.25} = -0.8 \Rightarrow \operatorname{cov}(x,y) = -4.2$

$\sum xy = 8(-4.2 + 4.5 \times 10) = 326.4$

Published values give $\sum y = 81.2$ (too large by 1.2)

and $\sum xy = 333.6$ (too large by 7.2, which is 6×1.2)

so when $x = 6$, y should be $9.8 - 1.2 = 8.6$ (it was printed upside down!)

Exercise 7.3 (page 139)

1 (i) $y = ax^3 + b$; y and x^3; $m = a, c = b$.

(ii) $y^2 = ax + b$; y^2 and x; $m = a, c = b$.

(iii) $\log y = -b \log x + \log a$; $\log y$ and $\log x$; $m = -b, c = \log a$

(iv) $\log y = b \log (x - 2) + \log a$; $\log y$ and $\log (x - 2)$; $m = b, c = \log a$

(v) $\log y = -x^2 \log b + \log a$; $\log y$ and x^2; $m = -\log b, c = \log a$

(vi) $\log (y - 4) = b \log x + \log a$; $\log (y - 4)$ and $\log x$; $m = b, c = \log a$

(vii) $\dfrac{y}{x} = a + bx^2$; $\dfrac{y}{x}$ and x^2; $m = b, c = a$

(viii) $x^3 y = bx + a$; $x^3 y$ and x; $m = b, c = a$.

2 For x and y, $r = \dfrac{4.152}{2.182 \times 2.796} = 0.68$

For $z = x^2$ and y, $r = \dfrac{16.123}{5.767 \times 2.796} = 0.99997$

This shows that x^2 and y (almost) satisfy a linear relationship $y = ax^2 + b$

L.o.r. of y on z is $y - 8.5 = \dfrac{16.123}{33.262} (z - 5.076)$

i.e. $y = 0.485x^2 + 6.040$; when $x = 3.2$, $y = 11.00$.

3 For $z = \log x$, $w = \log y$, $\sum z = -1.025, \sum z^2 = 0.390$

$\sum w = 8.031, \sum w^2 = 17.750, \sum zw = -2.513$

$r = \dfrac{-0.114}{0.179 \times 0.637} = -0.99997$; if $y = ax^b$, then $\log y = b \log x + \log a$

so $\log y$ and $\log x$ satisfy a linear relationship; this is confirmed by

$r \approx -1$. L.o.r. is $w - 2.008 = \dfrac{-0.114}{0.0319} (z + 0.256)$

i.e. $\log y = -3.567 \log x + 1.094$; thus $b = -3.567$ and $\log a = 1.094$, $a = 12.42$.

4 For $z = \log x$, $w = \log y$; $\sum z = 16.982, \sum z^2 = 50.581$

$\sum w = 11.734, \sum zw = 36.001$

L.o.r. is $w - 1.956 = \dfrac{0.4647}{0.4191} (z - 2.830)$

i.e. $\log y = 1.1088 \log x - 1.1826$; thus $b = 1.1088$, $\log a = -1.1826$, $a = 0.06568$

Hence $y = 0.06568x^{1.088}$

When (i) $x = 5000$, $y = 829.6$ s; should be fairly reliable

(ii) $x = 42200$, $y = 8830$ s; doubtful reliability since this is well beyond the given values of x.

5 If $w = \log y$, then $\sum x = 132, \sum x^2 = 2024, \sum w = 13.137, \sum xw = 125.694$

L.o.r. is $w - 1.0947 = \dfrac{-1.5675}{47.667} (x - 11)$

$\log y = -0.03288x + 1.4565$

If $y = ab^{-x}$, then $\log y = -x \log b + \log a$

Hence $a = 28.606$, $b = 1.0787$

x	0	2	4	6	8	10
Predicted $y = 28.6 \times 1.08^{-x}$	28.6	24.6	21.1	18.2	15.6	13.4

x	12	14	16	18	20	22
Predicted $y = 28.6 \times 1.08^{-x}$	11.5	9.9	8.5	7.3	6.3	5.4

For the year 2000, $x = 40$; formula gives $y = 1.4$ millions, but this is not a reliable prediction (extrapolating into the unknown; also the curve does not fit the given data particularly well).

6 Let $w = \dfrac{y-1}{x}$, $z = x^2$

z	1	1	4	9	16	25
w	4	4	3.55	3.167	2.4	1.46

$\sum z = 56, \sum z^2 = 980, \sum w = 18.577, \sum zw = 125.6$

L.o.r. is $w - 3.0961 = \dfrac{-7.9637}{76.222}(z - 9.333)$

i.e. $\dfrac{y-1}{x} = -0.1045x^2 + 4.071$

thus $y = 1 + 4.071x - 0.1045x^3$.

Exercise 7.4 (page 145)

1 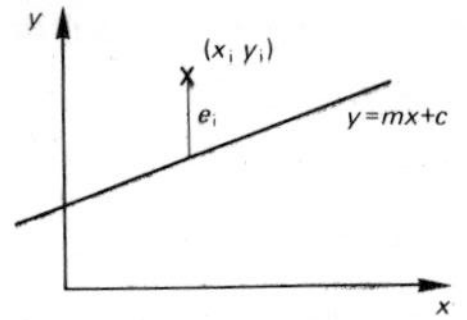

e_i is the difference between the actual value of y when $x = x_i$ (i.e. $y = y_i$) and the value of y predicted by the line $y = mx + c$ (i.e. $y = mx_i + c$), and so e_i is the 'error' in the prediction.

It is clearly desirable that the errors $e_1, e_2, e_3, \ldots$, should be as small as possible. If m and c are chosen so as to minimise $E = \sum e_i^2$ (i.e. the 'method of least squares') the corresponding line $y = mx + c$ is the line of regression of y on x.

2 $E = (3 - m - c)^2 + (7 - 4m - c)^2 = 17m^2 + 10mc + 2c^2 - 62m - 20c + 58$

E is minimised when $\dfrac{\partial E}{\partial m} = 0$, i.e. $34m + 10c - 62 = 0$ and $\dfrac{\partial E}{\partial c} = 0$, i.e. $10m + 4c - 20 = 0$ giving $m = \dfrac{4}{3}, c = \dfrac{5}{3}$

Minimum value of E = 0, showing that both points lie on the line $y = \dfrac{4}{3}x + \dfrac{5}{3}$

The l.o.r. is $y = \dfrac{4}{3}x + \dfrac{5}{3}$, which is the line joining (1, 3) to (4, 7).

3 $E = (4 - c)^2 + (2 - m - c)^2 + (8 - 4m - c)^2 = 17m^2 + 10mc + 3c^2 - 68m - 28c + 84$

E is minimised when $34m + 10c - 68 = 0$ and $10m + 6c - 28 = 0$ giving $m = \dfrac{16}{13}, c = \dfrac{34}{13}$

L.o.r. is $y = \dfrac{16}{13}x + \dfrac{34}{13}$.

4 $E = (10 + 4m - c)^2 + (-11m - c)^2 + (6 - 2m - c)^2 + (4 - 5m - c)^2$
$= 166m^2 + 28mc + 4c^2 + 16m - 40c + 152$

E is minimised when $332m + 28c + 16 = 0$ and $28m + 8c - 40 = 0$ giving $m = -\dfrac{2}{3}, c = \dfrac{22}{3}$

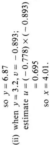

Minimum value of E = 0, because all 4 points lie on the line $y = -\dfrac{2}{3}x + \dfrac{22}{3}$ and hence $r = -1$.

5 $E = (1 + 5m - c)^2 + (4 - 2m - c)^2 + (7 - 3m - c)^2$
$= 38m^2 - 48m + 3c^2 - 24c + 66$
$= 38\left(m^2 - \dfrac{24}{19}m\right) + 3(c^2 - 8c) + 66 = 38\left(m - \dfrac{12}{19}\right)^2 + 3(c-4)^2 + \dfrac{54}{19}$

E is minimised when $m = \dfrac{12}{19}, c = 4$, so l.o.r. is $y = \dfrac{12}{19}x + 4$

RSS = minimum value of $E = \dfrac{54}{19}$.

6 $\bar{x} = 2.6889, s_x = 1.8947, \bar{y} = 5.6444, s_y = 2.7370$

u	0.96	0.16	1.33	1.22	−0.31	−1.42	−1.42	−0.79	0.27
v	−0.24	0.86	−1.77	−0.53	−0.31	0.64	1.37	1.08	−1.11

$r = \dfrac{\sum uv}{n} = \dfrac{-7.0045}{9} = -0.778$

(i) when $x = 1.6, u = -0.575$;
estimate $v = (-0.778) \times (-0.575)$
$= 0.447$
so $y = 6.87$

(ii) when $y = 3.2, v = -0.893$;
estimate $u = (-0.778) \times (-0.893)$
$= 0.695$
so $x = 4.01$.

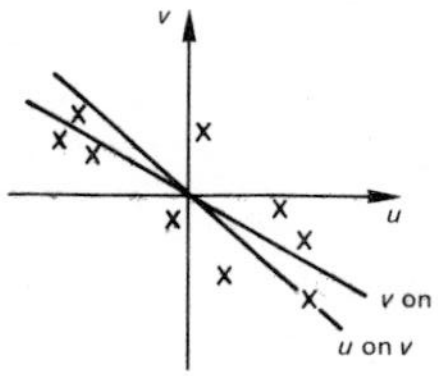

Exercise 8.1 (page 154)

1 (i) $\dfrac{7}{10}$ (ii) $\dfrac{7}{10}$ (iii) $\dfrac{3}{10}$ (iv) $\dfrac{7}{10} \times \dfrac{6}{9} = \dfrac{7}{15}$ (v) $\dfrac{7}{10} \times \dfrac{3}{9} = \dfrac{7}{30}$

(vi) P(both R) + P(both B) = $\left(\dfrac{7}{10} \times \dfrac{6}{9}\right) + \left(\dfrac{3}{10} \times \dfrac{2}{9}\right) = \dfrac{8}{15}$ (vii) $1 - \dfrac{8}{15} = \dfrac{7}{15}$

(viii) P(BBR) + P(BRB) + P(RBB) = $\dfrac{3}{10} \times \dfrac{2}{9} \times \dfrac{7}{8} \times 3 = \dfrac{7}{40}$

or Y transferred: P(6G,2Y)

$$Q(2G, 6Y) \quad \text{Prob} = \frac{3}{9} \times \left\{ \left(\frac{6}{8} \times \frac{2}{8}\right) + \left(\frac{2}{8} \times \frac{6}{8}\right) \right\} = \frac{2}{16}$$

$$\text{P(same colour)} = \frac{7}{16}.$$

13 If P is chosen, then P(same colour) $= \frac{7}{16}$ (see Question 12)

If Q is chosen, then P(same colour) $= \frac{46}{105}$ (using a similar argument)

$$\text{P(same colour)} = \frac{1}{2} \times \frac{7}{16} + \frac{1}{2} \times \frac{46}{105} = \frac{1471}{3360}.$$

Exercise 8.2 (page 159)

1 (i) $P(H < 188) = 0.9032$ (ii) $P(H > 180) = 0.3085$
(iii) $P(180 < H < 188) = 0.2117.$

2 $P(V > 240) = 0.7477$ (i) $P(V_1 > 240 \text{ and } V_2 > 240) = (0.7477)^2 = 0.5591$
(ii) $P(V_1 > 240, V_2 < 240) + P(V_1 < 240, V_2 > 240) = 0.7477 \times 0.2523 \times 2 = 0.3773.$

3 (i) $P(T < 45) = 0.7200$; P(all finish) $= (0.7200)^{10} = 0.0374$
(ii) If probability that an individual student finishes is p, then $p^{10} = 0.5$
Thus $p = 0.933$. Time corresponds to $z = 1.499$; time 56 minutes.

4 (i) $\left(\frac{1}{2}\right)^3 = \frac{1}{8}$

(ii) P(2 heads in 7 tosses *and* head on 8th) $= \binom{7}{2} \times \left(\frac{1}{2}\right)^7 \times \left(\frac{1}{2}\right)$
$$= \frac{21}{256}$$
$$= 0.003693.$$

(iii) P(less than 3 heads in 15 tosses) $= \left(\frac{1}{2}\right)^{15} + 15\left(\frac{1}{2}\right)^{15} + \binom{15}{2}\left(\frac{1}{2}\right)^{15}$
$$= 0.003693.$$

5 (i) $\frac{5}{8} + \left(\frac{3}{8} \times \frac{1}{5}\right) = 0.7$

(ii) Number of right answers is B(10, 0.7)
$$P(\geqslant 9 \text{ right}) = 10 \times (0.7)^9 \times 0.3 + (0.7)^{10}$$
$$= 0.1493.$$

(ix) $1 - P(RRR) = 1 - \frac{7}{10} \times \frac{6}{9} \times \frac{5}{8} = \frac{17}{24}$ (x) $P(RRRB) = \frac{7}{10} \times \frac{6}{9} \times \frac{5}{8} \times \frac{3}{7} = \frac{1}{8}$

(xi) P(last 3 discs BRR) $= \frac{3}{10} \times \frac{7}{9} \times \frac{6}{8} = \frac{7}{40}.$

2 (i) $0.4 \times 0.4 = 0.16$ (ii) $0.6 \times 0.1 = 0.06$ (iii) $0.16 + 0.06 = 0.22$
(iv) $\left.\begin{array}{l} P(RRN) = 0.4 \times 0.4 \times 0.6 = 0.096 \\ P(RNR) = 0.4 \times 0.6 \times 0.1 = 0.024 \\ P(NRR) = 0.6 \times 0.1 \times 0.4 = 0.024 \end{array}\right\}$ Probability $= 0.144.$

3 (i) $0.3 \times 0.5 \times 0.7 \times 0.9 = 0.0945$ (ii) $0.3 \times 0.5 \times 0.7 \times 0.1 = 0.0105$
(iii) $0.3 \times 0.5 \times 0.3 \times 0.8 = 0.036$ (iv) $0.3 \times 0.5 \times 0.6 \times 0.8 = 0.072$
(v) $0.7 \times 0.4 \times 0.6 \times 0.8 = 0.1344$ (vi) 0.2529 (vii) $0.3474.$

4 (i) 0.8 (ii) 0.7 (iii) $1 - 0.2 - 0.3 = 0.5$
(iv) $0.2 \times 0.2 = 0.04$ (v) $0.2 \times 0.3 \times 2 = 0.12.$

5 (i) $\left(\frac{5}{6}\right)^4 = \frac{625}{1296}$ (ii) $1 - \left(\frac{5}{6}\right)^4 = \frac{671}{1296}$
(iii) $1 \times \frac{5}{6} \times \frac{4}{6} \times \frac{3}{6} = \frac{5}{18}$ (iv) $1 \times \frac{1}{2} \times \frac{1}{2} \times \frac{1}{2} = \frac{1}{8}.$

6 (i) $\frac{4}{52} \times \frac{3}{51} = \frac{1}{221}$ (ii) $1 - \frac{39}{52} \times \frac{38}{51} = \frac{15}{34}$
(iii) $1 \times \frac{39}{51} = \frac{13}{17}$ (iv) $1 \times \frac{51}{52} \times \frac{50}{51} = \frac{1}{26}.$

7 (i) 0.8 (ii) $0.4 \times 0.2 = 0.08$ (iii) $0.4 \times 0.8 = 0.32$
P(one B & LH, other B & RH) $= 0.08 \times 0.32 \times 2 = 0.0512.$

8 (i) $\frac{8}{20} \times \frac{7}{19} = \frac{14}{95}$ (ii) $1 - \frac{14}{20} \times \frac{13}{19} = \frac{99}{190}$
(iii) $\left(\frac{12}{20} \times \frac{11}{19}\right) + \left(\frac{6}{20} \times \frac{5}{19}\right) + \left(\frac{2}{20} \times \frac{1}{19}\right) = \frac{41}{95}.$

9 (i) $\left(\frac{5}{10}\right)^4 = \frac{1}{16}$ (ii) $1 \times \frac{9}{10} \times \frac{8}{10} \times \frac{7}{10} = 0.504.$

10 P(loan granted) $= (0.2 \times 0.7) + (0.5 \times 0.9) + (0.3 \times 0.4) = 0.71.$

11 $P(GG) + P(YY) = \left(\frac{6}{9} \times \frac{2}{7}\right) + \left(\frac{3}{9} \times \frac{5}{7}\right) = \frac{3}{7}$

12 G transferred: P(5G, 3Y)
$$Q(3G, 5Y) \quad \text{Prob} = \frac{6}{9} \times \left\{ \left(\frac{5}{8} \times \frac{3}{8}\right) + \left(\frac{3}{8} \times \frac{5}{8}\right) \right\} = \frac{5}{16}$$

6 (i) $1 - P(\text{no link breaks}) = 1 - (0.99)^n$
(ii) $1 - (0.99)^n < 0.4 \Rightarrow n < 50.8$, maximum of 50 links.

7

X	0	1	2	3
Probability	$\frac{1}{8}$	$\frac{3}{8}$	$\frac{3}{8}$	$\frac{1}{8}$

$P(X_1 = X_2) = \left(\frac{1}{8} \times \frac{1}{8}\right) + \left(\frac{3}{8} \times \frac{3}{8}\right) + \left(\frac{3}{8} \times \frac{3}{8}\right) + \left(\frac{1}{8} \times \frac{1}{8}\right) = \frac{5}{16}$.

8 $P(Y > X) = P(X=2, Y>2) + P(X=3, Y>3) + P(X=4, Y>4) + P(X=5, Y>5) + P(X=6, Y>6)$
$= (0.1 \times 1) + (0.2 \times 0.9) + (0.3 \times 0.8) + (0.3 \times 0.6) + (0.1 \times 0.3)$
$= 0.73$.

9 (i) $k = \frac{6}{5}$, $P(2 < X < 3) = 0.1$
(ii) Number of shots with $2 < X < 3$ is B(4, 0.1)
$P(\text{exactly } 2) = 6 \times (0.1)^2 \times (0.9)^2 = 0.0486$.

10 Number of forward steps is B$\left(5, \frac{1}{3}\right)$

Forward steps	0	1	2	3	4	5
Finishing position	5 m back	3 m back	1 m back	1 m for.	3 m for.	5 m for.
Probability	$\frac{32}{243}$	$\frac{80}{243}$	$\frac{80}{243}$	$\frac{40}{243}$	$\frac{10}{243}$	$\frac{1}{243}$

(i) $\frac{40}{81}$ (ii) $\frac{17}{81}$.

11 $P(\text{score of exactly } 9) = \frac{1}{9}$
$P(A \text{ wins}) = \frac{1}{9} + \left(\frac{8}{9}\right)^3 \times \frac{1}{9} + \left(\frac{8}{9}\right)^6 \times \frac{1}{9} + \cdots = \frac{81}{217}$
$P(B \text{ wins}) = \frac{8}{9} \times \frac{1}{9} + \left(\frac{8}{9}\right)^4 \times \frac{1}{9} + \cdots = \frac{72}{217}$
$P(C \text{ wins}) = \left(\frac{8}{9}\right)^2 \times \frac{1}{9} + \left(\frac{8}{9}\right)^5 \times \frac{1}{9} + \cdots = \frac{64}{217}$.

12 $P(X_1 \leqslant 100 \text{ and } X_2 > 100) + P(100 < X_1 < 130 \text{ and } X_2 \text{ anything}) + P(X_1 \geqslant 130 \text{ and } X_2 < 130)$
$= (0.1056 \times 0.8944) + (0.6284 \times 1) + (0.2660 \times 0.7340)$
$= 0.9181$.

Exercise 8.3 (page 168)

1

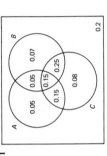

(i) 0.2
(ii) 0.32
(iii) 0.6
(iv) 0.73.

2

Total $= 116 - 2x = 100$
so $\quad x = 8$

(i) $\frac{39}{100}$

(ii) $\frac{52}{100}$

For 2 students
$P(\text{at least 1 taking all subjects})$
$= 1 - \frac{92}{100} \times \frac{91}{99} = \frac{382}{2475}$.

3

(i)

(ii) $P(A \mid B) = \dfrac{P(A \cap B)}{P(B)}$
$= \dfrac{0.1}{0.3} = \dfrac{1}{3}$

(iii) $P(B \mid A) = \dfrac{P(B \cap A)}{P(A)}$
$= \dfrac{0.1}{0.5} = 0.2$

(iv) $P(A \cap B') = 0.4$ (v) $P(A \mid B') = \dfrac{P(A \cap B')}{P(B')}$
$= \dfrac{0.4}{0.7} = \dfrac{4}{7}$

(vi) A, B not independent since e.g. $P(A \mid B) = \frac{1}{3}$ is not the same as $P(A) = 0.5$.

4 (i) $P(T \mid S) = 0.9$, but $P(T) = 0.8$, hence S, T are *not* independent.
(ii) $P(S \cap T) = P(S) \times P(T \mid S) = 0.6 \times 0.9 = 0.54$

(iii) D——G gives 5! arrangements; also G——D gives 5! arrangements hence the number of arrangements is 5! × 2 = 240
(iv) Consider as 'AB', C, D, E, F, G giving 6! arrangements; also consider as 'BA', C, D, E, F, G; hence the number of arrangements is 6! × 2 = 1440
(v) 'BEG', A, C, D, F gives 5! arrangements, similarly consider 'GEB', A, C, D, F; hence the number of arrangements is 5! × 2 = 240

The probabilities are (i) $\frac{720}{5040} = \frac{1}{7}$ (ii) $\frac{2}{7}$ (iii) $\frac{1}{21}$

(iv) $\frac{2}{7}$ (v) $\frac{1}{21}$.

2 Regard P as fixed throughout. The number of arrangements is 5! = 120
(i) 2 arrangements for U, then 4! for others; no of arrangements is 2 × 4! = 48
(ii) 4 choices for Q; 3 choices for opposite place; 3! for others
no of arrangements is 4 × 3 × 3! = 72
(iii) 2! arrangements of the men, 3! arrangements of the women; no of arrangements is 2! × 3! = 12.

3 8! = 40320 (i) Consider as 'PET', C, O, M, U, R; so 6! = 720 arrangements
(ii) 'OUE', C, M, P, T, R gives 6! arrangements; then OUE can be arranged in 3! ways; hence 6! × 3! = 4320 arrangements.

4 7 × 6 × 5 = 210 ways (i) R—— gives 6 × 5 = 30 ways: similarly ——R, —R— hence 30 × 3 = 90 ways
(ii) 5 choices for 3rd letter; then 3! arrangements; hence 5 × 3! = 30 ways.

5 (i) $\frac{8!}{2!} = 20160$ (ii) $\frac{8!}{2!2!} = 5040$ (iii) $\frac{13!}{2!3!2!2!} = 129\,729\,600$.

6 $\frac{10!}{3!3!2!} = 50400$ (i) $\frac{9!}{3!2!2!} = 15120$ (ii) $\frac{8!}{3!2!} = 3360$

(iii) No of arrangements with 'AC' or 'CA' is $\frac{9!}{3!3!2!} \times 2 = 10080$

P(A next to C) = $\frac{10080}{50400} = \frac{1}{5}$.

7 (i) 7 × 6 × 5 = 210 (ii) EE—, E–E, or –EE; so 6 × 3 = 18 (iii) 18
(vi) 1 (v) 247;
(a) $\frac{18+1}{247} = \frac{19}{247} = \frac{1}{13}$
(b) No of arrangements with no Ts; all letters different: 6 × 5 × 4 = 120
EE—, E–E, –EE: 5 × 3 = 15; EEE 1 Total 136 ways
P(no T) = $\frac{136}{247}$ (c) P(at least one T) = $1 - \frac{136}{247} = \frac{111}{247}$.

(iii)

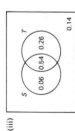

(iv) P(S∪T) = 0.86 (v) P(S|T) = $\frac{P(S\cap T)}{P(T)} = \frac{0.54}{0.80} = 0.675$.

5 (i) $\frac{11}{36}$ (ii) $\frac{5}{36}$ (iii) $\frac{2}{36}$ (iv) $\frac{11}{36} + \frac{5}{36} - \frac{2}{36} = \frac{14}{36}$

(v) $\frac{\frac{2}{36}}{\frac{5}{36}} = \frac{2}{5}$.

6 (i) $\left(\frac{3}{8} \times \frac{5}{7}\right) + \left(\frac{5}{8} \times \frac{3}{7}\right) = \frac{30}{56}$ (ii) $\frac{5}{8} \times \frac{3}{7} = \frac{15}{56}$

(iii) $\frac{P(1\text{st G \& two different})}{P(\text{two different})} = \frac{\frac{15}{56}}{\frac{30}{56}} = \frac{1}{2}$.

7 Number of bad eggs is B(12, 0.1)

$P(X = 2 \mid X \geqslant 1) = \frac{P(X=2)}{P(X\geqslant 1)} = \frac{0.2301}{0.7176} = 0.3207$.

8 $P(W < 1020 \mid W \geqslant 990) = \frac{P(990 \leqslant W < 1020)}{P(W \geqslant 990)} = \frac{0.5197}{0.8643} = 0.6013$.

9 (i) P(pass) = (0.7 × 0.6) + (0.3 × 0.1) = 0.45

(ii) P(dangerous | pass) = $\frac{P(\text{dangerous and pass})}{P(\text{pass})} = \frac{0.03}{0.45} = \frac{1}{15}$

(iii) P(dangerous | fail) = $\frac{P(\text{dangerous and fail})}{P(\text{fail})} = \frac{0.27}{0.55} = \frac{27}{55}$.

10 P(lives in town P | votes Labour) = $\frac{P(\text{lives in town P and votes Labour})}{P(\text{votes Labour})}$
$= \frac{0.13}{0.345} = 0.3768$.

Exercise 8.4 (page 174)

1 The total number of arrangements is 7! = 5040
(i) A fixed in the middle; remaining 6 can be arranged in 6! = 720 ways
(ii) Number of ways with B at one particular end is 6! = 720; there are 2 ends; hence no of ways is 6! × 2 = 1440 ways

(ix) $252 - \binom{8}{3} = 196$

(x) $252 - \binom{8}{4} = 182$.

3 The number with 6–1–4 is $\binom{8}{6} \times \binom{2}{1} \times \binom{7}{4} = 1960$

the number with 5–1–5 is $\binom{8}{5} \times \binom{2}{1} \times \binom{7}{5} = 2352$

$\left.\begin{array}{c}\\\\\end{array}\right\}$ Total number 4312

Total number of selections is $\binom{17}{11}$. P(team) $= \dfrac{4312}{\binom{17}{11}} = \dfrac{4312}{12376} = 0.3484$.

4 $\binom{52}{5} = 2598960$ (i) $\dfrac{\binom{4}{2} \times \binom{48}{3}}{\binom{52}{5}} = 0.0399$ (ii) $\dfrac{\binom{4}{2} \times \binom{4}{3}}{\binom{52}{5}} = \dfrac{1}{108290} = 9.23 \times 10^{-6}$

(iii) $\dfrac{\binom{4}{2} \times \binom{48}{11}}{\binom{52}{13}} = 0.2135$

5 (i) $\dfrac{\binom{48}{13}}{\binom{52}{13}} = 0.3038$ (ii) $\dfrac{\binom{4}{1} \times \binom{48}{12}}{\binom{52}{13}} = 0.4388$

(iv) $\dfrac{\binom{4}{3} \times \binom{48}{10}}{\binom{52}{13}} = 0.0412$ (v) $\dfrac{\binom{48}{9}}{\binom{52}{13}} = 0.0026$.

6 (i) $\dfrac{\binom{13}{5} \times \binom{39}{8}}{\binom{52}{13}} = 0.1247$

(ii) $1 - \dfrac{\binom{39}{13}}{\binom{52}{13}} - \dfrac{\binom{13}{1} \times \binom{39}{12}}{\binom{52}{13}} - \dfrac{\binom{13}{2} \times \binom{39}{11}}{\binom{52}{13}} = 1 - 0.0128 - 0.0801 - 0.2059 = 0.7013$

(iii) $\dfrac{\binom{13}{5} \times \binom{13}{3} \times \binom{13}{3} \times \binom{13}{2}}{\binom{52}{13}} = 0.01293$.

7 $\binom{10}{5} = 252$ ways.

8 (i) 3! = 6 arrangements; P(each arrangement) $= \frac{1}{6} \times \frac{1}{6} \times \frac{1}{6} = \frac{1}{216}$

P(a 4, a 5 and a 6) $= \frac{1}{216} \times 6 = \frac{1}{36}$

(ii) 663 can be arranged in $\frac{3!}{2!}$ = 3 ways

P(two 6s and a 3) $= \left(\frac{1}{6}\right)^3 \times 3 = \frac{3}{216} = \frac{1}{72}$.

9 (i) 6655 can be arranged in $\frac{4!}{2!2!}$ = 6 ways

P(two 6s and two 5s) $= \left(\frac{1}{6}\right)^4 \times 6 = \frac{1}{216}$

(ii) P(two 6s, a 5 and a 4) $= \left(\frac{1}{6}\right)^4 \times \frac{4!}{2!} \times \frac{12}{1296} = \frac{1}{108}$.

10 (i) 77752 can be arranged in $\frac{5!}{3!}$ = 20 ways

so P(three 7s, a 5 and a 2) $= \left(\frac{1}{10}\right)^5 \times 20 = 0.0002$.

(ii) P(two 9s, two 6s and a 0) $= \left(\frac{1}{10}\right)^5 \times \frac{5!}{2!2!} = 0.0003$.

Exercise 8.5 (page 179)

1 (i) $\binom{8}{3} = 56$ (ii) $\binom{10}{2} = 45$ (iii) $\binom{15}{11} = 1365$ (iv) $\binom{10}{4} = 210$

(v) $\binom{52}{5} = 2598960$ (vi) $\binom{20}{7} = 77520$ (vii) $\binom{9}{5} = 126$

(viii) $\binom{80}{6} = 300500200$.

2 $\binom{10}{5} = 252$ (i) $\binom{9}{4} = 126$ (ii) $\binom{7}{2} = 21$ (iii) $\binom{7}{4} \times \binom{3}{1} = 105$

(iv) $\binom{7}{3} \times \binom{3}{2} = 105$ (v) $\binom{8}{4} = 70$ (vi) $\binom{6}{2} \times \binom{3}{2} = 45$

(vii) $252 - \binom{7}{5} = 231$ (viii) $252 - \binom{8}{5} = 196$

8 $\binom{12}{5} \times \binom{7}{4} \times \binom{3}{3} = 27\,720$ ways

Number of ways
with J and K both in BM group is $\binom{10}{3} \times \binom{7}{4} \times \binom{3}{3} = 4200$
both in SM group is $\binom{10}{5} \times \binom{5}{2} \times \binom{3}{3} = 2520$ } 7980
both in HP group is $\binom{10}{5} \times \binom{5}{4} \times \binom{1}{1} = 1260$

$P(\text{J and K in same group}) = \dfrac{7980}{27720} = \dfrac{19}{66}$.

9 The flavour of the three sweets can be chosen in $\binom{8}{1}$ ways

The two flavours of the two lots of 2 sweets can be chosen in $\binom{7}{2}$ ways

The three flavours which have 1 sweet each can be chosen in $\binom{5}{3}$ ways

The number of different packets is $\binom{8}{1} \times \binom{7}{2} \times \binom{5}{3} = 1680$.

10 2 + 4: $\binom{8}{2} \times \binom{6}{4} = 420$
3 + 3: $\binom{8}{3} \times \binom{6}{3} = 1120$ } Total 2590 ways
4 + 2: $\binom{8}{4} \times \binom{6}{2} = 1050$

Number of ways of choosing 6 questions at random from 14 is $\binom{14}{6} = 3003$

$P(\text{satisfies rubric}) = \dfrac{2590}{3003} = 0.8625$.

Exercise 8.6 (page 182)

1 (i) P(three 6s and two 5s) $= \left(\frac{1}{6}\right)^5 \times \dfrac{5!}{3!2!}$

P(three of one, two of another) $= \left(\frac{1}{6}\right)^5 \times \dfrac{5!}{3!2!} \times 6 \times 5 = \dfrac{25}{648}$

(ii) P(two 6s, two 5s, and one 4) $= \left(\frac{1}{6}\right)^5 \times \dfrac{5!}{2!2!}$

P(two of one, two of another, one of a third) $= \left(\frac{1}{6}\right)^5 \times \dfrac{5!}{2!2!} \times \binom{6}{2} \times 4 = \dfrac{25}{108}$.

2 P(5 spades, 3 hearts, 3 diamonds, 2 clubs) $= \dfrac{\binom{13}{5} \times \binom{13}{3} \times \binom{13}{3} \times \binom{13}{2}}{\binom{52}{13}} = 0.01293$

P(5 of one suit, 3 of another, 3 of a third, 2 of a fourth)
$= 0.01293 \times 4 \times \binom{3}{2} \times 1 = 0.1552$.

3 P(two 1s, two 2s, two 3s) $= \left(\frac{1}{10}\right)^6 \times \dfrac{6!}{2!2!2!}$

P(two of one, two of another, two of a third) $= \left(\frac{1}{10}\right)^6 \times \dfrac{6!}{2!2!2!} \times \binom{10}{3} = 0.0108$.

4 P(two Gs, one R, one B) $= \frac{5}{10} \times \frac{4}{9} \times \frac{3}{8} \times \frac{2}{7} \times \frac{4!}{2!}$

P(one G, two Rs, one B) $= \frac{5}{10} \times \frac{3}{9} \times \frac{2}{8} \times \frac{4}{7} \times \frac{4!}{2!}$ } P(at least one of each) $= \frac{1}{2}$.

P(one G, one R, two Bs) $= \frac{5}{10} \times \frac{3}{9} \times \frac{2}{8} \times \frac{4}{7} \times \frac{4!}{2!}$

5 (i) $\dfrac{10!}{5!3!2!} = 2520$

(ii) $2520 \times (0.5)^5 \times (0.4)^3 \times (0.1)^2 = 0.0504$

(iii) $\dfrac{10!}{6!2!2!} \times (0.5)^6 \times (0.4)^2 \times (0.1)^2 = 0.0315$.

Exercise 8.7 (page 185)

1

Score	1	2	3	4	5	6
Prob.	2p	p	p	p	p	2p

Total $8p = 1$, so $p = \dfrac{1}{8}$.

P(total score $\geqslant 10$) $=$ P(4 + 6) + P(5 + 5 or 6) + P(6 + 4 or 5 or 6)
$= \frac{1}{8} \times \frac{2}{8} + \frac{1}{8} \times \frac{3}{8} + \frac{2}{8} \times \frac{4}{8} = \frac{13}{64}$

2 (i) 0.4 (ii) 0.22 (iii) $0.22 \times 0.4 + 0.78 \times 0.1 = 0.166$
(iv) $0.166 \times 0.4 + 0.834 \times 0.1 = 0.1498$.

3 (i) $\left(\frac{9}{10}\right)^5$ (ii) $\left(\frac{8}{10}\right)^5$ (iii) $\left(\frac{9}{10}\right)^5 - \left(\frac{8}{10}\right)^5 = 0.26281$.

4 $\binom{36}{13} \Big/ \binom{52}{13} = 0.003639$.

5 (i) $(0.7)^4 \times 0.3 = 0.07203$
(ii) P(discovered once in first 6 and discovered in 7th) $= 6 \times (0.7)^5 \times 0.3 \times 0.3$
$= 0.09076$
(iii) P(discovered at most once in 10 competitions) $= (0.7)^{10} + 10 \times (0.7)^9 \times 0.3$
$= 0.1493$.

6 $0.6 \times P(T < 25) + 0.3 \times P(T < 24) + 0.1 \times P(T < 23)$
$= 0.6 \times 0.5 + 0.3 \times 0.3085 + 0.1 \times 0.1587 = 0.4084$.

7

A(6R 4N)

B(4R 4N) — $\frac{4}{8}$ R — $\frac{4}{8}$ N — A(6R 4N) $\frac{6}{10}\times\frac{5}{9}$ RR / A(5R 5N) $\frac{5}{10}\times\frac{4}{9}$ RR

B(3R 5N) — $\frac{3}{8}$ R — $\frac{5}{8}$ N — A(7R 3N) $\frac{7}{10}\times\frac{6}{9}$ RR / A(6R 4N) $\frac{6}{10}\times\frac{5}{9}$ RR

P(both red) = $\frac{8}{25}$.

8 $\dfrac{\binom{10}{2} \times \binom{20}{3}}{\binom{30}{5}} = 0.35998$.

9 $\binom{5}{2} \times \left(\frac{1}{3}\right)^2 \times \left(\frac{2}{3}\right)^3 = \frac{80}{243}$.

10 (i) $0.7 \times 0.8 + 0.1 \times 1 + 0.2 \times 0.4 = 0.74$
(ii) P(H | satis) $= \dfrac{\text{P(H and satis)}}{\text{P(satis)}} = \dfrac{0.08}{0.74} = \dfrac{4}{37}$.
(iii) $(0.74)^2 = 0.5476$
(iv) $0.7 \times (0.8)^2 + 0.1 \times 1^2 + 0.2 \times (0.4)^2 = 0.58$.

11 (i) P(A wins 7 out of first 11 and A wins 12th) + P(B wins 7 out of first 11 and B wins 12th)
$= \binom{11}{7} \times \left(\frac{2}{3}\right)^7 \times \left(\frac{1}{3}\right)^4 \times \left(\frac{2}{3}\right) + \binom{11}{7} \times \left(\frac{1}{3}\right)^7 \times \left(\frac{2}{3}\right)^4 \times \left(\frac{1}{3}\right)$
$= 0.15896 + 0.00994 = 0.16890$.
(ii) (a) $\dfrac{0.15896}{0.16890} = \dfrac{16}{17}$ (b) $\dfrac{0.00994}{0.16890} = \dfrac{1}{17}$.

12 (i) $\frac{3}{8} \times \frac{2}{7} = \frac{3}{28}$ (ii) $\frac{1}{2}$
(iii) Consider just {R, L, L, L}; P(R first of these) $= \frac{1}{4}$
(iv) Total number of arrangements is $\dfrac{8!}{2!3!} = 3360$
No of arrangements with 'AA' together is $\dfrac{7!}{3!} = 840$
Probability $= \dfrac{840}{3360} = \dfrac{1}{4}$
(v) $\dfrac{1}{3360}$.

13 P(both at 1st table) + P(both at 2nd table) $= \dfrac{8}{14} \times \dfrac{7}{13} + \dfrac{6}{14} \times \dfrac{5}{13} = \dfrac{43}{91}$.

14

C   0.1 0.4 0.2   M   0.3

(i) $P(C \cap M) = 0.4$
(ii) $P(C \cap M') = 0.1$
(iii) $P(C \,|\, M') = \dfrac{P(C \cap M')}{P(M')} = \dfrac{0.1}{0.4} = 0.25$.

15 Events overlap (e.g. For 666, all 3 events occur)
so P(at least one six) $\neq$ P(1st is six) + P(2nd is six) + P(3rd is six)
P(at least one six) = 1 − P(no sixes) $= 1 - \left(\frac{5}{6}\right)^3 = \dfrac{91}{216}$.

16 Correct. (The 3 events do not overlap.)

17 The given outcomes are not equally likely.
There are 4 equally likely outcomes HH, HT, TH, TT, so P(1 of each) $= \frac{2}{4} = \frac{1}{2}$.

18 Events overlap (e.g. For 6460, both $6***$ and $**6*$ occur.)
It is possible to adapt the given method by considering one 6, two 6s, three 6s and four 6s separately, but easier calculated as

$$P(\text{highest is a } 6) = P(\text{all} \leqslant 6) - P(\text{all} \leqslant 5) = \left(\frac{7}{10}\right)^4 - \left(\frac{6}{10}\right)^4 = 0.1105.$$

19 Events are not independent: $P(X < 60 \text{ and } X > 20) = P(X < 60) \times P(X > 20 \mid X < 60)$
and $P(X > 20 \mid X < 60)$ is not the same as $P(X > 20)$
$P(X < 60 \text{ and } X > 20) = P(20 < X < 60) = P(X < 60) - P(X \leqslant 20)$
$= 0.84 - 0.28 = 0.56.$

20 (i) Events are not independent.

$$P(\text{total is 8 and differ by 2}) = P(3 + 5 \text{ or } 5 + 3) = \frac{2}{36} = \frac{1}{18}$$

(ii) Events overlap.

$$P(\text{total is 8 or differ by 2}) = \frac{5}{36} + \frac{8}{36} - \frac{2}{36} = \frac{11}{36}.$$

21 Correct.

22 Argument would be correct if we were given that the *first* child is a boy, but the actual information given is weaker than this.

$$P(\text{all boys} \mid \text{at least one boy}) = \frac{P(\text{all boys})}{P(\text{at least one boy})} = \frac{\frac18}{\frac78} = \frac17.$$

23 Correct.

Confirmed by $P(\text{no fives} \mid \text{no sixes}) = \dfrac{P(\text{no fives and no sixes})}{P(\text{no sixes})} = \dfrac{\left(\frac46\right)^3}{\left(\frac56\right)^3} = \left(\dfrac45\right)^3.$

24 The given probability includes cases where the 4 helpers are obtained after asking *less* than 10 people.
We require $\Pr(3 \text{ agree out of first 9 and 10th person agrees}) = \binom{9}{3} \times \left(\frac14\right)^3 \times \left(\frac34\right)^6 \times \frac14$

$\approx 0.0584.$

25 The number of choices for the suits is not 4×3, but $\binom{4}{2}$
(e.g. '2 spades and 2 hearts' is the same as '2 hearts and 2 spades').

$$P(\text{2 of one suit, 2 of another}) = \frac{468}{20825} \times \binom{4}{2} \approx 0.1348.$$

Exercise 9.1 (page 196)

1 (i)

X＼Y	0	1	2
0	$\frac14$	$\frac14$	0
1	0	$\frac14$	$\frac14$
	$\frac14$	$\frac12$	$\frac14$

(ii) $E[X] = \frac12,\quad \text{var}(X) = \frac14$
$E[Y] = 1,\quad \text{var}(Y) = \frac12$

(iii) e.g. $P(X = 0 \text{ and } Y = 2) = 0$
and $P(X = 0) \times P(Y = 2) = \frac12 \times \frac14 = \frac18$

These are not equal, so X and Y are *not* independent.

(iv) $E[XY] = 0 \times \frac14 + \cdots + 1 \times \frac14 + 2 \times \frac14 \times 1 = \frac34$

(v) $\text{cov}(X,Y) = \frac34 - \frac12 \times 1 = \frac14$

(vi) $\rho = \dfrac{\frac14}{\sqrt{\frac14}\,\sqrt{\frac12}} = \dfrac{1}{\sqrt2}.$

2 (i)

X＼Y	0	1	2	
0	$\frac{16}{36}$	$\frac{8}{36}$	$\frac{1}{36}$	$\frac{25}{36}$
1	$\frac{8}{36}$	$\frac{2}{36}$	0	$\frac{10}{36}$
2	$\frac{1}{36}$	0	0	$\frac{1}{36}$
	$\frac{25}{36}$	$\frac{10}{36}$	$\frac{1}{36}$	

(ii) e.g. $P(X = 2 \text{ and } Y = 1) = 0$
$P(X = 2) \times P(Y = 1) = \frac{1}{36} \times \frac{10}{36}$
These are not equal so X and Y are *not* independent.

(iii) $E[X] = \frac{12}{36},\; E[Y] = \frac{12}{36},\; E[XY] = \frac{2}{36},\; \text{cov}(X,Y) = -\frac{1}{18}$

(iv) $\text{var}(X) = \frac{5}{18},\; \text{var}(Y) = \frac{5}{18},\; \rho = -\frac15.$

3

X＼Y	1	2	3	4	5	6	
1	$\frac{1}{36}$	0	0	0	0	0	$\frac{1}{36}$
2	$\frac{2}{36}$	$\frac{1}{36}$	0	0	0	0	$\frac{3}{36}$
3	$\frac{2}{36}$	$\frac{2}{36}$	$\frac{1}{36}$	0	0	0	$\frac{5}{36}$
4	$\frac{2}{36}$	$\frac{2}{36}$	$\frac{2}{36}$	$\frac{1}{36}$	0	0	$\frac{7}{36}$
5	$\frac{2}{36}$	$\frac{2}{36}$	$\frac{2}{36}$	$\frac{2}{36}$	$\frac{1}{36}$	0	$\frac{9}{36}$
6	$\frac{2}{36}$	$\frac{2}{36}$	$\frac{2}{36}$	$\frac{2}{36}$	$\frac{2}{36}$	$\frac{1}{36}$	$\frac{11}{36}$
	$\frac{11}{36}$	$\frac{9}{36}$	$\frac{7}{36}$	$\frac{5}{36}$	$\frac{3}{36}$	$\frac{1}{36}$	

$E[X] = \dfrac{161}{36},\qquad E[Y] = \dfrac{91}{36}$

$E[XY] = \dfrac{441}{36}$

$\text{cov}(X,Y) = \dfrac{1225}{1296}.$

Exercise 9.2 (page 201)

1 (i) $\mathrm{var}(X + 3Y) = \mathrm{var}(X) + 6\,\mathrm{cov}(X, Y) + 9\,\mathrm{var}(Y)$
 (ii) $\mathrm{var}(2X - 5Y) = 4\,\mathrm{var}(X) - 20\,\mathrm{cov}(X, Y) + 25\,\mathrm{var}(Y)$.

2 $\mathrm{cov}(X, Y) = \tfrac{1}{4} \times 7 \times 12 = 21$
 (i) $E[X + Y] = 60$
   $\mathrm{var}(X + Y) = 7^2 + 2 \times 21 + 12^2 = 235; \ \sigma = 15.33$
 (ii) $E[X - Y] = -10$
   $\mathrm{var}(X - Y) = 7^2 - 2 \times 21 + 12^2 = 151; \ \sigma = 12.29$
 (iii) $E[3X + 5Y] = 3 \times 25 + 5 \times 35 = 250$
   $\mathrm{var}(3X + 5Y) = 9 \times 7^2 + 30 \times 21 + 25 \times 12^2 = 4671; \ \sigma = 68.34$.

3 (i) $\mathrm{var}(X + Y) = 6^2 + 2\rho \times 6 \times 10 + 10^2 = 136 + 120\rho$
 max $\sigma = 16$ when $\rho = 1$; min $\sigma = 4$ when $\rho = -1$
 (ii) X, Y independent $\Rightarrow \rho = 0 \Rightarrow \sigma = \sqrt{136} = 11.66$
 (iii)

4 (i) $E[X + Y] = 104$; $\mathrm{var}(X + Y) = 15^2 + 2 \times 0.7 \times 15 \times 12 + 12^2$; $\sigma = 24.92$
 (a) $P(X + Y > 130) = P(Z > 1.043) = 0.1485$
 (b) $P(X + Y < 90) = 0.2870$
 (ii) $E[X - Y] = -8$; $\mathrm{var}(X - Y) = 15^2 - 2 \times 0.7 \times 15 \times 12 + 12^2$; $\sigma = 10.82$
 $P(X > Y) = P(X - Y > 0) = P(Z > 0.740) = 0.2296$.

5 Let May rainfall be X cm, June rainfall be Y cm
 (i) $P(X > 20) = 0.1056$
 (ii) $P(Y > 20) = 0.3446$
 (iii) $E[X + Y] = 33$; $\mathrm{var}(X + Y) = 4^2 + 2 \times (-0.3) \times 4 \times 5 + 5^2$; $\sigma = 5.385$
 $P(X + Y > 40) = 0.0968$
 (iv) $E[X - Y] = -3$; $\mathrm{var}(X - Y) = 4^2 - 2 \times (-0.3) \times 4 \times 5 + 5^2$; $\sigma = 7.280$
 $P(X > Y) = P(X - Y > 0) = 0.3402$.

6 Let my arrival time be X, boss' time Y; boss arrives first if $Y < X$
 $E[Y - X] = 9$, $\mathrm{var}(Y - X) = 6^2 - 2 \times 0.4 \times 6 \times 4 + 4^2$; $\sigma = 5.727$
 $P(Y < X) = P(Y - X < 0) = 0.0581$.

7 Let weight of passenger be X kg and weight of his baggage be Y kg
 (i) $\dfrac{85 - 74}{\sigma_X} = 1.282$ and $\dfrac{24 - 20}{\sigma_Y} = 0.8416$
   $\sigma_X = 8.580$ kg $\sigma_Y = 4.753$ kg
 (ii) $E[X + Y] = 94$ kg
   If standard deviation of $X + Y$ is σ, then $\dfrac{108 - 94}{\sigma} = 1.282$
   $\sigma = 10.920$
   Thus $10.920^2 = 8.580^2 + 2\rho \times 8.580 \times 4.753 + 4.753^2$
   $\rho = 0.283$.

---

4 (i)

$X \backslash Y$	1	2	3	
1	0.12	0.15	0.03	0.3
2	0.28	0.35	0.07	0.7
	0.4	0.5	0.1	

All 6 entries are the product of the row and column totals
e.g. $0.15 = 0.3 \times 0.5$ so X and Y are independent.

(ii) $E[XY] = 1 \times 0.12 + 2 \times 0.15 + \cdots + 6 \times 0.07 = 2.89$
 $E[X] = 1.7$, $E[Y] = 1.7$
(iii) $\mathrm{cov}(X, Y) = 0$.

5 (i)

$X \backslash Y$	0	1	2	3	4	
0	0.04	0.12	0.16	0.06	0.02	0.4
1	0.03	0.09	0.12	0.045	0.015	0.3
2	0.02	0.06	0.08	0.03	0.01	0.2
3	0.01	0.03	0.04	0.015	0.005	0.1
	0.1	0.3	0.4	0.15	0.05	

$\rho = 0$
(ii) The positions in the table for which $X > Y$ are enclosed by the dotted lines.
 $P(X > Y) = 0.19$.

6 (i) $P(X = 1 \text{ and } Y = 1) = 0$, $P(X = 1) \times P(Y = 1) = \dfrac{2}{8} \times \dfrac{2}{8}$.
 These are not equal so X and Y are *not* independent.
(ii) $E[XY] = 1$, $E[X] = 1$, $E[Y] = 1$, $\mathrm{cov}(X, Y) = 0$
 If $\mathrm{cov}(X, Y) = 0$, it does not necessarily follow that X and Y are independent.

7 $\rho = \dfrac{\mathrm{cov}(X, Y)}{\sigma_X \sigma_Y} = \dfrac{E[XY] - E[X]E[Y]}{\sigma_X \sigma_Y}$
(i) Yes (ii) Yes (iii) No.

8 (i) $E[X + Y] = 0 \times \tfrac{1}{4} + 1 \times \tfrac{1}{4} + 2 \times \tfrac{1}{4} + 3 \times \tfrac{1}{4} = \tfrac{3}{2}$
 $E[X] = \tfrac{1}{2}$, $E[Y] = 1$
(ii) $E[(X + Y)^2] = 0 \times \tfrac{1}{4} + 1 \times \tfrac{1}{4} + 4 \times \tfrac{1}{4} + 9 \times \tfrac{1}{4} = \tfrac{14}{4}$
 $\mathrm{var}(X + Y) = \dfrac{14}{4} - \left(\dfrac{3}{2}\right)^2 = \dfrac{5}{4}$
(iii) $\mathrm{var}(X) + 2\,\mathrm{cov}(X, Y) + \mathrm{var}(Y) = \tfrac{1}{4} + 2 \times \tfrac{1}{4} + \tfrac{1}{2} = \tfrac{5}{4}$

8 Using $\text{var}(X + Y) = \text{var}(X) + 2\,\text{cov}(X, Y) + \text{var}(Y)$
$\text{cov}(X, Y) = -3$
$\text{var}(X - Y) = 20 - 2 \times (-3) + 12 = 38$.

9 (i) $s_x = 2.608$, $s_y = 2.417$
Values of $(x + y)$ are 5, 12, 10, 18, 17; standard deviation is 4.758
(ii) $4.758^2 = 2.608^2 + 2r \times 2.608 \times 2.417 + 2.417^2$
$r = 0.793$
(iii) $\bar{x} = 6$, $\bar{y} = 6.4$, $\sum xy = 217$
$\text{cov}(x, y) = \dfrac{217}{5} - 6 \times 6.4 = 5$
$r = \dfrac{5}{2.608 \times 2.417} = 0.793$.

Exercise 9.3 (page 208)

1 (i) $E[X + Y] = 13$ (ii) $E[X - Y] = 3$ (iii) $E[2X + 3Y] = 31$
$\text{var}(X + Y) = 13$ $\text{var}(X - Y) = 13$ $\text{var}(2X + 3Y) = 4 \times 4 + 9 \times 9 = 97$
(iv) $E[X - 4Y] = -12$ (v) $E[\frac{1}{2}(X + Y)] = 6.5$
$\text{var}(X - 4Y) = 4 + 16 \times 9 = 148$ $\text{var}(\frac{1}{2}(X + Y)) = \frac{1}{4} \times 13 = 3.25$.

2 The score on a single die has mean 3.5 and variance $\dfrac{35}{12}$
$E[S] = E[R + 3W - 2B] = 3.5 + 3 \times 3.5 - 2 \times 3.5 = 7$
$\text{var}(S) = \text{var}(R + 3W - 2B) = \dfrac{35}{12} + 9 \times \dfrac{35}{12} + 4 \times \dfrac{35}{12} = \dfrac{245}{6}$.

3 (i) $E[X] = 2$
$E[X^2] = \displaystyle\int_1^3 x^2(\tfrac{1}{2})\,dx = \dfrac{13}{3}$, $\text{var}(X) = \frac{1}{3}$, $\sigma = 0.5774$
(ii) $E[X + Y] = E[X] + E[Y] = 2 + 2 = 4$
$\text{var}(X + Y) = \text{var}(X) + \text{var}(Y) = \frac{1}{3} + \frac{1}{3} = \frac{2}{3}$, $\sigma = 0.8165$
(iii) $E[X - Y] = 0$
$\text{var}(X - Y) = \frac{2}{3}$, $\sigma = 0.8165$.

4 Let weight of man be X kg, and weight of woman be Y kg.
(i) $X + Y$ has mean 148 kg and standard deviation $\sqrt{12^2 + 8^2} = 14.42$ kg
(a) $P(X + Y > 160) = P(Z > 0.832) = 0.2027$
(b) $P(\frac{1}{2}(X + Y) < 72) = P(X + Y < 144) = 0.3909$
(ii) $X - Y$ has mean 8 kg and standard deviation 14.42 kg
(a) $P(X > Y + 5) = P(X - Y > 5) = 0.5824$
(b) $P(Y > X) = P(X - Y < 0) = 0.2895$
(iii) Total weight is $W = X_1 + X_2 + \dots + X_5 + Y_1 + Y_2 + Y_3$
$E[W] = 600$
$\text{var}(W) = 12^2 + 12^2 + \dots + 12^2 + 8^2 + 8^2 + 8^2 = 912$, $\sigma_W = 30.20$
$P(W < 560) = 0.0926$.

5 Let thickness of bar be X mm, and width of slot be Y mm.
$Y - X$ has mean 2 and standard deviation $\sqrt{2^2 + 1.5^2} = 2.5$
(i) $P(X > Y) = P(Y - X < 0) = P(Z < -0.8) = 0.2119$
(ii) $P(Y - X > 3) = 0.3446$.

6 (i) $X - Y$ has mean 16 and standard deviation $\sqrt{2^2 + 5^2} = 5.385$
(ii) $P(10 + Y < X) = P(X - Y > 10) = 0.8673$
(iii) If I leave home at t minutes past 8, we require $P(t + Y < X) = 0.95$
$P(X - Y > t) = 0.95$
t corresponds to $z = -1.645$; $t = 7.14$ minutes past 8.

7 (i) Total consumption is $T = H_1 + H_2 + \dots + H_{250} + F_1 + F_2 + F_3 + F_4 + G$
$E[T] = 250 \times 2.8 + 4 \times 15.5 + 8.0 = 770$ kW
$\text{var}(T) = 250 \times 0.8^2 + 4 \times 3.6^2 + 1.5^2 = 214.09$, $\sigma_T = 14.63$ kW
(ii) $P(T > 800) = 0.0202$
(iii) Required consumption t corresponds to $z = -2.326$; $t = 736.0$ kW.

8 Distribution is

X	0	1	2	...	9
Probability	$\frac{1}{10}$	$\frac{1}{10}$	$\frac{1}{10}$		$\frac{1}{10}$

$E[X] = 4.5$, $\text{var}(X) = 8.25$
$S = X_1 + X_2 + \dots + X_{40}$ $E[S] = 180$, $\text{var}(S) = 330$
$P(180 < S < 200) \approx P(180.5 < S < 199.5$ in Normal$) = P(0.028 < Z < 1.073)$
$\qquad\qquad = 0.3472$.

9 X is B(30, 0.62), $E[X] = 30 \times 0.62 = 18.6$, $\text{var}(X) = 30 \times 0.62 \times 0.38 = 7.068$
Y is B(70, 0.24), $E[Y] = 16.8$, $\text{var}(Y) = 12.768$
$X + Y$ has mean 35.4 and standard deviation $\sqrt{7.068 + 12.768} = 4.454$
$P(X + Y > 40) \approx P(X + Y > 40.5$ in Normal$) = 0.1261$.

10 (i) Yes   (ii) Yes   (iii) No   (iv) Yes.

Exercise 10.1 (page 214)

1 (i) (a) A pupil who makes regular use of the tuckshop has a greater chance of being selected than one who does not.
(b) A pupil who is interested has a greater chance of being selected than one who is not interested.
(c) Systematic sample.
(d) Stratified sample.
(e) A pupil in a small form has a greater chance of being selected than one in a large form.
(ii) Number the pupils from 1 to 1120, and use random number tables.

3 $\bar{X} \pm 2.576 \times \dfrac{\sigma}{\sqrt{n}} = £87.14, £162.86.$

4 (i) $\bar{X} + 1.645 \times \dfrac{\sigma}{\sqrt{n}}$ (ii) $\bar{X} - 2.326 \times \dfrac{\sigma}{\sqrt{n}}$ (iii) $\bar{X} + 2.878 \times \dfrac{\sigma}{\sqrt{n}}$.

5 $202.0 - 1.645 \times \dfrac{18}{\sqrt{10}} = 192.64$ kN.
One can be '95% confident' that the mean breaking tension is greater than 192.64 kN, in the sense that this procedure will lead to a correct conclusion for 95% of samples.

6 $\bar{X} = 126.472; \ s \approx 3.1153;$ estimate $\sigma \approx 3.1153$
90% upper confidence limit for μ is $\bar{X} + 1.282 \times \dfrac{\sigma}{\sqrt{n}} = 126.72$ minutes.

7 (a) 95% confidence limits are $4.2 \pm 1.96 \times \dfrac{2.7}{\sqrt{180}}$ i.e. $3.81 < \mu < 4.59$
  (b) (i) 95% confidence limits are $\bar{X} \pm 1.96 \times \dfrac{\sigma}{\sqrt{n}}$; we require $1.96 \times \dfrac{\sigma}{\sqrt{n}} = 0.1$
    i.e. $n = 2801$.
    (ii) We require $1.96 \times \dfrac{\sigma}{\sqrt{n}} = 0.05$, $n = 11203$.
  (c) 99% confidence limits are $\bar{X} \pm 2.576 \times \dfrac{\sigma}{\sqrt{n}}$;
    we require $2.576 \times \dfrac{\sigma}{\sqrt{n}} = 0.1$, $n = 4838$.

8 95% confidence limits are $\bar{X} \pm 1.96 \times \dfrac{\sigma}{\sqrt{n}}$; we require $1.96 \times \dfrac{\sigma}{\sqrt{n}} = 0.05$, $n = 192757$.

9 $\dfrac{\sigma}{\sqrt{n}} = 0.011$
  (i) Confidence limits 2.06 and 2.10 are $2.08 \pm 0.02 = \bar{X} \pm 1.818 \times \dfrac{\sigma}{\sqrt{n}}$
    i.e. these are 93.1% confidence limits.

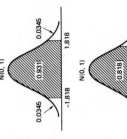

  (ii) 2.07 is $2.08 - 0.01 = \bar{X} - 0.909 \times \dfrac{\sigma}{\sqrt{n}}$
    i.e. this is an 81.8% lower confidence limit.

2 (i) $\mu = 19.6$ (ii) $\sigma = 8.2$ (iii) $\bar{X} = 19.12$
  (iv) $s = 4.934$ (v) $Z = -0.1309$ (vi) $t = -0.1946$.

3 $\mu = 172.1$ cm, $\sigma = 10.00$ cm.

Exercise 10.2 (page 218)

1 $\bar{X}$ is approximately Normal, mean 78.5, standard deviation $\dfrac{11.2}{\sqrt{20}} = 2.504$

(i) $P(\bar{X} > 82.0) = P(Z > 1.398) = 0.0810$
(ii) For 95% of samples, $\bar{X}$ lies between $78.5 \pm 1.96 \times 2.504$, i.e. $73.59 < \bar{X} < 83.41$.

2 $\bar{X}$ is approximately Normal, mean 52 and standard deviation $\dfrac{18}{\sqrt{150}} = 1.470$

$P(\bar{X} > 54) + P(\bar{X} < 50) = P(Z > 1.361) + P(Z < -1.361) = 0.1734.$

3 $\bar{X}$ (and $\bar{Y}$) is approximately Normal, mean 125, standard deviation $\dfrac{12}{\sqrt{60}} = 1.549$

Hence $(\bar{X} - \bar{Y})$ is approximately Normal, mean $125 - 125 = 0$, standard deviation $\sqrt{1.549^2 + 1.549^2} = 2.191$
$P(\bar{X} - \bar{Y} > 2.5) + P(\bar{X} - \bar{Y} < -2.5) = P(Z > 1.141) + P(Z < -1.141) = 0.2538.$

4 $\bar{X}$ has mean 179, standard deviation $\dfrac{12}{\sqrt{32}}$, $\bar{Y}$ has mean 177, standard deviation $\dfrac{8}{\sqrt{75}}$,

Hence $(\bar{X} - \bar{Y})$ has mean $179 - 177 = 2$, and standard deviation
$$\sqrt{\dfrac{144}{32} + \dfrac{64}{75}} = 2.314$$

$P(\bar{Y} > \bar{X}) = P(\bar{X} - \bar{Y} < 0) = P(Z < -0.864) = 0.1938.$

Exercise 10.3 (page 223)

1 (i) $\bar{X} \pm 1.96 \times \dfrac{\sigma}{\sqrt{n}} = \bar{X} \pm 0.7$ (ii) $\bar{X} \pm 2.326 \times \dfrac{\sigma}{\sqrt{n}} = \bar{X} \pm 0.831$

(iii) $\bar{X} \pm 2.576 \times \dfrac{\sigma}{\sqrt{n}} = \bar{X} \pm 0.92$ (iv) $\bar{X} \pm 3.09 \times \dfrac{\sigma}{\sqrt{n}} = \bar{X} \pm 1.104$

95% confidence interval is $57.5 < \mu < 58.9$.

2 $\bar{X} \pm 1.96 \times \dfrac{\sigma}{\sqrt{n}} = 210.66, 220.14$ kg.

10 $\alpha = -2.326$, $\beta = 1.751$

With probability 0.95, $-2.326 < \dfrac{\bar{X} - \mu}{\sigma/\sqrt{n}} < 1.751$

i.e. $\mu - 2.326 \times \dfrac{\sigma}{\sqrt{n}} < \bar{X} < \mu + 1.751 \times \dfrac{\sigma}{\sqrt{n}}$

i.e. $\bar{X} - 1.751 \times \dfrac{\sigma}{\sqrt{n}} < \mu < \bar{X} + 2.326 \times \dfrac{\sigma}{\sqrt{n}}$

When $n = 50$, $\sigma = 1.4$, $\bar{X} = 57.2$, this is $56.85 < \mu < 57.66$ (difference 0.81).
Usual limits $\bar{X} \pm 1.96 \times \dfrac{\sigma}{\sqrt{n}}$ give $56.81 < \mu < 57.59$ (difference 0.78).

Prefer 2nd interval because it is shorter.

Exercise 10.4 (page 231)

1 (i) (a) Normal, mean μ, standard deviation $\dfrac{\sigma}{\sqrt{n}}$ (b) Standard Normal

(ii) $Z = \dfrac{\bar{X} - 60}{\sigma/\sqrt{n}}$ Accept H_0 if $-1.96 \leqslant Z \leqslant 1.96$

Reject H_0 if $Z < -1.96$ or $Z > 1.96$

(a) $Z = 0.686$; accept H_0
(b) $Z = -2.527$; reject H_0
(c) $Z = 2.921$; reject H_0.

(iii) Reject H_0 if $|Z| > 2.326$
(iv) Reject H_0 if $|Z| > 2.576$
(v) Reject H_0 if $|Z| > 3.09$.

2 $H_0: \mu = 78.36$ against $H_1: \mu \neq 78.36$

$Z = \dfrac{\bar{X} - 78.36}{\sigma/\sqrt{n}}$ (5% rejection region $|Z| > 1.96$)

$= 3.215$

Reject H_0. Strong evidence that the mean length has changed.

3 $Z = \dfrac{\bar{X} - 50}{\sigma/\sqrt{n}}$ (5% rejection region $|Z| > 1.96$)

$= -1.431$

Accept H_0.

4 $H_0: \mu = 0.7$ against $H_1: \mu \neq 0.7$

$Z = \dfrac{\bar{X} - 0.7}{\sigma/\sqrt{n}}$ (5% rejection region $|Z| > 1.96$)

$= \dfrac{0.6837 - 0.7}{\dfrac{0.012}{\sqrt{10}}}$

$= -4.295$

Reject H_0. Very strong evidence that mean is not 0.7 ℓ

5 $H_0: \mu = 650$ against $H_1: \mu \neq 650$

$Z = \dfrac{\bar{X} - 650}{\sigma/\sqrt{n}}$ (1% rejection region $|Z| > 2.576$)

$= 2.559$

Accept H_0. Not quite enough evidence (at 1% level) to say that the mean is different.

6 $H_0: \mu = 176$ against $H_1: \mu \neq 176$

$Z = \dfrac{\bar{X} - 176}{\sigma/\sqrt{n}}$ (5% rejection region $|Z| > 1.96$)

$= 2.113$

Reject H_0. There is some evidence of a difference.

7 $H_0: \mu = 1$ against $H_1: \mu \neq 1$

$Z = \dfrac{\bar{X} - 1}{\sigma/\sqrt{n}}$ (5% rejection region: $|Z| > 1.96$)

$= -1.905$

Accept H_0.

8 $H_0: \mu = 48.7$ against $H_1: \mu \neq 48.7$

$Z = \dfrac{\bar{X} - 48.7}{\sigma/\sqrt{n}}$ (5% rejection region: $|Z| > 1.96$)

$= -3.039$

Reject H_0. Strong evidence that he is marking differently.

9 $H_0: \mu = 2.4$ against $H_1: \mu \neq 2.4$

$Z = \dfrac{\bar{X} - 2.4}{\sigma/\sqrt{n}} = \dfrac{2.2 - 2.4}{0.7/\sqrt{30}}$

$= -1.565$

A significance level of 11.8%

N(0, 1)

5.9%   5.9%

−1.565   1.565

16 $H_0: \mu = 850$ against $H_1: \mu > 850$

$Z = \frac{\bar{X} - 850}{\sigma/\sqrt{n}}$ (1% rejection region: $Z > 2.326$)

$= 2.454$

Reject H_0. There is evidence of longer life.

Exercise 11.1   (page 238)

1   (i) $B\left(20, \frac{1}{6}\right)$ (ii) $B(5, \frac{1}{2})$ (iii) No   (iv) $B(20, \frac{1}{4})$ (v) No

(vi) $B(12, p)$ where $p = P(\text{left-handed})$ (vii) No   (viii) $B(150, 0.01)$ (ix) No.

2   The number X, which fail, is $B(9, 0.3)$

(i) $P(X = 0) = 0.0404$ (ii) $P(3 \leqslant X \leqslant 5) = 0.5119$ (iii) $P(X \geqslant 7) = 0.0043$.

3   The number X, which grow, is $B(12, 0.6)$

$P(X > 0) = 1$, $P(X \geqslant 1) \approx 1$, $P(X \geqslant 2) = 0.9997$, $P(X \geqslant 3) = 0.9972$

$P(X \geqslant 4) = 0.9847$, $P(X \geqslant 5) = 0.9427$; hence $N = 4$.

4   $P(\text{at least one will work}) = 1 - (\frac{1}{4})^n = 0.99$ when $n = \frac{\ln(0.01)}{\ln(0.75)} \approx 16$.

5   Amount collected, X pounds, is $B(8, 0.7)$; $\mu = 5.6$, $\sigma^2 = 1.68$

$P(\mu - \sigma < X < \mu + \sigma) = P(4.3 < X < 6.9) = P(X = 5) + P(X = 6) = 0.5506$.

6   $np = 7.2$, $npq = 1.44 \Rightarrow q = 0.2$, $p = 0.8$, $n = 9$

X is $B(9, 0.8)$; $P(X = 6) = 0.1762$.

7   $\bar{X} = 3$, estimate $5p \approx 3$, i.e. $p \approx 0.6$.

X	0	1	2	3	4	5
Probability	0.0102	0.0768	0.2304	0.3456	0.2592	0.0778
Expected frequency	1.6	12.0	35.9	53.9	40.4	12.1

Not a good fit

8   $\bar{X} = 1.2$; if we have $B(8, p)$, we estimate that $8p = 1.2$, $p = 0.15$,

X	0	1	2	3	4	5	6	7	8
Probability	0.2725	0.3847	0.2376	0.0839	0.0185	0.0026	0.0002	0.0000	0.0000
Expected frequency	27.2	38.5	23.8	8.4	1.8	0.3	0.0	0.0	0.0

Very good fit

10   (i) Let $Z = \frac{\bar{X} - 80}{\sigma/\sqrt{n}}$ Reject H_0 if $Z > 1.645$

(a) $Z = 1.392$; accept H_0
(b) $Z = 1.657$; reject H_0
(c) $Z = -3.394$; accept H_0

(ii) Let $Z = \frac{\bar{X} - 45}{\sigma/\sqrt{n}}$ Reject H_0 if $Z < -1.645$

(iii) Let $Z = \frac{\bar{X} - 120}{\sigma/\sqrt{n}}$ Reject H_0 if $Z > 2.326$

(iv) Let $Z = \frac{\bar{X} - 2.8}{\sigma/\sqrt{n}}$ Reject H_0 if $Z < -2.878$.

11   $H_0: \mu = 75$ against $H_1: \mu > 75$

$Z = \frac{\bar{X} - 75}{\sigma/\sqrt{n}}$ (5% rejection region: $Z > 1.645$)

$= 1.791$

Reject H_0. Some evidence that the new diet is effective.

12   $H_0: \mu = 25$ against $H_1: \mu < 25$

$Z = \frac{\bar{X} - 25}{\sigma/\sqrt{n}}$ (5% rejection region: $Z < -1.645$)

$= -5.443$

Reject H_0. Very strong evidence that the sacks are underweight.

13   $H_0: \mu = 192$ against $H_1: \mu \neq 192$

$Z = \frac{\bar{X} - 192}{\sigma/\sqrt{n}}$ (5% rejection region: $|Z| > 1.96$)

$= -1.75$

Accept H_0. No evidence that his performance has altered.

14   $H_0: \mu = 5.2$ against $H_1: \mu > 5.2$

$Z = \frac{\bar{X} - 5.2}{\sigma/\sqrt{n}}$ (5% rejection region: $Z > 1.645$)

$= 1.418$

Accept H_0. Insufficient evidence to justify the claim.

15   $H_0: \mu = 176$ against $H_1: \mu \neq 176$

$Z = \frac{\bar{X} - 176}{\sigma/\sqrt{n}}$ (5% rejection region: $|Z| > 1.96$)

$= -1.831$

Accept H_0. The claim is not justified.

9 B(200, 0.3) ~ N(60, 6.481²)
P(50 ≤ X ≤ 65) = P(49.5 < X < 65.5 in Normal) = P($-1.620 < Z < 0.849$) = 0.7494.

10 B$\left(150, \frac{1}{6}\right)$ ~ N(25, 4.564²)
P(20 ≤ X ≤ 30) = P(19.5 < X < 30.5 in Normal) = 0.7720.

11 Number X travelling with T-A is B$(10000, \frac{1}{2})$ ~ N(5000, 50²)
If they provide N seats, then P($X > N$) = 0.01
($N + 0.5$) corresponds to $z = 2.326$
$N = 5115.8$; they should provide 5116 seats.
When number X travelling is B$\left(10\,000, \frac{3}{5}\right)$ ~ N(6000, 48.990²)
P($X > 6100.5$) = 0.0202.

12 (i) $(n + 1)p = 10.4$, mode is 10   (ii) $(n + 1)p = 2.96$; mode is 2
(ii) $(n + 1)p = 9$; modes are 8, 9   (iv) $(n + 1)p = 32.94$; mode is 32

13 $(n + 1)p = 301 \times 0.032 = 9.632$; most likely number is 9.

Exercise 11.2 (page 243)

1 B(16, 0.2); P(4) = 0.2001.

2 (i) B$(10, \frac{1}{4})$ (ii) B$(10, \frac{1}{4})$ (iii) B$(20, \frac{1}{4})$

(a) P(4 tickets) = $\binom{20}{4}(\frac{1}{4})^4(\frac{3}{4})^{16}$ = 0.1897

(b) Number X who obtain tickets for both matches is B$\left(10, \frac{1}{16}\right)$
P($X \geqslant 2$) = 0.1260.

3 (i) B(8, 0.7)   (ii) B(5, 0.7)   (iii) B(2, 0.7)   (iv) B(15, 0.7)
P($X \geqslant 13$) = 0.1268.

4 (i)

X	0	1
Probability	$\frac{4}{5}$	$\frac{1}{5}$

Y	0	1	2
Probability	$\frac{16}{25}$	$\frac{8}{25}$	$\frac{1}{25}$

$X + Y$	0	1	2	3
Probability	$\frac{64}{125}$	$\frac{48}{125}$	$\frac{12}{125}$	$\frac{1}{125}$

(ii)

X	0	1
Probability	$\frac{1}{2}$	$\frac{1}{2}$

Y	0	1	2
Probability	$\frac{16}{25}$	$\frac{8}{25}$	$\frac{1}{25}$

$X + Y$	0	1	2	3
Probability	0.32	0.48	0.18	0.02

If ($X + Y$) is B(3, p), $p^3 = 0.02 \Rightarrow p = 0.2714$
$(1 - p)^3 = 0.32 \Rightarrow p = 0.3160$,
hence ($X + Y$) is *not* Binomial.

5 (i) $\mu = 60$, $\sigma = 6.928$ (ii) $\mu = 30$, $\sigma = 5.050$
(iii) $\mu = 90$, $\sigma = 8.573$; P($X > 100.5$) = 0.1103.

Exercise 11.3 (page 247)

1 $\frac{X}{n}$ has mean 0.3, standard deviation $\sqrt{\dfrac{0.3 \times 0.7}{n}}$

(a) If $n = 400$, $\dfrac{X}{n} \sim$ N(0.3, 0.0229²)

(i) P$\left(\dfrac{X}{n} > 0.33\right) = 0.0954$

(ii) With probability 0.95, $0.2551 < \dfrac{X}{n} < 0.3449$

(b) Between $0.3 \pm 1.96\sqrt{\dfrac{0.3 \times 0.7}{n}}$; when $1.96\sqrt{\dfrac{0.3 \times 0.7}{n}} = 0.02$, $n = 2017$.

2 95% confidence limits are $\dfrac{X}{n} \pm 1.96\sqrt{\dfrac{\theta(1 - \theta)}{n}} = \dfrac{42}{200} \pm 1.96\sqrt{\dfrac{0.21 \times 0.79}{200}}$
$= 0.154, 0.266$.

3 99% confidence limits are $\dfrac{X}{n} \pm 2.576\sqrt{\dfrac{\theta(1 - \theta)}{n}} \approx 0.06, 0.12$
$2.576\sqrt{\dfrac{\theta(1 - \theta)}{n}} = 0.01 \Rightarrow n = 5435$.

4 95% upper confidence limit is $\dfrac{X}{n} + 1.645\sqrt{\dfrac{\theta(1 - \theta)}{n}} = 0.113$.

5 With 95% confidence, error bounds are $\pm 1.96\sqrt{\dfrac{\theta(1-\theta)}{n}}$.

Now $\theta(1-\theta) \leqslant \frac{1}{4}$ for all θ, so error bounds can be given as $\pm 1.96\sqrt{\dfrac{1}{4n}}$.

(i) When $n = 1000$, error bounds are ± 0.031

(ii) $1.96\sqrt{\dfrac{1}{4n}} = 0.005$ gives $n = 38416$.

Exercise 12.1 (page 253)

1 (i) $P(X=0) = e^{-3} = 0.0498$ (ii) 0.1494 (iii) 0.2240 (iv) 0.2240
(v) 0.1680 (vi) $P(X<2) = 0.1991$ (vii) $P(X \geqslant 3) = 0.5768$.

2 (i) $P(X=2) = e^{-1} \times \dfrac{1^2}{2!} = 0.1839$ (ii) $P(X \geqslant 2) = 0.2642$

3 (i) $P(X=0) = e^{-1.2} = 0.3012$ (ii) $P(X=3) = 0.0867$
P(can satisfy demand) = $P(X \leqslant 3) = 0.9662$
$P(X \leqslant 0) = 0.3012$, $P(X \leqslant 1) = 0.6626$, $P(X \leqslant 2) = 0.8795$, $P(X \leqslant 3) = 0.9662$
$P(X \leqslant 4) = 0.9923$, $P(X \leqslant 5) = 0.9985$, $P(X \leqslant 6) = 0.9997$
Hence 6 sets should be stocked.

4 In B(50, 0.08), $P(X=3) = \binom{50}{3}(0.08)^3(0.92)^{47} = 0.1993$
Similarly the other probabilities are 0.1973, 0.1959, 0.1955, 0.1954
All have mean 4; In Poisson(4), $P(X=3) = e^{-4} \times \dfrac{4^3}{3!} = 0.1954$.

5 Number of claims is B(750, 0.0032), which is approximately Poisson(2.4).
(i) $P(X=0) = e^{-2.4} = 0.0907$ (ii) $P(X=2) = 0.2613$.

6 Number of defectives is B(250, 0.02), which is approximately Poisson(5)
(i) $P(X=2) = 0.0842$ (ii) $P(X<3) = 0.1247$.

7 Number with birthday on that date is $B\left(1151, \dfrac{1}{365}\right)$, which is approximately Poisson(3.1534). $P(X \geqslant 5) = 0.2111$.

8 Number away is B(158, 0.02), which is approximately Poisson(3.16)
(i) $P(X=4) = 0.1763$ (ii) $P(2 \leqslant X \leqslant 5) = 0.7226$.

9 (i) In B(60, 0.05), $P(X=4) = \binom{60}{4}(0.05)^4(0.95)^{56} = 0.1724$
(ii) In Poisson(3), $P(X=4) = e^{-3} \times \dfrac{3^4}{4!} = 0.1680$.

Exercise 12.2 (page 256)

1 (i) Poisson($\frac{1}{2}$); $P(X>1) = 0.0902$
(ii) Poisson(3.5); $P(X<3) = 0.3208$.

2 Poisson(2.8); $P(X>4) = 0.3081$

3 (i) Poisson($\frac{2}{7}$); $P(X=0) = 0.7515$
(ii) Poisson($\frac{4}{7}$); $P(X>2) = 0.0204$.

4 (i) Poisson($\frac{1}{3}$); $P(X=3) = 0.0254$
(ii) Number of customers entering during 5 minutes is Poisson$\left(\dfrac{10}{3}\right)$.
$P(X=0) = 0.0357$.

5 Poisson(4.2); $P(X<2) = 0.0780$.

6 (i) Poisson(2.4); $P(X=0) = 0.0907$
(ii) Poisson(1.6); $P(X>3) = 0.0788$.

7 (i) Poisson(6); $P(X=6) = 0.1606$
(ii) (a) $2t$ (b) In Poisson($2t$), $P(X \geqslant 1) = 1 - e^{-2t} = 0.99$ when $t = 2.303$ minutes.

Exercise 12.3 (page 260)

1 (i) $\mu = 6$, $\sigma = \sqrt{6}$; $\mu \pm \sigma = 3.55, 8.45$
$P(3.55 < X < 8.45) = P(X=4, 5, 6, 7 \text{ or } 8) = 0.6960$
(ii) $P(\mu - \sigma < X < \mu + \sigma) = P(0.28 < X < 2.72) = P(X = 1 \text{ or } 2) = 0.5857$.

2 (i) 7 (ii) 3 (iii) 4 and 5.

3 (i) Poisson(4.57), most likely number is 4
(ii) Poisson(141.71), most likely number is 141.

4 Mean 2.4 goals

x	0	1	2	3	4	5	6	$\geqslant 7$
Probability from Poisson (2.4)	0.0907	0.2177	0.2613	0.2090	0.1254	0.0602	0.0241	0.0116
Expected frequency	7.3	17.4	20.9	16.7	10.0	4.8	1.9	0.9

5 $\bar{x} = 4.5$, $s^2 = 4.242$; mean ≈ variance is consistent with Poisson

X	0	1	2	3	4	5
Probability from Poisson (4.5)	0.0111	0.0500	0.1125	0.1687	0.1898	0.1708
Expected frequency	2.8	12.5	28.1	42.2	47.5	42.7

X	6	7	8	9	≥10
Probability from Poisson (4.5)	0.1281	0.0824	0.0463	0.0232	0.0171
Expected frequency	32.0	20.6	11.6	5.8	4.3

Exercise 12.4 (page 262)

1 Poisson(5); P(X < 3) = 0.1247

2 Poisson(2), Poisson(0.1), Poisson(0.25); total number is Poisson(2.35) P(X ≥ 4) = 0.2109

3 (i) Poisson(0.02 n) (ii) P(X ≥ 1) = $1 - e^{-0.02n}$
$1 - e^{-0.02n} = 0.95 \Rightarrow n = 149.8$; so 150 oysters should be taken.
When $n = 150$, number of pearls is Poisson(3); P(X > 4) = 0.1847.

Exercise 12.5 (page 264)

1 $\mu = 26$, $\sigma = \sqrt{26} = 5.0990$
(i) P(X > 30) ≈ P(X > 30.5 in Normal) = P(Z > 0.883) = 0.1886
(ii) P(20 ≤ X ≤ 30) ≈ P(19.5 < X < 30.5 in Normal) = 0.7103.

2 (i) $P(4 \leqslant X \leqslant 7) = e^{-6}\frac{6^4}{4!} + \cdots + e^{-6}\times\frac{6^7}{7!} = 0.5928$

(ii) P(4 ≤ X ≤ 7) ≈ P(3.5 < X < 7.5 in Normal) = P(-1.021 < Z < 0.612) = 0.5760.

3 (i) Poisson(144) ~ N(144, 12²); P(X > 160) ≈ P(X > 160.5 in Normal) = 0.0845
(ii) Poisson(12) ~ N(12, 3.4641²); P(X < 10) ≈ P(X < 9.5 in Normal) = 0.2352.

4 Poisson(40) ~ N(40, 6.3246²); P(X ≥ 55) ≈ P(X > 54.5 in Normal) = 0.0109.

Exercise 12.6 (page 265)

1 $P(X = 0) = e^{-\mu} = e^{-\mu}\frac{41}{250} \Rightarrow \mu = 1.8079$; In Poisson(5.4237), P(X = 2) = 0.0649.

2

Demand	0	1	2	3	≥4
Probability	0.0608	0.1703	0.2384	0.2225	0.3081

 0.5305

(i) Profit (3 cars owned) | -24 | -4 | 16 | 36 | | E[X] = £20.8
(ii) Profit (4 cars owned) | -32 | -12 | 8 | 28 | 48 | E[X] = £18.9.

3
(i) In Poisson ($\frac{1}{3}$), P(X ≥ 1) = 0.2835
(ii) Number of days is B(5, 0.2835); P(X ≥ 3) = 0.1419
(iii) For each day, P (report) = P(≥2 incidents) = 0.0446
Number of reports sent during a month is B(22, 0.0446); μ = 0.982, σ = 0.968.

4 P(2 sixes in first 7 throws) × P(six on 8th throw) = $\binom{7}{2}\left(\frac{1}{6}\right)^2\left(\frac{5}{6}\right)^5 \times \frac{1}{6} = 0.0391$

P(k − 1 successes in first n − 1 trials) × P(success on nth trial)
$= \binom{n-1}{k-1}p^{k-1}(1-p)^{(n-1)-(k-1)}p = \binom{n-1}{k-1}p^k(1-p)^{n-k}$.

5 Add up to $1 - e^{-1.5} = 0.7769$; divide by 0.7769; $P(X=1) = \frac{0.3347}{0.7769} = 0.4308$
P(X=2) = 0.3231, P(X=3) = 0.1616, P(X=4) = 0.0606.

6

X	2	3	4	5
All plants	0.2240	0.2240	0.1680	0.1008
Plants sold	0.3125	0.3125	0.2344	0.1406

E[X] = 3.203, E[X²] = 11.328, σ = 1.033.

Exercise 13.1 (page 270)

1 (i)

mode is 1

(ii) $1 = \int_1^4 \frac{k}{x^2}\, dx \Rightarrow k = \frac{4}{3}$

(iii) $E[X] = \int_1^4 x \cdot \frac{4}{3x^2} \, dx = \frac{4}{3} \ln 4$

$E[X^2] = \int_1^4 x^2 \cdot \frac{4}{3x^2} \, dx = 4$, var$(X) = 4 - \left(\frac{4}{3} \ln 4\right)^2 = 0.5834$

(iv) $P(2 < X < 3) = \int_2^3 \frac{4}{3x^2} \, dx = \frac{2}{9}$

(v) $P(X > 2.5) = \int_{2.5}^4 \frac{4}{3x^2} \, dx = \frac{1}{5}$

(vi) $\int_1^M \frac{4}{3x^2} \, dx = \frac{1}{2} \Rightarrow M = 1.6$.

2 $f(x) \geq 0$ and $\int_1^\infty f(x) \, dx = \int_1^\infty \frac{2}{x^3} \, dx = 1$

(i) $P(X < 2) = \int_1^2 \frac{2}{x^3} \, dx = \frac{3}{4}$

(ii) $P(X \geq 5) = \int_5^\infty \frac{2}{x^3} \, dx = \frac{1}{25}$

(iii) $E[X] = \int_1^\infty x \frac{2}{x^3} \, dx = 2$.

3

(i) $\int_0^\infty f(x) \, dx = \int_0^\infty e^{-x} \, dx = \left[-e^{-x} \right]_0^\infty = 1$

(ii) $P(X > 2) = \int_2^\infty e^{-x} \, dx = e^{-2} \approx 0.1353$

(iii) $0.95 = \int_T^\infty e^{-x} \, dx = e^{-T}$, so $T = 0.0513$ minutes

(iv) $0.5 = \int_M^\infty e^{-x} \, dx = e^{-M}$, so $M = 0.693$ minutes

(v) $E[X] = \int_0^\infty x e^{-x} \, dx = \left[-x e^{-x} - e^{-x} \right]_0^\infty = 1$ minute

$E[X^2] = \int_0^\infty x^2 e^{-x} \, dx = \left[-x^2 e^{-x} - 2x e^{-x} - 2e^{-x} \right]_0^\infty = 2$

var$(X) = 1$, $\sigma = 1$ minute.

4 (i) $P\left(0 < X < \frac{2}{\lambda}\right) = \int_0^{2/\lambda} \lambda e^{-\lambda x} \, dx = 1 - e^{-2} = 0.8647$

(ii) $0.1 = \int_a^\infty \lambda e^{-\lambda x} \, dx = e^{-\lambda a}$, so $a = \frac{\ln 10}{\lambda}$.

---

5 $E[X] = \int_0^a x \cdot \frac{1}{a} \, dx = \frac{1}{2} a$

$E[X^2] = \int_0^a x^2 \cdot \frac{1}{a} \, dx = \frac{1}{3} a^2$, var$(X) = \frac{1}{3} a^2 - \left(\frac{1}{2} a\right)^2 = \frac{1}{12} a^2$.

6 (a) $h = 1$ **(b)** $E[X] = 0$ **(c)** $f(x) = \begin{cases} 1 + x & \text{if } -1 \leq x \leq 0 \\ 1 - x & \text{if } 0 \leq x \leq 1 \end{cases}$

(d) $E[X^2] = \int_{-1}^0 x^2(1+x) \, dx + \int_0^1 x^2(1-x) \, dx = \frac{1}{6}$, so var$(X) = \frac{1}{6}$

(e) $P(X < -0.8165) + P(X > 0.8165)$
$= (1 - 0.8165)^2$
$= 0.0337$.

7

(i) $1 = \int_0^1 A x \, dx + 2A = \frac{8}{3} A$, so $A = \frac{3}{8}$

(ii) For median M, $(3 - M) \times \frac{3}{8} = \frac{1}{2}$

so $M = \frac{5}{3}$

(iii) $E[X] = \int_0^1 x \frac{3}{8} \sqrt{x} \, dx + \int_1^3 x \frac{3}{8} \, dx = 1.65$

$E[X^2] = \int_0^1 x^2 \cdot \frac{3}{8} \sqrt{x} \, dx + \int_1^3 x^2 \cdot \frac{3}{8} \, dx = \frac{47}{14}$

var$(X) = \frac{1777}{2800} \approx 0.6346$.

Exercise 13.2   (page 276)

1 (i) 0   **(ii)** $\frac{1}{2}x$ **(iii)** 1.

380

The semi-interquartile range, $\frac{1}{2}(Q_3 - Q_1) = \frac{1}{4}\ln 3 \approx 0.2747$

Mean $= \frac{1}{\lambda} = 0.5$; standard deviation $= \frac{1}{\lambda} = 0.5$.

7 $F(x) = \begin{cases} 0 & \text{if } x < 2 \\ 28 - 36x + 15x^2 - 2x^3 & \text{if } 2 \leq x \leq 3 \\ 1 & \text{if } x > 3. \end{cases}$

8 (i) $h = \frac{1}{2}$

(ii) $f(x) = \begin{cases} \frac{1}{2} + \frac{1}{4}x & \text{if } -2 \leq x \leq 0 \\ \frac{1}{2} - \frac{1}{4}x & \text{if } 0 \leq x \leq 2. \end{cases}$

(iii) If $-2 \leq x \leq 0$, $F(x) = \frac{1}{2}x + \frac{1}{8}x^2 + C$
$F(-2) = 0 \Rightarrow C = \frac{1}{2}$; $F(x) = \frac{1}{2} + \frac{1}{2}x + \frac{1}{8}x^2$
$F(0) = \frac{1}{2}$

(iv) If $0 \leq x \leq 2$, $F(x) = \frac{1}{2}x - \frac{1}{8}x^2 + K$
$F(0) = \frac{1}{2} \Rightarrow K = \frac{1}{2}$; $F(x) = \frac{1}{2} + \frac{1}{2}x - \frac{1}{8}x^2$.

9 If $0 \leq x < 1$, $F(x) = \frac{1}{3}x + C_1$; $F(0) = 0 \Rightarrow C_1 = 0$; $F(1) = \frac{1}{3}$
If $1 \leq x < 2$, $F(x) = \frac{1}{2}x + C_2$; $F(1) = \frac{1}{3} \Rightarrow C_2 = -\frac{1}{6}$; $F(2) = \frac{5}{6}$
If $2 \leq x < 3$, $F(x) = \frac{1}{6}x + C_3$; $F(2) = \frac{5}{6} \Rightarrow C_3 = \frac{1}{2}$;

$F(x) = \begin{cases} 0 & \text{if } x < 0 \\ \frac{1}{3}x & \text{if } 0 \leq x < 1 \\ \frac{1}{2}x - \frac{1}{6} & \text{if } 1 \leq x < 2 \\ \frac{1}{6}x + \frac{1}{2} & \text{if } 2 \leq x < 3 \\ 1 & \text{if } x > 3. \end{cases}$

10 $\bar{x} = 50$ hours, $\lambda = 0.02$

x	25	50	75	100	125	150
$F(x) = 1 - e^{-0.02x}$	0.3935	0.6321	0.7769	0.8647	0.9179	0.9502

x	175	200	225	250
$F(x)$	0.9698	0.9817	0.9889	0.9933

2 $F(x) = \begin{cases} 0 & \text{if } x < 3 \\ x - 3 & \text{if } 3 \leq x \leq 4 \\ 1 & \text{if } x > 4. \end{cases}$

3 $F(x) = \begin{cases} 0 & \text{if } x < 0 \\ x^2 & \text{if } 0 \leq x \leq 1 \\ 1 & \text{if } x > 1. \end{cases}$

4 For $1 \leq x \leq 2$, $F(x) = \frac{1}{7}x^3 + C$
$F(1) = 0 \Rightarrow C = -\frac{1}{7}$
Thus $F(x) = \frac{1}{7}x^3 - \frac{1}{7}$ if $1 \leq x \leq 2$
(i) if $x < 1$, $F(x) = 0$
(ii) if $x > 2$, $F(x) = 1$.

5 $F(x) = \begin{cases} 0 & \text{if } x < 1 \\ -\frac{2}{x} + 2 & \text{if } 1 \leq x \leq 2 \\ 1 & \text{if } x > 2 \end{cases}$

(i) $P\left(X < \frac{3}{2}\right) = F\left(\frac{3}{2}\right) = \frac{2}{3}$ (ii) $P(X > 1.8) = 1 - F(1.8) = \frac{1}{9}$

(iii) $P(1.4 < X < 1.6) = F(1.6) - F(1.4) = \frac{5}{28}$

6 $F(x) = \begin{cases} 0 & \text{if } x < 0 \\ 1 - e^{-2x} & \text{if } x \geq 0 \end{cases}$

For the median M, $F(M) = \frac{1}{2} \Rightarrow M = \frac{1}{2}\ln 2 = 0.3466$
$F(Q_3) = \frac{3}{4} \Rightarrow Q_3 = \ln 2 = 0.6931$
$F(Q_1) = \frac{1}{4} \Rightarrow Q_1 = \frac{1}{2}\ln\frac{4}{3} = 0.1438$

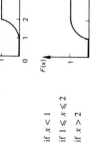
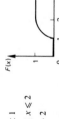

	0 < X < 25	25 < X < 50	50 < X < 75	75 < X < 100	100 < X < 125
Prob.	0.3935	0.2386	0.1448	0.0878	0.0532
Exp. freq.	295.1	179.0	108.6	65.9	39.9

	125 < X < 150	150 < X < 175	175 < X < 200	200 < X < 225	X > 225
Prob.	0.0323	0.0196	0.0119	0.0072	0.0067
Exp. freq.	24.2	14.7	8.9	5.4	5.0

11 $f(x) = F'(x) = \begin{cases} 1 - \frac{1}{2}x & \text{if } 0 \leqslant x \leqslant 2 \\ 0 & \text{otherwise} \end{cases}$

$E[X] = \int_0^2 x(1 - \tfrac{1}{2}x)\,dx = \frac{2}{3}$

$E[X^2] = \int_0^2 x^2(1 - \tfrac{1}{2}x)\,dx = \frac{2}{3}$; var$(X) = \frac{2}{9}$

12 cdf is $F(x) = \sin\frac{1}{2}x$ (i) $P(1 < X < 2) = F(2) - F(1) = 0.3620$

(ii) $f(x) = \frac{1}{2}\cos\frac{1}{2}x$ for $0 \leqslant x \leqslant \pi$ (iii) $E[X] = \int_0^\pi x\,\tfrac{1}{2}\cos\tfrac{1}{2}x\,dx = \pi - 2.$

13 (i) cdf is $F(x) = 1 - \left(\frac{10}{x} - 1\right) = 2 - \frac{10}{x}$ (for $5 \leqslant x \leqslant 10$)

(ii) $f(x) = \frac{10}{x^2}$ (for $5 \leqslant x \leqslant 10$)

(iii) $E[X] = \int_5^{10} x \cdot \frac{10}{x^2}\,dx = 10\ln 2.$

14 (i) $P\left(\frac{3}{2} < X < \frac{5}{2}\right) = F\left(\frac{5}{2}\right) - F\left(\frac{3}{2}\right) = 0.775 - 0.4 = 0.375$

(ii)

$f(x) = \begin{cases} 0 & \text{if } x < 0 \\ \frac{2}{5}x & \text{if } 0 \leqslant x < 1 \\ \frac{2}{5} & \text{if } 1 \leqslant x < 2 \\ \frac{4}{5} - \frac{1}{5}x & \text{if } 2 \leqslant x \leqslant 4 \\ 0 & \text{if } x > 4. \end{cases}$

Graph: $f(x)$ against x, with $\frac{2}{5}$ marked on the vertical axis and $1, 2, 3, 4$ on the horizontal axis.

15 When $x < 0$, $F(x) = 0$; when $x > 3$, $F(x) = 1$

$F(3) = 1 \Rightarrow k \times 3^3 = 1 \Rightarrow k = \frac{1}{27}$

$f(x) = \frac{1}{9}x^2$ (for $0 \leqslant x \leqslant 3$), $E[X] = \int_0^3 x \cdot \frac{1}{9}x^2\,dx = 2.25$

$E[X^2] = \int_0^3 x^2 \cdot \frac{1}{9}x^2\,dx = 5.4$

var$(X) = 0.3375.$

16 $P(X \leqslant x) = \dfrac{\frac{4}{3}\pi x^3}{\frac{4}{3}\pi a^3} = \dfrac{x^3}{a^3}$

cdf is $F(x) = \dfrac{x^3}{a^3}$; pdf $f(x) = \dfrac{3x^2}{a^3}$ (for $0 \leqslant x \leqslant a$)

$E[X] = \int_0^a x \cdot \frac{3x^2}{a^3}\,dx = \frac{3}{4}a$, $E[X^2] = \int_0^a x^2 \cdot \frac{3x^2}{a^3}\,dx = \frac{3}{5}a^2$

var$(X) = \frac{3}{80}a^2$, $\sigma = 0.1936a.$

17 $P(T > t) = \dfrac{k}{(t+2)^2}$; $P(T > 0) = 1 \Rightarrow k = 4$

cdf $F(t) = P(T \leqslant t) = 1 - \dfrac{4}{(t+2)^2}$

(i) $P(2 < T < 3) = F(3) - F(2) = 0.09$

(ii) pdf is $f(t) = \dfrac{8}{(t+2)^3}$ (for $t \geqslant 0$)

$E[T] = \int_0^\infty t\,\frac{8}{(t+2)^3}\,dt = \int_2^\infty \frac{8(u-2)}{u^3}\,du$ (where $t + 2 = u$)

= 2 minutes.

Exercise 13.3   (page 282)

1 (i) $E[X] = \int_0^1 x \cdot 2x\,dx = \frac{2}{3}$ (ii) $E[X^2] = \int_0^1 x^2 \cdot 2x\,dx = \frac{1}{2}$

(iii) $E[X^3] = \int_0^1 x^3 \cdot 2x\,dx = \frac{2}{5}$ (iv) $E[2X - \bar{X}] = \int_0^1 (2x - \bar{x})2x\,dx = \frac{8}{15}$

(v) $E\left[\dfrac{1}{X^2+1}\right] = \int_0^1 \dfrac{2x}{x^2+1}\,dx = \ln 2.$

2 (i) $E[X^4] = \int_1^3 x^4 \cdot \frac{1}{2} \, dx = 24.2$ (ii) $E[X(3-X)] = \int_1^3 x(3-x) \cdot \frac{1}{2} \, dx = \frac{5}{3}$

(iii) $E\left[\frac{1}{X}\right] = \int_1^3 \frac{1}{x} \cdot \frac{1}{2} \, dx = \frac{1}{2} \ln 3$ (iv) $E\left[\frac{1}{X^2}\right] = \int_1^3 \frac{1}{x^2} \cdot \frac{1}{2} \, dx = \frac{1}{3}$

(v) $E[\sin \frac{1}{3}\pi X] = \int_1^3 (\sin \frac{1}{3}\pi x) \cdot \frac{1}{2} \, dx = \frac{9}{4\pi}$.

3 $E[Y] = E[2X^2] = \int_{-1}^1 2x^2 \cdot \frac{3}{4}(1-x^2) \, dx = \frac{2}{5}$

$E[Y^2] = E[4X^4] = \int_{-1}^1 4x^4 \cdot \frac{3}{4}(1-x^2) \, dx = \frac{12}{35}$

$\text{var}(Y) = \frac{12}{35} - \left(\frac{2}{5}\right)^2 = \frac{32}{175}$.

4 If $Y = \frac{1}{X}$, $E[Y] = \int_1^2 \frac{1}{x} \, dx = \ln 2$; $E[Y^2] = \int_1^2 \frac{1}{x^2} \, dx = \frac{1}{2}$

$\text{var}(Y) = \frac{1}{2} - (\ln 2)^2$.

5 If $Y = e^{-X}$, $E[Y] = \int_0^\infty e^{-x} \cdot 2e^{-2x} \, dx = \frac{2}{3}$; $E[Y^2] = \int_0^\infty e^{-2x} \cdot 2e^{-2x} \, dx = \frac{1}{2}$

$\text{var}(Y) = \frac{1}{2} - (\frac{2}{3})^2 = \frac{1}{18}$; $\sigma \approx 0.2357$.

6 $E[A] = \int_0^a \pi x^2 \cdot \frac{1}{a} \, dx = \frac{1}{3}\pi a^2$; $E[A^2] = \int_0^a \pi^2 x^4 \cdot \frac{1}{a} \, dx = \frac{1}{5}\pi^2 a^4$

$\text{var}(A) = \frac{4}{45}\pi^2 a^4$, $\sigma = \frac{2}{3\sqrt{5}}\pi a^2$.

7 (i) $k = \frac{12}{625}$ (ii) $E[X] = \int_0^5 x \cdot \frac{12}{625} x^2(5-x) \, dx = 3$

(iii) $E[V] = E\left[\frac{4}{3}\pi X^3\right] = \int_0^5 \frac{4}{3}\pi x^3 \cdot \frac{12}{625} x^2(5-x) \, dx = \frac{1000}{21}\pi$.

8 $E[A] = \int_0^{10} x(10-x) \cdot \frac{1}{10} \, dx = \frac{50}{3} \text{ cm}^2$; $E[A^2] = \int_0^{10} x^2(10-x)^2 \cdot \frac{1}{10} \, dx = \frac{1000}{3}$

$\text{var}(A) = \frac{500}{9}$; $\sigma = \frac{10}{3}\sqrt{5} \text{ cm}^2$.

9 $P(6 < Y < 7) = P(6 < 3X + 5 < 7) = P(\frac{1}{3} < X < \frac{2}{3}) = \int_{1/3}^{2/3} 6x(1-x) \, dx = \frac{13}{27}$.

10 (i) $P(0.5 < Y < 1.5) = P\left(\frac{1}{16} < X < \frac{9}{16}\right) = \int_{1/16}^{9/16} 3x^2 \, dx = \frac{91}{512}$

(ii) $P(Y > 1) = P(X > \frac{1}{4}) = \int_{1/4}^1 3x^2 \, dx = \frac{63}{64}$.

11 (i) $P(2 < Y < 18) = P(1 < X < 3) = \int_1^3 e^{-x} \, dx = e^{-1} - e^{-3} \approx 0.3181$

(ii) $P(Y < 4) = P(X < \sqrt{2}) = \int_0^2 e^{-x} \, dx = 1 - e^{-2} \approx 0.7569$.

12 $P(Y < 4) = P(X^2 < 4) = P(-2 < X < 2) = 0.4$.

13 $P(S < 620) = P(4\pi R^2 < 620) = P(R < 7.024) = \frac{1}{4} \times 1.024 = 0.2560$.

14 $P(4 < T < 6) = P\left(4 < \frac{160}{V} < 6\right) = P\left(\frac{80}{3} < V < 40\right) = P(-1.067 < Z < 1.6) = 0.8022$.

15 Area $Y = \frac{1}{2} 5 \times 8 \times \sin \theta = 20 \sin \theta$

(i) $E[Y] = \int_0^\pi 20 \sin \theta \cdot \frac{1}{\pi} \, d\theta = \frac{40}{\pi} = 12.73 \text{ cm}^2$

$E[Y^2] = \int_0^\pi 400 \sin^2 \theta \cdot \frac{1}{\pi} \, d\theta = 200$; $\text{var}(Y) = 37.89$, $\sigma = 6.16 \text{ cm}^2$

(ii) $P(Y > 10) = P(\sin \theta > \frac{1}{2}) = P\left(\frac{1}{6}\pi < \theta < \frac{5}{6}\pi\right) = \frac{2}{3}$.

16

$L = 2R \cos \theta$

(i) $E[L] = \int_{-\pi/2}^{\pi/2} 2R \cos \theta \cdot \frac{1}{\pi} \, d\theta = \frac{4R}{\pi}$

(ii) $P(L > R\sqrt{2}) = P\left(\cos \theta > \frac{\sqrt{2}}{2}\right)$
$= P(-\frac{1}{4}\pi < \theta < \frac{1}{4}\pi)$
$= \frac{1}{2}$.

(iii) Since $P(L > R\sqrt{2}) = \frac{1}{2}$, the median length is $R\sqrt{2}$.

17 (i) For time T, the mean is 8 minutes with standard deviation $\frac{4}{\sqrt{12}} \approx 1.155$ minutes

The median is 8 minutes.

(ii) $V = \frac{600}{60T} = \frac{10}{T}$.

(iii) $E[V] = \int_6^{10} \frac{10}{t} \cdot \frac{1}{4} dt = \frac{5}{2} \ln \frac{10}{6} = 1.277$ m/s

$E[V^2] = \int_6^{10} \frac{100}{t^2} \cdot \frac{1}{4} dt = \frac{5}{3}$

var $(V) = 0.0358$, $\qquad \sigma = 0.189$ m/s

(iv) $\frac{1}{2} = P(T < 8) = P\left(\frac{10}{V} < 8\right) = P(V > 1.25)$; so the median speed is 1.25 m/s.

Exercise 13.4 (page 286)

1 $\int_0^4 e^{-x^2/2} dx \approx \frac{1}{3} \times 0.5 \times \{1.0000 + 4 \times 0.8825 + 2 \times 0.6065 + 4 \times 0.3247$
$\qquad + 2 \times 0.1353 + 4 \times 0.0439 + 2 \times 0.0111$
$\qquad + 4 \times 0.0022 + 0.0003\}$
$\qquad = 1.2532.$

Thus $\int_{-4}^4 e^{-x^2/2} dx \approx 2 \times 1.2532 = 2.5065$; and $\sqrt{2\pi} = 2.5066.$

2 $\int_1^2 \frac{1}{\sqrt{2\pi}} e^{-x^2/2} dx \approx \frac{1}{3} \times 0.5 \times \{0.2420 + 4 \times 0.1295 + 0.0540\}$
$\qquad = 0.1357$
$P(1 < Z < 2) = \Phi(2) - \Phi(1) = 0.1359.$

3 $P(1.4 < Z < 1.6) =$ area under pdf between 1.4 and 1.6
$\qquad \approx$ (height at mid-point) $\times$ (width)
$\qquad = \phi(1.5) \times 0.2$
$\qquad = 0.0259$
$P(1.4 < Z < 1.6) = \Phi(1.6) - \Phi(1.4) = 0.0260.$

4 $\phi'(x) = \frac{-1}{\sqrt{2\pi}} x e^{-x^2/2}$, $\phi''(x) = \frac{1}{\sqrt{2\pi}} (x^2 - 1) e^{-x^2/2}$

Points of inflexion occur when $\phi''(x) = 0$, i.e. when $x = \pm 1$.

5 $e^{-x^2/2} = 1 + (-\frac{1}{2}x^2) + \frac{(-\frac{1}{2}x^2)^2}{2!} + \cdots$
$\qquad = 1 - \frac{x^2}{2} + \frac{x^4}{8} + \frac{x^6}{48} + \frac{x^8}{384} - \cdots$
$\Phi(x) = \int \frac{1}{\sqrt{2\pi}} e^{-x^2/2} dx$
$\qquad = \frac{1}{\sqrt{2\pi}} \left\{x - \frac{x^3}{6} + \frac{x^5}{40} - \frac{x^7}{336} + \frac{x^9}{3456} - \cdots \right\} + C,$

and $\Phi(0) = \frac{1}{2} \Rightarrow C = \frac{1}{2}$.
$\Phi(0.3) = 0.617911$

6 (i) $E[Z^3] = \int_{-\infty}^\infty x^3 \frac{1}{\sqrt{2\pi}} e^{-x^2/2} dx = \frac{1}{\sqrt{2\pi}} \left[-x^2 e^{-x^2/2} - 2e^{-x^2/2} \right]_{-\infty}^\infty = 0$

(ii) $E[Z^4] = \int_{-\infty}^\infty x^4 \frac{1}{\sqrt{2\pi}} e^{-x^2/2} dx = \frac{1}{\sqrt{2\pi}} \left[-x^3 e^{-x^2/2} - 3xe^{-x^2/2} \right]_{-\infty}^\infty$
$\qquad + \frac{3}{\sqrt{2\pi}} \int_{-\infty}^\infty e^{-x^2/2} dx$
$\qquad = 0 + 3 = 3$

(iii) $E[Z^2] = 1$; var $(Z^2) = E[Z^4] - (E[Z^2])^2 = 2$

7

$f(x) = \frac{2}{\sqrt{2\pi}} e^{\frac{1}{2}x^2}$

$E[X] = \int_0^\infty x \frac{2}{\sqrt{2\pi}} e^{-x^2/2} dx = \frac{2}{\sqrt{2\pi}} \left[-e^{-x^2/2} \right]_0^\infty$
$\qquad = \frac{2}{\sqrt{2\pi}} = 0.7979$

$E[X^2] = \int_0^\infty x^2 \frac{2}{\sqrt{2\pi}} e^{-x^2/2} dx$
$\qquad = \frac{2}{\sqrt{2\pi}} \left[-xe^{-x^2/2} \right]_0^\infty + \frac{2}{\sqrt{2\pi}} \int_0^\infty e^{-x^2/2} dx$
$\qquad = 0 + \frac{2}{\sqrt{2\pi}} \times \frac{1}{2} \times \sqrt{2\pi}$
$\qquad = 1$

var $(X) = 1 - \frac{2}{\pi}$; $\sigma \approx 0.6028.$

8 (i) $P(1 < Z < 2) = 0.1359$

(ii) $\Phi(M) = \Phi(1) + \frac{1}{2} \times 0.1359 = 0.9093$
$M = 1.337$

Area under $\frac{1}{\sqrt{2\pi}} e^{-x^2/2}$ for $1 \le x \le 2$ is 0.1359

To make the area one, pdf is

$$f(x) = \begin{cases} \dfrac{1}{0.1359} \times \dfrac{1}{\sqrt{2\pi}} e^{-x^2/2} & \text{if } 1 \le x \le 2 \\ 0 & \text{otherwise} \end{cases}$$

(iii)

(iv) mean $= \int_1^2 xf(x) dx = \frac{1}{0.1359} \times \frac{1}{\sqrt{2\pi}} \left[-e^{-x^2/2} \right]_1^2 = 1.383.$

Exercise 13.5 (page 293)

1 (i) $\dfrac{1}{3\sqrt{2\pi}} e^{-(x-2)^2/18}$ (ii) $\dfrac{1}{5\sqrt{2\pi}} e^{-x^2/50}$

(iii) $\dfrac{1}{2.4\sqrt{2\pi}} e^{-(x-57.6)^2/11.52}$ (iv) $\dfrac{1}{2\sqrt{2\pi}} e^{-(x+10)^2/8}$

(v) $\dfrac{1}{4\sqrt{\pi}} e^{-(x-12)^2/16}$

2 (i) $\mu = 0, \sigma = 4$ (ii) $\mu = 25, \sigma = \sqrt{5}$

(iii) $\mu = -6, \sigma = \dfrac{1}{\sqrt{2}}$.

3 (i) $\mu = 4, \sigma^2 = 1.5$ (ii) $A = \dfrac{1}{\sqrt{3\pi}}$

(iii) $P(4 < X < 5) = 0.2928$.

4 (i) $P(X > 1000) = 0.6554$

(iii)

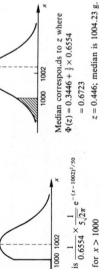

pdf is $\dfrac{1}{0.6554} \times \dfrac{1}{5\sqrt{2\pi}} e^{-(x-1002)^2/50}$
for $x > 1000$

(ii)

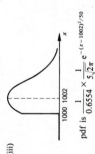

Median corresponds to z where
$\Phi(z) = 0.3446 + \tfrac{1}{2} \times 0.6554$
$= 0.6723$
$z = 0.446$; median is 1004.23 g.

5

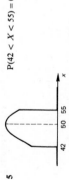

$f(x) = \begin{cases} \dfrac{1}{0.2541} \times \dfrac{1}{20\sqrt{2\pi}} e^{-(x-50)^2/800} = 0.0785\, e^{-(x-50)^2/800} & \text{if } 42 < x < 55 \\ 0 & \text{otherwise} \end{cases}$

$P(42 < X < 55) = 0.2541$

mean is $0.0785 \displaystyle\int_{42}^{55} x\, e^{-(x-50)^2/800}\, dx$.

6 $\bar{x} = 176.1, s = 9.372$

x	<155	155–160	160–165	165–170	170–175
Probability	0.0122	0.0307	0.0753	0.1393	0.1959
expected frequency	3.1	7.7	18.8	34.8	49.0

x	175–180	180–185	185–190	190–195	> 195
Probability	0.2079	0.1676	0.1021	0.0471	0.0219
expected frequency	52.0	41.9	25.5	11.8	5.5

7

x	155	160	165	170	175	180	185	190	195
Cumulative probability	0.012	0.036	0.116	0.268	0.456	0.660	0.828	0.932	0.972
z	-2.26	-1.80	-1.20	-0.62	-0.11	0.41	0.95	1.49	1.91

Graph of z against x is approximately a straight line.
When z = 0; mean $\mu \approx 176.2$
When z = 1, x = 185.5; standard deviation $\sigma \approx 185.5 - 176.2 = 9.3$.

8 $\mu \approx 176.2, \sigma \approx 9.3$

Exercise 13.6 (page 298)

These samples were obtained using the random digits on p 294; in questions 5, 7–10 cumulative probabilities are taken to 3 decimal places.

1 Allocating 0–4: H, 5–9: T, digits 33975... give
HHTTT TTTTH THTHT HHTTH.

2 Reading in threes, 001–400: select, 401–800: subtract 400,
000, 801–999: ignore,

digits 456 109 ... give

56, 109, 303, 125, 393, 175, 209, 181, 327, 19, 185, 353.

3

X	0	1	2	3	4	5	6
Probability	$\frac{1}{64}$	$\frac{6}{64}$	$\frac{15}{64}$	$\frac{20}{64}$	$\frac{15}{64}$	$\frac{6}{64}$	$\frac{1}{64}$
Cum prob.	$\frac{1}{64}$	$\frac{7}{64}$	$\frac{22}{64}$	$\frac{42}{64}$	$\frac{57}{64}$	$\frac{63}{64}$	$\frac{64}{64}$

Reading in pairs, and ignoring 00, 65–99,
digits 93 40 94 ... give 3, 4, 2, 1, 4, 4.

4

X	0	1	2	3
probability	$\frac{8}{729}$	$\frac{84}{729}$	$\frac{294}{729}$	$\frac{343}{729}$
Cum prob.	$\frac{8}{729}$	$\frac{92}{729}$	$\frac{386}{729}$	$\frac{729}{729}$

Reading in threes, and ignoring 000, 730–999,
digits 265 456 679 ... give 2, 3, 3, 3, 2, 3, 2, 3.

5

X	0	1	2	3	4	5	6	7
Probability	0.135	0.271	0.271	0.180	0.090	0.036	0.012	0.003
Cum prob.	0.135	0.406	0.677	0.857	0.947	0.983	0.995	0.999

Reading in threes (and ignoring 000), digits .642 .466 .540 ... give
2, 2, 2, 1, 1, 0, 1, 3.

6

X	1	2	3	4	5
Cum prob.	0.08	0.33	0.67	0.87	1.00

Reading in pairs, allocating 01–08 : 1, 09–33 : 2, ... and 88–99, 00 : 5,
digits 64 39 44 ... give 3, 3, 3, 1, 3, 4, 3, 3, 1, 3.

7 $x = 3p$, and digits .410 .610 .307 ... give
1.230, 1.830, 0.921, 2.502, 1.431, 0.600, 1.335, 2.118.

8 cdf $F(x) = x^3$. If $F(x) = p$, then $x = p^{1/3}$
digits .034 .532 .119 ... give
0.324, 0.810, 0.492, 0.914, 0.745.

9 $F(x) = 1 - e^{-0.2x}$ If $F(x) = p$ then $x = -5\ln(1-p)$
digits .863 .655 .011 ... give
9.94, 5.32, 0.06, 3.16, 7.73, 10.60.

10 Using digits N(0, 1):
Sample from N(0, 1):
-0.782, -0.423, 0.619, 0.410, -0.942, -1.572, -0.212, 2.054,
-1.675, -2.097
Sample from N(50, 20^2):
34.4, 41.5, 62.4, 58.2, 31.2, 18.6, 45.8, 91.1, 16.5, 8.1
$\bar{x} = 40.8$, $s = 23.7$.

Exercise 14.1 (page 304)

1 $H_0 : \mu = 7.5$ against $H_1 : \mu < 7.5$
Reject H_0 if $\bar{X} < 7.4$

Let $Z = \dfrac{\bar{X} - 7.5}{\sigma/\sqrt{n}} \sim$ N(0, 1) if H_0 is true.

(i) Reject H_0 if $Z < -1.265$. Significance level 10.3%
If the mean weight is actually 7.5 kg, there is a 10.3% chance of stopping the machine.

(ii) Reject H_0 if $Z < -1.645$ gives $n \approx 17$.

2 Let $Z = \dfrac{\bar{X} - 600}{\sigma/\sqrt{n}} \sim$ N(0, 1) if H_0 is true.

(i) H_0 accepted if $-1.96 < Z < 1.96$
i.e. $596.605 < \bar{X} < 603.395$

If $\mu = 605$, then $\bar{X}$ has mean 605, standard deviation $\dfrac{15}{\sqrt{75}} = 1.732$

P(H_0 accepted) = P($596.605 < \bar{X} < 603.395$) = 0.1770

(ii) H_0 accepted if $597.60 < \bar{X} < 602.40$

If $\mu = 605$, then $\bar{X}$ has mean 605, standard deviation $\dfrac{15}{\sqrt{150}}$

P(H_0 accepted) = 0.0169
Increasing the sample size has greatly reduced this probability.

3 $H_0 : \mu_1 = \mu_2$ against $H_1 : \mu_1 \neq \mu_2$

Let $Z = \dfrac{\bar{X}_1 - \bar{X}_2}{\sqrt{\sigma_1^2/m_1 + \sigma_2^2/m_2}} \sim$ N(0, 1)   (5% rejection region: $|Z| > 1.96$)
$= -1.890$

Accept H_0. Not quite enough evidence to suggest a difference.

4 $H_0: \mu_1 = \mu_2$ against $H_1: \mu_1 > \mu_2$
$\bar{X}_1 = 15.125, \sigma_1 = 3.488, \bar{X}_2 = 13.95, \sigma_2 \approx 4.003$

Let $Z = \dfrac{\bar{X}_1 - \bar{X}_2}{\sqrt{\sigma_1^2/n_1 + \sigma_2^2/n_2}} \sim N(0,1)$ (5% rejection region: $Z > 1.645$)
$= 1.979$

Reject H_0. There is sufficient evidence.

5 $H_0: \mu_1 = \mu_2$ against $H_1: \mu_1 \neq \mu_2$
$\bar{X}_1 = 13.65, \bar{X}_2 = 12.1$

Let $Z = \dfrac{\bar{X}_1 - \bar{X}_2}{\sqrt{\sigma_1^2/n_1 + \sigma_2^2/n_2}} \sim N(0,1)$ (5% rejection region: $|Z| > 1.96$)
$= 1.510$

Accept H_0. No significant difference.

6 $(\bar{X}_1 - \bar{X}_2)$ has mean $(\mu_1 - \mu_2)$, standard deviation $\sqrt{\dfrac{\sigma_1^2}{n_1} + \dfrac{\sigma_2^2}{n_2}}$

For 95% of samples, $(\bar{X}_1 - \bar{X}_2)$ lies between

$$(\mu_1 - \mu_2) \pm 1.96 \sqrt{\dfrac{\sigma_1^2}{n_1} + \dfrac{\sigma_2^2}{n_2}}$$

and hence $(\mu_1 - \mu_2)$ lies between

$$(\bar{X}_1 - \bar{X}_2) \pm 1.96 \sqrt{\dfrac{\sigma_1^2}{n_1} + \dfrac{\sigma_2^2}{n_2}}$$

95% confidence limits are 182.52, 189.48 seconds.

7 $H_0: \mu_1 = \mu_2$ against $H_1: \mu_1 > \mu_2$

Let $Z = \dfrac{\bar{X}_1 - \bar{X}_2}{\sqrt{\sigma_1^2/n_1 + \sigma_2^2/n_2}} \sim N(0,1)$ (5% rejection region: $Z > 1.645$)
$= 3.637$

Reject H_0. Very strong evidence that $\mu_1 > \mu_2$.
99% lower confidence limit for $(\mu_1 - \mu_2)$ is $(\bar{X}_1 - \bar{X}_2) - 2.326 \sqrt{\dfrac{\sigma_1^2}{n_1} + \dfrac{\sigma_2^2}{n_2}}$
$= 0.58$ m.

8 $\bar{X}$ has mean μ, standard deviation $\dfrac{\sigma}{\sqrt{n}} = \dfrac{\mu}{\sqrt{n}}$

thus $\dfrac{\bar{X} - \mu}{\mu/\sqrt{n}} \sim N(0,1)$

$H_0: \mu = 75$ against $H_1: \mu > 75$

Let $Z = \dfrac{\bar{X} - \mu}{\mu/\sqrt{n}} \sim N(0,1)$ (5% rejection region: $Z > 1.645$)
$= 1.678$

Reject H_0. There is some evidence.

9 $H_0: \mu = 5.8$ against $H_1: \mu \neq 5.8$

Let $Z = \dfrac{\bar{X} - \mu}{\mu/\sqrt{n}} \sim N(0,1)$ (1% rejection region: $|Z| > 2.576$)
$= 2.395$

Accept H_0. Insufficient evidence of a change.

Exercise 14.2   (page 309)

1 $H_0: \theta = \frac{1}{2}$ against $H_1: \theta \neq \frac{1}{2}$

Let $Z = \dfrac{X - n \times \frac{1}{2}}{\sqrt{n \times \frac{1}{2} \times \frac{1}{2}}} \sim N(0,1)$ (5% rejection region : $|Z| > 1.96$)
$= 1.697$

Accept H_0. No evidence of bias.

2 $H_0: \theta = \frac{1}{6}$ against $H_1: \theta \neq \frac{1}{6}$

If H_0 is true, $X \sim B\left(120, \frac{1}{6}\right)$ and $P(X \geq 30) = 0.0100$
Significance level is 2.0%.
95% confidence limits for θ are $\dfrac{X}{n} \pm 1.96 \sqrt{\dfrac{\theta(1 - \theta)}{n}}$

Putting $\theta \approx \dfrac{X}{n} = \dfrac{30}{120}$ gives
approximate 95% confidence interval for θ is $0.173 < \theta < 0.327$.

3 $H_0: \theta = 0.8$ against $H_1: \theta < 0.8$

Let $Z = \dfrac{X - n \times 0.8}{\sqrt{n \times 0.8 \times 0.2}} \sim N(0,1)$ (5% rejection region : $Z < -1.645$)
$= -2.239$

Reject H_0. This sample confirms his belief.

4 $H_0: \theta = 0.24$ against $H_1: \theta \neq 0.24$

Let $Z = \dfrac{X - n \times 0.24}{\sqrt{n \times 0.24 \times 0.76}} \sim N(0,1)$ (5% rejection region : $|Z| > 1.96$)
$= 1.777$

Accept H_0. No evidence of difference.

387

2. $f'(x) = \frac{1}{2}Cx^{m/2-2}e^{-x/2}(n-2-x)$
   $f'(x) = 0$ when $x = n-2$

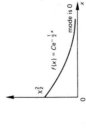

χ_3^2: $f(x) = Cx^{1/2}e^{-x/2}$
$f'(x) = \frac{1}{2}Cx^{-1/2}e^{-x/2}(1-x)$

χ_4^2: $f(x) = Cxe^{-x/2}$
$f'(x) = \frac{1}{2}Ce^{-x/2}(2-x)$

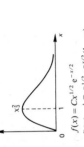

χ_5^2: $f(x) = Cx^{3/2}e^{-x/2}$
$f'(x) = \frac{1}{2}Cx^{1/2}e^{-x/2}(3-x)$

χ_1^2: $f(x) = Cx^{-\frac{1}{2}}e^{-\frac{1}{2}x}$ — mode is 0

χ_2^2: $f(x) = Ce^{-\frac{1}{2}x}$ — mode is 0

3. (i) $P(Z^2 < 2.706) = P(-1.645 < Z < 1.645) = 0.9000$
   (ii) If $P(Z^2 > a) = 0.01$, then $P(Z > \sqrt{a}) = 0.005$
        $a = (2.576)^2 = 6.636$.

4. (i) $\int_0^\infty Ce^{-x/2}\,dx = 1 \Rightarrow C = \frac{1}{2}$
   (ii) $P(Y^2 > 6) = \int_6^\infty \frac{1}{2}e^{-x/2}\,dx = e^{-3} = 0.0498$
   (iii) $\int_a^\infty \frac{1}{2}e^{-x/2}\,dx = 0.9$ gives $a = -2\ln(0.9) = 0.2107$
   (iv) $\int_M^\infty \frac{1}{2}e^{-x/2}\,dx = 0.5$ gives $M = 2\ln 2 = 1.3863$.

5. $H_0: \theta_1 = \theta_2$ against $H_1: \theta_1 \neq \theta_2$ (5% rejection region: $|Z| > 1.96$)

Let $Z = \dfrac{X_1/n_1 - X_2/n_2}{\sqrt{\dfrac{p(1-p)}{n_1} + \dfrac{p(1-p)}{n_2}}} \sim N(0,1)$

$p = \dfrac{528}{1500}$, and $Z = 2.752$

Reject H_0. Strong evidence of a change.

6. $H_0: \theta_1 = \theta_2$ against $H_1: \theta_1 < \theta_2$ (1% rejection region: $Z < -2.326$)

Let $Z = \dfrac{X_1/n_1 - X_2/n_2}{\sqrt{\dfrac{p(1-p)}{n_1} + \dfrac{p(1-p)}{n_2}}} \sim N(0,1)$

$p = \dfrac{85}{120}$, and $Z = -2.209$

Accept H_0. Insufficient evidence.

7. $H_0: \theta = 0.1$ against $H_1: \theta > 0.1$

If H_0 is true, $X \sim B(20, 0.1)$; mean (2) is much less than 5

$P(X \geq 5) = 0.0432$

Significance level is 4.32%, so we would usually reject H_0.

8. $\bar{X} = 1997.69$; estimate $\sigma = 18.276$

   (i) $H_0: \mu = 2000$ against $H_1: \mu < 2000$ (5% rejection region: $Z < -1.645$)

   Let $Z = \dfrac{\bar{X} - 2000}{\sigma/\sqrt{n}} \sim N(0,1)$
   $= -2.826$

   Reject H_0. Strong evidence that (A) is not being met.

   (ii) $H_0: \theta = \dfrac{1}{40}$ against $H_1: \theta > \dfrac{1}{40}$

   Let $Z = \dfrac{X - n \times \dfrac{1}{40}}{\sqrt{n \times \dfrac{1}{40} \times \dfrac{39}{40}}} \sim N(0,1)$ (5% rejection region: $Z > 1.645$)

   $X = 17$, and $Z = 1.289$

   Accept H_0. Condition (B) appears satisfactory.

Exercise 14.3   (page 320)

1 (i) $a = 11.07$ (ii) $b = 2.558$ (iii) 0.01   (iv) 0.94
  (v) $n = 6$ (vi) $c = 10.60$ (vii) $d = 34.76$.

388

5 $1 = \int_0^\infty Cxe^{-x/2}\,dx = C\left[-2xe^{-x/2} - 4e^{-x/2}\right]_0^\infty$, giving $C = \frac{1}{4}$

$F(x) = \int \frac{1}{4}xe^{-x/2}\,dx = -\frac{1}{2}(x+2)e^{-x/2} + k,$

and $F(0) = 0 \Rightarrow k = 1.$
$P(Y^2 < 1) = F(1) = 0.0902$
$P(Y^2 > 8) = 1 - F(8) = 0.0916.$

6 $E[Y^2] = E[Z_1^2] + \cdots + E[Z_n^2] = n$
$\operatorname{var}(Y^2) = \operatorname{var}(Z_1^2) + \cdots + \operatorname{var}(Z_n^2) = 2n$
Y^2 is the sum of n independent random variables, each having the same distribution; by the Central Limit Theorem, Y^2 is approximately Normal for large enough n.
$Y^2 \sim \chi_{30}^2$ is approximately Normal, mean 30, variance 60
a and b are $30 \pm 1.645\sqrt{60}$ i.e. $a = 17.26$, $b = 42.74$
(True values are $a = 18.49$, $b = 43.77$).

7 (i) $\sqrt{2a} = \sqrt{59} - 1.645$ $\sqrt{2b} = \sqrt{59} + 1.645$
$\qquad a = 18.22$ $b = 43.49$
(ii) $\sqrt{2c} = \sqrt{639} + 2.326$
$\qquad c = 381.0$

8 H_0: die is fair
Expected frequencies are all $\dfrac{100}{6}$
Let $Y^2 = \sum \dfrac{(O-E)^2}{E} \sim \chi_5^2$ (5% rejection region: $Y^2 > 11.07$)
$\quad = 15.56$
Reject H_0. Strong evidence that the die is biased (too many 1s and 6s, too few 4s).

9 H_0: ratios are $1:2:2:1$
Expected frequencies are 25, 50, 50, 25
Let $Y^2 = \sum \dfrac{(O-E)^2}{E} \sim \chi_3^2$ (5% rejection region: $Y^2 > 7.815$)
$\quad = 8.00$
Reject H_0. Not consistent with the theory (too many Type B).

10 H_0: all digits are equally likely.
(i) Expected frequencies are all 40
Let $Y^2 = \sum \dfrac{(O-E)^2}{E} \sim \chi_9^2$ (5% rejection region: $Y^2 > 16.92$)
$\quad = 8.85$
Accept H_0.

(ii) Expected frequencies are all 80
$Y^2 = 5.125$
Accept H_0.

11 H_0: P(girl) $= \frac{1}{2}$
Using B(6, $\frac{1}{2}$)

X	0	1	2	3	4	5	6
E	0.8	4.7	11.7	15.6	11.7	4.7	0.8

($0.8 + 4.7 = 5.5$; $4.7 + 0.8 = 5.5$)

Let $Y^2 = \sum \dfrac{(O-E)^2}{E} \sim \chi_4^2$ (5% rejection region: $Y^2 > 9.488$)
$\quad = 3.69$
Accept H_0. A good fit.

12 H_0: distribution is Binomial.
$\bar{x} = 1.4$, so using B(7, 0.2)

X	0	1	2	3	4	5	6	7
E	75.5	132.1	99.0	41.3	10.3	1.5	0.1	0.0

($10.3 + 1.5 + 0.1 + 0.0 = 11.9$)

Let $Y^2 = \sum \dfrac{(O-E)^2}{E} \sim \chi_3^2$ (5% rejection region: $Y^2 > 7.815$)
$\quad = 9.04$
Reject H_0. Not a good fit; too much at the extremes, and not enough in the middle (e.g. $X = 3$).

13 H_0: distribution is Poisson (2.5).

X	0	1	2	3	4	5	$\geqslant 6$
E	11.5	28.7	35.9	29.9	18.7	9.4	5.9

Let $Y^2 = \sum \dfrac{(O-E)^2}{E} \sim \chi_6^2$ (5% rejection region: $Y^2 > 12.59$)
$\quad = 15.72$
Reject H_0. Not consistent; too many high values.

14 H_0: distribution is Poisson
$\bar{x} = 1.723$, so using Poisson (1.723)

X	0	1	2	3	4	$\geqslant 5$
E	65.1	112.3	96.7	55.6	23.9	11.4

Let $Y^2 = \sum \dfrac{(O-E)^2}{E} \sim \chi^2_4$ (5% rejection region: $Y^2 > 9.488$)

$\quad = 4.54$

Accept H_0. A good fit.

15 H_0: distribution is Normal
$\bar{x} = 1997.69$, $s = 18.276$, so using N($1997.69, 18.276^2$)

X	<1960	1960–1970	1970–1980	1980–1990	1990–2000
E	9.8	22.6	50.9	85.2	106.6

X	2000–2010	2010–2020	2020–2030	2030–2040	$\geqslant 2040$
E	99.8	69.6	36.2	14.1	5.2

Let $Y^2 = \sum \dfrac{(O-E)^2}{E} \sim \chi^2_7$ (5% rejection region: $Y^2 > 14.07$)

$\quad = 78.86$

Reject H_0. Extremely strong evidence that this data does not come from a Normal distribution.

16 H_0: there is no association.

E	fair	dark	ginger
blue	8.4	14.3	5.3
brown	21.6	36.7	13.7

Let $Y^2 = \sum \dfrac{(O-E)^2}{E} \sim \chi^2_2$ (5% rejection region: $Y^2 > 5.991$)

$\quad = 10.38$

Reject H_0. Strong evidence of association; a relatively high proportion of fair-haired people have blue eyes.

17 H_0: there is no association.

E	small	family	luxury
< 25	21.5	37.3	11.2
25 – 40	32.2	56.0	16.8
> 40	38.3	66.7	20.0

Let $Y^2 = \sum \dfrac{(O-E)^2}{E} \sim \chi^2_4$ (5% rejection region: $Y^2 > 9.488$)

$\quad = 25.85$

Reject H_0. Very strong evidence of association; young drivers have more small cars and fewer luxury cars than expected.

18 H_0: there is no association.

E	sf	g	o	e
2 bed	88.4	163.2	27.2	61.2
3 bed	110.5	204.0	34.0	76.5
4/5 bed	22.1	40.8	6.8	15.3

Let $Y^2 = \sum \dfrac{(O-E)^2}{E} \sim \chi^2_6$ (5% rejection region: $Y^2 > 12.59$)

$\quad = 10.95$

Accept H_0. No evidence of association.

19 (i) H_0: there is no association.

O	died	survived
A	55	35
B	28	32

E	died	survived
A	49.8	40.2
B	33.2	26.8

Let $Y^2 = \sum \dfrac{(O-E)^2}{E} \sim \chi^2_1$ (5% rejection region: $Y^2 > 3.841$)

$\quad = 3.039$

Accept H_0. There is no evidence of association.

(ii) $H_0: \theta_1 = \theta_2$ against $H_1: \theta_1 \neq \theta_2$

Let $Z = \dfrac{X_1/n_1 - X_2/n_2}{\sqrt{\dfrac{p(1-p)}{n_1} + \dfrac{p(1-p)}{n_2}}} \sim \text{N}(0,1)$ (5% rejection region: $|Z| > 1.96$)

$p = \frac{83}{150}$ and $Z = 1.743$

Accept H_0. No evidence of a difference.
This test is: Reject H_0 if $Z^2 > (1.96)^2 = 3.84$
and $Z^2 = (1.743)^2 = 3.04$, i.e. the test is equivalent to that in (i).

20 $Y^2 = \frac{(4.7)^2}{49.8} + \frac{(4.7)^2}{40.2} + \frac{(4.7)^2}{33.2} + \frac{(4.7)^2}{26.8} = 2.483$.

Exercise 14.4 (page 327)

1 For 95% of samples, $2.70 < \frac{10S^2}{7.2^2} < 19.02$

i.e. $3.74 < S < 9.93$.

2 For 98% of samples, $10.86 < \frac{25S^2}{36.0^2} < 42.98$

i.e. $23.73 < S < 47.20$.

3 For 5% of samples, $\frac{15S^2}{10.0^2} > 23.68$, i.e. $S > 12.56$.

4 $H_0: \sigma = 2.5$ against $H_1: \sigma \neq 2.5$.

Let $Y^2 = \frac{16S^2}{2.5^2} \sim \chi^2_{15}$ (5% rejection region: $Y^2 < 6.262$ or $Y^2 > 27.49$)

$= 26.21$

Accept H_0. No evidence of a difference.

5 $H_0: \sigma = 8$ against $H_1: \sigma > 8$

Let $Y^2 = \frac{11S^2}{8^2} \sim \chi^2_{10}$ (5% rejection region: $Y^2 > 18.31$)

$S = 11.26$ and $Y^2 = 21.79$
Reject H_0. There is evidence that $\sigma > 8$.
We have assumed that the weights are Normally distributed.

6 $H_0: \sigma = 6$ against $H_1: \sigma \neq 6$

Let $Y^2 = \frac{20S^2}{6^2} \sim \chi^2_{19}$ (5% rejection region: $Y^2 < 8.907$ or $Y^2 > 32.85$)

$S = 3.961$ and $Y^2 = 8.715$
Reject H_0. It does appear that σ has been altered.

7 For 95% of samples, $0.484 < \frac{5S^2}{\sigma^2} < 11.14$

i.e. $\frac{5S^2}{11.14} < \sigma^2 < \frac{5S^2}{0.484}$

$S = 0.3493$ gives 95% confidence limits for σ as 0.23, 1.12.

8 For 98% of samples, $14.26 < \frac{30S^2}{\sigma^2} < 49.59$

$S = 11.4$ gives 98% confidence limits for σ as 8.87, 16.54 cm.

9 $\sqrt{2a} = \sqrt{997} - 1.96$ $\sqrt{2b} = \sqrt{997} + 1.96$
$a = 438.53$ $b = 562.31$

(i) $H_0: \sigma = 25.0$ against $H_1: \sigma \neq 25.0$

Let $Y^2 = \frac{500S^2}{25.0^2} \sim \chi^2_{499}$ (5% rejection region: $Y^2 < 438.53$ or $Y^2 > 562.31$)

$= 574.59$

Reject H_0.

(ii) For 95% of samples, $438.53 < \frac{500S^2}{\sigma^2} < 562.31$

95% confidence limits for σ are 25.27, 28.62.

10 (i) $E[S^2] = \frac{(n-1)}{n}\sigma^2$; $\text{var}\left(\frac{nS^2}{\sigma^2}\right) = 2(n-1)$

hence $\text{var}(S^2) = 2(n-1) \times \frac{\sigma^4}{n^2}$

(ii) $\sqrt{\frac{2nS^2}{\sigma^2}}$ is approximately Normal, with mean $\sqrt{2(n-1)} - 1 = \sqrt{2n-3}$

and standard deviation 1
hence

S is approximately Normal, with mean $\sqrt{2n-3} \times \frac{\sigma}{\sqrt{2n}}$

and standard deviation $1 \times \frac{\sigma}{\sqrt{2n}}$

Exercise 14.5 (page 337)

1 (i) $a = 2.776$ (ii) $b = 2.896$ (iii) $c = -1.725$ (iv) 0.045
(v) 0.9   (vi) $n = 13$ (vii) $d = 2.571$ (viii) $e = 63.66$.

2 $f(x) = C(1 + x^2)^{-1}$; $1 = C \int_{-\infty}^{\infty} (1 + x^2)^{-1} \, dx = C \left[\tan^{-1} x \right]_{-\infty}^{\infty}$ gives $C = \frac{1}{\pi}$

$F(x) = \int_{-\infty}^{x} \frac{1}{\pi} (1 + x^2)^{-1} \, dx = \frac{1}{\pi} \tan^{-1} x + k$, and $F(0) = \frac{1}{2} \Rightarrow k = \frac{1}{2}$

(i) $P(-12.7 < T < 12.7) = F(12.7) - F(-12.7) = 0.9500$
(ii) $F(a) = 0.99$, giving $a = 31.82$.

3 $f(x) = (2 + x^2)^{-3/2} = 2^{-3/2} \left(1 + \frac{x^2}{2}\right)^{-3/2}$ which is of the form $C\left(1 + \frac{x^2}{n}\right)^{-\frac{1}{2}(n+1)/2}$

with $n = 2$; hence $T \sim t_2$

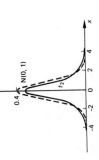

(i) $P(T < 4) = F(4) = 0.9714$
(ii) $F(a) = 0.9$, giving $a = 1.8856$.

4 $H_0 : \mu = 12.00$ against $H_1 : \mu \neq 12.00$

Let $T = \frac{\bar{X} - 12.00}{S/\sqrt{7}} \sim t_7$ (5% rejection region: $|T| > 2.365$)

$\bar{X} = 12.471$, $S = 0.6022$ and $T = 2.070$
Accept H_0. Insufficient evidence.

5 $H_0 : \mu = 6.46$ against $H_1 : \mu > 6.46$

Let $T = \frac{\bar{X} - 6.46}{S/\sqrt{5}} \sim t_5$ (5% rejection region: $T > 2.015$)

$\bar{X} = 6.5517$, $S = 0.1313$ and $T = 1.561$
Accept H_0. No evidence of an improvement.

6 $H_0 : \mu = 25$ against $H_1 : \mu \neq 25$

Let $T = \frac{\bar{X} - 25}{S/\sqrt{11}} \sim t_{11}$ (5% rejection region: $|T| > 2.201$)

$\bar{X} = 30.833$, $S = 7.988$ and $T = 2.422$
Reject H_0. It is significantly different.

7 $H_0 : \mu = 1.5$ against $H_1 : \mu < 1.5$

Let $T = \frac{\bar{X} - 1.5}{S/\sqrt{14}} \sim t_{14}$ (1% rejection region: $T < -2.624$)
$= -2.673$

Reject H_0. There is evidence that $\mu < 1.5$.

8 $H_0 : \mu = 93$ against $H_1 : \mu \neq 93$

Let $T = \frac{\bar{X} - 93}{S/\sqrt{20}} \sim t_{20}$ (5% rejection region: $|T| > 2.086$)

$\bar{X} = 85.714$, $S = 15.997$ and $T = -2.036$
Accept H_0. Not quite enough evidence to say average speed has changed.

9 $\bar{X} = 2.63$, $S = 0.4791$

99% confidence limits are $\bar{X} \pm 3.182 \times \frac{S}{\sqrt{3}} = 1.75, 3.51$ m.

10 $\bar{X} = 16.8$, $S = 3.8$

99% lower confidence limit is $\bar{X} - 2.602 \times \frac{S}{\sqrt{15}} = 14.25$ kN.

11 $\bar{X} = 144.33$; σ is known, so
95% confidence limits are $\bar{X} \pm 1.96 \times \frac{\sigma}{\sqrt{n}} = 126.73, 161.94$ g.

12 $H_0 : \mu_1 = \mu_2$ against $H_1 : \mu_1 \neq \mu_2$

Let $T = \frac{\bar{D}}{S/\sqrt{6}} \sim t_6$ (5% rejection region: $|T| > 2.447$)

$\bar{D} = 2.9857$, $S = 3.2436$ and $T = 2.255$
Accept H_0. No evidence of a difference.
Assumptions: The 7 days are a random sample of days.
The differences are Normally distributed.

13 $H_0 : \mu_1 = \mu_2$ against $H_1 : \mu_1 > \mu_2$

Let $T = \frac{\bar{D}}{S/\sqrt{7}} \sim t_7$ (5% rejection region: $T > 1.895$)

$\bar{D} = 9$, $S = 8.5$ and $T = 2.801$
Reject H_0. The treatment does appear to be effective.

14 $H_0 : \mu_1 = \mu_2$ against $H_1 : \mu_1 \neq \mu_2$ (5% rejection region: $|T| > 2.145$)

Let $T = \frac{\bar{X}_1 - \bar{X}_2}{S\sqrt{\frac{1}{9} + \frac{1}{7}}} \sim t_{14}$

$\bar{X}_1 = 211.111$, $S_1 = 5.065$, $\bar{X}_2 = 206.714$, $S_2 = 6.902$, $S = 6.349$ and $T = 1.374$
Accept H_0. No evidence that the gestation periods are different.

15 $H_0: \mu_1 = \mu_2$ against $H_1: \mu_1 < \mu_2$

Let $T = \dfrac{\bar{X}_1 - \bar{X}_2}{S\sqrt{\frac{1}{20}+\frac{1}{12}}} \sim t_{30}$ (5% rejection region: $T < -1.697$)

$S = 1.919$ and $T = -1.470$

Accept H_0. No evidence that exercise increases the mean weight loss.

16 $H_0: \mu_1 = \mu_2$ against $H_1: \mu_1 \neq \mu_2$

Let $T = \dfrac{\bar{D}}{S/\sqrt{9}} \sim t_9$ (5% rejection region: $|T| > 2.262$)

$\bar{D} = -4.3$, $S = 4.7127$ and $T = -2.737$

Reject H_0. This does indicate a difference in the mean times.

17 $\bar{D} = -14.667$, $S = 7.616$

95% confidence limits for $(\mu_1 - \mu_2)$ are $\bar{D} \pm 2.306 \times \dfrac{S}{\sqrt{8}}$

giving $8.46 < \mu_2 - \mu_1 < 20.88$.

18 $\bar{X}_1 = 90.2$, $S_1 = 12.221$, $\bar{X}_2 = 57.2$, $S_2 = 7.332$; $S = 11.267$

90% confidence limits for $(\mu_1 - \mu_2)$ are $(\bar{X}_1 - \bar{X}_2) \pm 1.860 \times S\sqrt{\frac{1}{5}+\frac{1}{5}}$

i.e. $19.75, 46.25$ mm

Assumptions: The samples are random, and independent of each other. The populations are Normal, with the same standard deviation.

19 Test A scores and Test B scores are not independent, so we must consider the differences, having mean $(\mu_1 - \mu_2)$ and standard deviation σ

$H_0: \mu_1 = \mu_2$ against $H_1: \mu_1 \neq \mu_2$

Let $Z = \dfrac{\bar{D}}{\sigma/\sqrt{n}} \sim N(0,1)$ (5% rejection region: $|Z| > 1.96$)

$\sigma \approx 8.5$ and $Z = -3.17$

(or $T = \dfrac{\bar{D}}{S/\sqrt{149}} \sim t_{149} \sim N(0,1)$

$= -3.16$)

Reject H_0. Very strong evidence that people perform differently in the two tests.

20 $\bar{X} = 20.25$, $S = 5.044$

(i) 95% confidence limits for μ are $\bar{X} \pm 2.201 \times \dfrac{S}{\sqrt{11}} = 16.90, 23.60$

(ii) $\dfrac{12S^2}{\sigma^2} \sim \chi^2_{11}$. For 95% of samples, $3.816 < \dfrac{12S^2}{\sigma^2} < 21.92$

95% confidence limits for σ are $3.732, 8.944$.

Exercise 14.6 (page 344)

1 $H_0: m = 10$ against $H_1: m \neq 10$

$U = 12$. In $B(15, \frac{1}{2})$, $P(U \geq 12) = 0.02$

Significant at about 4%

Reject H_0.

2 $H_0: m = 144$ against $H_1: m < 144$

$U = 1$. In $B(7, \frac{1}{2})$, $P(U \leq 1) = \dfrac{1}{16}$

Significant at 6.25%.

3 $H_0: m = 165$ against $H_1: m \neq 165$

$U = 82$. In $B(185, \frac{1}{2})$, $P(U \leq 82) = 0.07$

Significant at about 14%

Accept H_0. This is consistent with the claim.

4 H_0: distributions are identical against H_1: distributions differ

$U = 11$. In $B(13, \frac{1}{2})$, $P(U \geq 11) = 0.01$

Significant at about 2%.

Reject H_0. The run-times of the two computers do appear to differ.

5 (i) H_0: distributions are identical against H_1: bottom is hotter.

$U = 6$. In $B(8, \frac{1}{2})$, $P(U \geq 6) = 0.14$

Significant at about 14%

Accept H_0. No evidence that the bottom is hotter.

(ii) $H_0: \mu_1 = \mu_2$ against $H_1: \mu_1 > \mu_2$

Let $T = \dfrac{\bar{D}}{S/\sqrt{9}} \sim t_9$ (5% rejection region: $T > 1.833$)

Differences are 4, -2, 4, 0, 0, 4, -2, 4, 4, 4

$\bar{D} = 2$, $S = 2.530$ and $T = 2.372$

Reject H_0. Evidence that the mean temperature at the bottom is greater.

Differences must have a Normal distribution; this does not seem a reasonable assumption for the above differences.

6 Ranks (town houses): 11, 6, 1, 4, 21, 7, 9, 17, 15, 3, 10, 12

$R_1 = 116$

H_0: distributions are identical against H_1: distributions differ

If H_0 is true, $E[R_1] = 138$, var $(R_1) = 230$

Let $Z = \dfrac{R_1 - 138}{\sqrt{230}} \sim N(0, 1)$ (5% rejection region: $|Z| > 1.96$)

$= -1.451$

Accept H_0. No evidence of any difference.

7 Ranks (standard batteries): 8, 3, 11, 1.5, 6, 9.5, 1.5, 5, 13

$R_1 = 58.5$

H_0: distributions are identical against H_1: long-life last longer.

Let $Z = \dfrac{R_1 - 85.5}{\sqrt{128.25}} \sim N(0, 1)$ (5% rejection region: $Z < -1.645$)

$= -2.384$

Reject H_0. Strong evidence that long-life batteries do last longer.

8 H_0: distributions are identical against H_1: distributions differ

Let $Z = \dfrac{R_1 - 460}{\sqrt{1916.7}} \sim N(0, 1)$ (5% rejection region: $|Z| > 1.96$)

$= -1.94$

Accept H_0. Not quite enough evidence to suggest a difference.

9 Ranks (girls): 5, 1, 2. $R = 8$

Sets with $R \leqslant 8$: {1,2,3}, {1,2,4}, {1,2,5}, {1,3,4}

Total number of sets is $\dbinom{11}{3} = 165$

If distributions are identical, then $P(R \leqslant 8) = \dfrac{4}{165} = 0.02424$

Level of significance is 4.85%.

10

	A	B	C	D	E	F	G	H	I	J	K	L
Sign	+	−	+	+	+	+	−	+	+	−	+	+
Difference	1.2	0.2	0.7	1.8	1.3	0.1	0.9	2.4	1.5	0.6	1.6	0.4
Rank	7	2	5	11	8	1	6	12	9	4	10	3

H_0: distributions are identical against H_1: distributions differ.

(i) $U = 3$. In B$(12, \frac{1}{2})$, $P(U \leqslant 3) \approx 0.075$

Significant at about 15%

Accept H_0. No evidence of any difference.

(ii) $R = 66$

If H_0 is true, $E[R] = 39$, var $(R) = 162.5$

Let $Z = \dfrac{R - 39}{\sqrt{162.5}} \sim N(0, 1)$ (5% rejection region: $|Z| > 1.96$)

$= 2.118$

Reject H_0. Some evidence of a difference.

11 (i) $H_0: \mu = 80$ against $H_1: \mu \neq 80$

Let $T = \dfrac{\bar{X} - 80}{S/\sqrt{5}} \sim t_5$ (5% rejection region: $|T| > 2.571$)

$\bar{X} = 38.5$, $S = 28.4$ and $T = -3.267$

Reject H_0. Some evidence that the mean is not 80 years.

(ii) $H_0: \lambda = \dfrac{1}{80}$ against $H_1: \lambda \neq \dfrac{1}{80}$

Let $Y^2 = 12 \times \dfrac{1}{80} \times \bar{X} \sim \chi^2_{12}$ (5% rejection region: $Y^2 < 4.404$ or $Y^2 > 23.34$)

$= 5.775$

Accept H_0. No evidence that the mean is different from 80 years.

(iii) $H_0: m = 80$ against $H_1: m \neq 80$

$U = 1$. In B$(6, \frac{1}{2})$, $P(U \leqslant 1) = 0.11$

Significant at about 22%

Accept H_0. No evidence that the median is not 80 years.

Statistical Tables

The Normal distribution – values of $\Phi(z) = p$

The table gives the probability p of a random variable distributed as N(0, 1) being less than z.

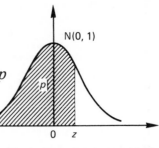

z	.00	.01	.02	.03	.04	.05	.06	.07	.08	.09	1	2	3	4	5	6	7	8	9
0.0	.5000	5040	5080	5120	5160	5199	5239	5279	5319	5359	4	8	12	16	20	24	28	32	36
0.1	.5398	5438	5478	5517	5557	5596	5636	5675	5714	5753	4	8	12	16	20	24	28	32	35
0.2	.5793	5832	5871	5910	5948	5987	6026	6064	6103	6141	4	8	12	15	19	23	27	31	35
0.3	.6179	6217	6255	6293	6331	6368	6406	6443	6480	6517	4	8	11	15	19	23	26	30	34
0.4	.6554	6591	6628	6664	6700	6736	6772	6808	6844	6879	4	7	11	14	18	22	25	29	32
0.5	.6915	6950	6985	7019	7054	7088	7123	7157	7190	7224	3	7	10	14	17	21	24	27	31
0.6	.7257	7291	7324	7357	7389	7422	7454	7486	7517	7549	3	6	10	13	16	19	23	26	29
0.7	.7580	7611	7642	7673	7704	7734	7764	7794	7823	7852	3	6	9	12	15	18	21	24	27
0.8	.7881	7910	7939	7967	7995	8023	8051	8078	8106	8133	3	6	8	11	14	17	19	22	25
0.9	.8159	8186	8212	8238	8264	8289	8315	8340	8365	8389	3	5	8	10	13	15	18	20	23
1.0	.8413	8438	8461	8485	8508	8531	8554	8577	8599	8621	2	5	7	9	12	14	16	18	21
1.1	.8643	8665	8686	8708	8729	8749	8770	8790	8810	8830	2	4	6	8	10	12	14	16	19
1.2	.8849	8869	8888	8907	8925	8944	8962	8980	8997	9015	2	4	6	7	9	11	13	15	16
1.3	.9032	9049	9066	9082	9099	9115	9131	9147	9162	9177	2	3	5	6	8	10	11	13	14
1.4	.9192	9207	9222	9236	9251	9265	9279	9292	9306	9319	1	3	4	6	7	8	10	11	13
1.5	.9332	9345	9357	9370	9382	9394	9406	9418	9429	9441	1	2	4	5	6	7	8	10	11
1.6	.9452	9463	9474	9484	9495	9505	9515	9525	9535	9545	1	2	3	4	5	6	7	8	9
1.7	.9554	9564	9573	9582	9591	9599	9608	9616	9625	9633	1	2	3	3	4	5	6	7	8
1.8	.9641	9649	9656	9664	9671	9678	9686	9693	9699	9706	1	1	2	3	4	4	5	6	6
1.9	.9713	9719	9726	9732	9738	9744	9750	9756	9761	9767	1	1	2	2	3	4	4	5	5
2.0	.9772	9778	9783	9788	9793	9798	9803	9808	9812	9817	0	1	1	2	2	3	3	4	4
2.1	.9821	9826	9830	9834	9838	9842	9846	9850	9854	9857	0	1	1	2	2	2	3	3	4
2.2	.9861	9864	9868	9871	9875	9878	9881	9884	9887	9890	0	1	1	1	2	2	2	3	3
2.3	.9893	9896	9898	9901	9904	9906	9909	9911	9913	9916	0	1	1	1	1	2	2	2	2
2.4	.9918	9920	9922	9925	9927	9929	9931	9932	9934	9936	0	0	1	1	1	1	1	2	2
2.5	.9938	9940	9941	9943	9945	9946	9948	9949	9951	9952									
2.6	.9953	9955	9956	9957	9959	9960	9961	9962	9963	9964									
2.7	.9965	9966	9967	9968	9969	9970	9971	9972	9973	9974									
2.8	.9974	9975	9976	9977	9977	9978	9979	9979	9980	9981									
2.9	.9981	9982	9982	9983	9984	9984	9985	9985	9986	9986									
3.0	.9987	9987	9987	9988	9988	9989	9989	9989	9990	9990									
3.1	.9990	9991	9991	9991	9992	9992	9992	9992	9993	9993									
3.2	.9993	9993	9994	9994	9994	9994	9994	9995	9995	9995									
3.3	.9995	9995	9996	9996	9996	9996	9996	9996	9996	9997									
3.4	.9997	9997	9997	9997	9997	9997	9997	9997	9997	9998									

differences untrustworthy

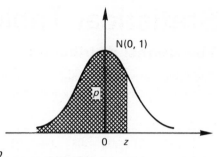

The inverse Normal function

The table gives the value of z such that $\Phi(z) = p$

p	.000	.001	.002	.003	.004	.005	.006	.007	.008	.009
.50	.0000	.0025	.0050	.0075	.0100	.0125	.0150	.0175	.0201	.0226
.51	.0251	.0276	.0301	.0326	.0351	.0376	.0401	.0426	.0451	.0476
.52	.0502	.0527	.0552	.0577	.0602	.0627	.0652	.0677	.0702	.0728
.53	.0753	.0778	.0803	.0828	.0853	.0878	.0904	.0929	.0954	.0979
.54	.1004	.1030	.1055	.1080	.1105	.1130	.1156	.1181	.1206	.1231
.55	.1257	.1282	.1307	.1332	.1358	.1383	.1408	.1434	.1459	.1484
.56	.1510	.1535	.1560	.1586	.1611	.1637	.1662	.1687	.1713	.1738
.57	.1764	.1789	.1815	.1840	.1866	.1891	.1917	.1942	.1968	.1993
.58	.2019	.2045	.2070	.2096	.2121	.2147	.2173	.2198	.2224	.2250
.59	.2275	.2301	.2327	.2353	.2378	.2404	.2430	.2456	.2482	.2508
.60	.2533	.2559	.2585	.2611	.2637	.2663	.2689	.2715	.2741	.2767
.61	.2793	.2819	.2845	.2871	.2898	.2924	.2950	.2976	.3002	.3029
.62	.3055	.3081	.3107	.3134	.3160	.3186	.3213	.3239	.3266	.3292
.63	.3319	.3345	.3372	.3398	.3425	.3451	.3478	.3505	.3531	.3558
.64	.3585	.3611	.3638	.3665	.3692	.3719	.3745	.3772	.3799	.3826
.65	.3853	.3880	.3907	.3934	.3961	.3989	.4016	.4043	.4070	.4097
.66	.4125	.4152	.4179	.4207	.4234	.4261	.4289	.4316	.4344	.4372
.67	.4399	.4427	.4454	.4482	.4510	.4538	.4565	.4593	.4621	.4649
.68	.4677	.4705	.4733	.4761	.4789	.4817	.4845	.4874	.4902	.4930
.69	.4959	.4987	.5015	.5044	.5072	.5101	.5129	.5158	.5187	.5215
.70	.5244	.5273	.5302	.5330	.5359	.5388	.5417	.5446	.5476	.5505
.71	.5534	.5563	.5592	.5622	.5651	.5681	.5710	.5740	.5769	.5799
.72	.5828	.5858	.5888	.5918	.5948	.5978	.6008	.6038	.6068	.6098
.73	.6128	.6158	.6189	.6219	.6250	.6280	.6311	.6341	.6372	.6403
.74	.6433	.6464	.6495	.6526	.6557	.6588	.6620	.6651	.6682	.6713
.75	.6745	.6776	.6808	.6840	.6871	.6903	.6935	.6967	.6999	.7031
.76	.7063	.7095	.7128	.7160	.7192	.7225	.7257	.7290	.7323	.7356
.77	.7388	.7421	.7454	.7488	.7521	.7554	.7588	.7621	.7655	.7688
.78	.7722	.7756	.7790	.7824	.7858	.7892	.7926	.7961	.7995	.8030
.79	.8064	.8099	.8134	.8169	.8204	.8239	.8274	.8310	.8345	.8381

p	.000	.001	.002	.003	.004	.005	.006	.007	.008	.009
.80	.8416	.8452	.8488	.8524	.8560	.8596	.8633	.8669	.8705	.8742
.81	.8779	.8816	.8853	.8890	.8927	.8965	.9002	.9040	.9078	.9116
.82	.9154	.9192	.9230	.9269	.9307	.9346	.9385	.9424	.9463	.9502
.83	.9542	.9581	.9621	.9661	.9701	.9741	.9782	.9822	.9863	.9904
.84	.9945	.9986	1.003	1.007	1.011	1.015	1.019	1.024	1.028	1.032
.85	1.036	1.041	1.045	1.049	1.054	1.058	1.063	1.067	1.071	1.076
.86	1.080	1.085	1.089	1.094	1.099	1.103	1.108	1.112	1.117	1.122
.87	1.126	1.131	1.136	1.141	1.146	1.150	1.155	1.160	1.165	1.170
.88	1.175	1.180	1.185	1.190	1.195	1.200	1.206	1.211	1.216	1.221
.89	1.227	1.232	1.237	1.243	1.248	1.254	1.259	1.265	1.270	1.276
.90	1.282	1.287	1.293	1.299	1.305	1.311	1.317	1.323	1.329	1.335
.91	1.341	1.347	1.353	1.360	1.366	1.372	1.379	1.385	1.392	1.398
.92	1.405	1.412	1.419	1.426	1.433	1.440	1.447	1.454	1.461	1.468
.93	1.476	1.483	1.491	1.499	1.506	1.514	1.522	1.530	1.538	1.546
.94	1.555	1.563	1.572	1.581	1.589	1.598	1.607	1.616	1.626	1.635
.95	1.645	1.655	1.665	1.675	1.685	1.695	1.706	1.717	1.728	1.739
.96	1.751	1.762	1.774	1.787	1.799	1.812	1.825	1.838	1.852	1.866
.97	1.881	1.896	1.911	1.927	1.943	1.960	1.977	1.995	2.014	2.034
.98	2.054	2.075	2.097	2.120	2.144	2.170	2.197	2.226	2.257	2.290
.99	2.326	2.366	2.409	2.457	2.512	2.576	2.652	2.748	2.878	3.090

Percentage points of the χ^2 (chi-squared) distribution

$p\%$	99	97.5	95	90	10	5.0	2.5	1.0	0.5
$n = 1$	.0002	.0010	.0039	.0158	2.706	3.841	5.024	6.635	7.879
2	.0201	.0506	0.103	0.211	4.605	5.991	7.378	9.210	10.60
3	0.115	0.216	0.352	0.584	6.251	7.815	9.348	11.34	12.84
4	0.297	0.484	0.711	1.064	7.779	9.488	11.14	13.28	14.86
5	0.554	0.831	1.145	1.610	9.236	11.07	12.83	15.09	16.75
6	0.872	1.237	1.635	2.204	10.64	12.59	14.45	16.81	18.55
7	1.239	1.690	2.167	2.833	12.02	14.07	16.01	18.48	20.28
8	1.646	2.180	2.733	3.490	13.36	15.51	17.53	20.09	21.95
9	2.088	2.700	3.325	4.168	14.68	16.92	19.02	21.67	23.59
10	2.558	3.247	3.940	4.865	15.99	18.31	20.48	23.21	25.19
11	3.053	3.816	4.575	5.578	17.28	19.68	21.92	24.72	26.76
12	3.571	4.404	5.226	6.304	18.55	21.03	23.34	26.22	28.30
13	4.107	5.009	5.892	7.042	19.81	22.36	24.74	27.69	29.82
14	4.660	5.629	6.571	7.790	21.06	23.68	26.12	29.14	31.32
15	5.229	6.262	7.261	8.547	22.31	25.00	27.49	30.58	32.80
16	5.812	6.908	7.962	9.312	23.54	26.30	28.85	32.00	34.27
17	6.408	7.564	8.672	10.09	24.77	27.59	30.19	33.41	35.72
18	7.015	8.231	9.390	10.86	25.99	28.87	31.53	34.81	37.16
19	7.633	8.907	10.12	11.65	27.20	30.14	32.85	36.19	38.58
20	8.260	9.591	10.85	12.44	28.41	31.41	34.17	37.57	40.00
21	8.897	10.28	11.59	13.24	29.62	32.67	35.48	38.93	41.40
22	9.542	10.98	12.34	14.04	30.81	33.92	36.78	40.29	42.80
23	10.20	11.69	13.09	14.85	32.01	35.17	38.08	41.64	44.18
24	10.86	12.40	13.85	15.66	33.20	36.42	39.36	42.98	45.56
25	11.52	13.12	14.61	16.47	34.38	37.65	40.65	44.31	46.93
26	12.20	13.84	15.38	17.29	35.56	38.89	41.92	45.64	48.29
27	12.88	14.57	16.15	18.11	36.74	40.11	43.19	46.96	49.64
28	13.56	15.31	16.93	18.94	37.92	41.34	44.46	48.28	50.99
29	14.26	16.05	17.71	19.77	39.09	42.56	45.72	49.59	52.34

$p\%$	99	97.5	95	90	10	5.0	2.5	1.0	0.5
30	14.95	16.79	18.49	20.60	40.26	43.77	46.98	50.89	53.67
35	18.51	20.57	22.47	24.80	46.06	49.80	53.20	57.34	60.27
40	22.16	24.43	26.51	29.05	51.81	55.76	59.34	63.69	66.77
50	29.71	32.36	34.76	37.69	63.17	67.50	71.42	76.15	79.49
100	70.06	74.22	77.93	82.36	118.5	124.3	129.6	135.8	140.2

Percentage points of the *t*-distribution

$p\%$	10	5	2	1
$n = 1$	6.314	12.71	31.82	63.66
2	2.920	4.303	6.965	9.925
3	2.353	3.182	4.541	5.841
4	2.132	2.776	3.747	4.604
5	2.015	2.571	3.365	4.032
6	1.943	2.447	3.143	3.707
7	1.895	2.365	2.998	3.499
8	1.860	2.306	2.896	3.355
9	1.833	2.262	2.821	3.250
10	1.812	2.228	2.764	3.169
11	1.796	2.201	2.718	3.106
12	1.782	2.179	2.681	3.055
13	1.771	2.160	2.650	3.012
14	1.761	2.145	2.624	2.977
15	1.753	2.131	2.602	2.947
20	1.725	2.086	2.528	2.845
30	1.697	2.042	2.457	2.750
50	1.676	2.009	2.403	2.678
100	1.660	1.984	2.364	2.626
∞	1.645	1.960	2.326	2.576

Percentage points of the
Normal distribution N(0, 1)

Index